Principles of Automotive Engines

Principles of Automotive Engines

Edited by
Lincoln Burns

Larsen & Keller
www.larsen-keller.com

Principles of Automotive Engines
Edited by Lincoln Burns
ISBN: 978-1-63549-104-3 (Hardback)

© 2017 Larsen & Keller

☰ Larsen & Keller

Published by Larsen and Keller Education,
5 Penn Plaza,
19th Floor,
New York, NY 10001, USA

Cataloging-in-Publication Data

Principles of automotive engines / edited by Lincoln Burns.
     p. cm.
Includes bibliographical references and index.
ISBN 978-1-63549-104-3
1. Automobiles--Motors. 2. Automobiles--Motors--Maintenance and repair.
3. Automobiles--Design andconstruction. 4. Motor vehicles--Design and construction.
I. Burns, Lincoln.
TL210 .F74 2017
629.25--dc23

The publisher's policy is to use permanent paper from mills that operate a sustainable forestry policy. Furthermore, the publisher ensures that the text paper and cover boards used have met acceptable environmental accreditation standards.

Printed and bound in the United States of America.

For more information regarding Larsen and Keller Education and its products, please visit the publisher's website www.larsen-keller.com

Table of Contents

Permissions

Index

Preface

This book elucidates new techniques and applications of automotive engine in a multidisciplinary approach. An engine is a machine used to convert chemical energy into mechanical energy through combustion of fuel. It is characterized by a high power-to-weight ratio. The different types of engines are heat engines, electric motors, clockwork motors, molecular motors, pneumatic motors, fire engines, etc. This book is designed to discuss in detail the various motors and their uses and functionality. Most of the topics introduced in it cover new techniques and the applications of engines. It studies, analyses and uphold the pillars of this subject area and its utmost significance in modern times. The various subfields of engines along with technological progress that have future implications are glanced at in it. This textbook is a complete source of knowledge on the present status of this important field.

Given below is the chapter wise description of the book:

Chapter 1- Engines are machines that are designed to convert energy into mechanical energy. Heat engines are majorly used to burn a fuel and then this fuel is used to create heat. This chapter will provide an integrated understanding of automotive engines.

Chapter 2- Vehicle engineering includes fields such as automotive engineering, vehicle dynamics, rolling stock and noise, vibration and harshness. The components of vehicle dynamics are automobile layout, electronics stability control, steering and suspension. This section is an overview of the subject matter incorporating all the major aspects of vehicle engineering.

Chapter 3- Some of the technologies of automotive engines that have been explained in the section are antifreeze, engine coolant temperature sensor, block heater, cold air intake, choke valve and radiator. Antifreeze helps in lowering the freezing point of any water based liquid. It is used to achieve freezing-point depression for environments that are cold.

Chapter 4- Internal combustion engines are heat engines that help in the combustion of a fuel. In these engines, the expansion of the high-temperature and high-pressure gases is produced by combustion; this combustion directly applies force to a particular part of the engine. The section on internal combustion engine offers an insightful focus, keeping in mind the subject matter.

Chapter 5- The aspects of automotive engines are vehicle emissions control, automotive electronics, automatic transmission, g-force, fuel efficiency and revolutions per minute. The study of reducing the emissions produced by motor vehicles is known as vehicle emissions control whereas automotive electronics is the system that is used in road vehicles. The major components of automotive engines are discussed in this chapter.

Chapter 6- Propulsion is a way of generating force leading to movement. The features related to propulsion are ground propulsion, maglev, marine propulsion, spacecraft propulsion and jet propulsion. The major components of propulsion are discussed in this chapter.

At the end, I would like to thank all those who dedicated their time and efforts for the successful completion of this book. I also wish to convey my gratitude towards my friends and family who supported me at every step.

Editor

Introduction to Automotive Engines

Engines are machines that are designed to convert energy into mechanical energy. Heat engines are majorly used to burn a fuel and then this fuel is used to create heat. This chapter will provide an integrated understanding of automotive engines.

Engine

A V6 internal combustion engine from a Mercedes car

An engine or motor is a machine designed to convert one form of energy into mechanical energy. Heat engines burn a fuel to create heat, which then creates a force. Electric motors convert electrical energy into mechanical motion; pneumatic motors use compressed air and clockwork motors in wind-up toys use elastic energy. In biological systems, molecular motors, like myosins in muscles, use chemical energy to create forces and eventually motion.

Terminology

The word "engine" derives from Old French engin, from the Latin ingenium—the root of the word ingenious. Pre-industrial weapons of war, such as catapults, trebuchets and battering rams, were called "siege engines", and knowledge of how to construct them was often treated as a military secret. The word "gin", as in "cotton gin", is short for "engine". Most mechanical devices invented during the industrial revolution were described as engines—the steam engine being a notable example. However, the original steam engines, such as those by Thomas Savery, were not mechanical engines but pumps. In this manner, a fire engine in its original form was merely a water pump, with the engine being transported to the fire by horses.

In modern usage, the term engine typically describes devices, like steam engines and internal combustion engines, that burn or otherwise consume fuel to perform mechanical work by exerting a torque or linear force (usually in the form of thrust). Examples of engines which exert a torque include the familiar automobile gasoline and diesel engines, as well as turboshafts. Examples of engines which produce thrust include turbofans and rockets.

When the internal combustion engine was invented, the term "motor" was initially used to distinguish it from the steam engine—which was in wide use at the time, powering locomotives and other vehicles such as steam rollers. "Motor" and "engine" later came to be used interchangeably in casual discourse. However, technically, the two words have different meanings. An engine is a device that burns or otherwise consumes fuel, changing its chemical composition, whereas a motor is a device driven by electricity, air, or hydraulic pressure, which does not change the chemical composition of its energy source. However, rocketry uses the term rocket motor, even though they consume fuel.

A heat engine may also serve as a prime mover—a component that transforms the flow or changes in pressure of a fluid into mechanical energy. An automobile powered by an internal combustion engine may make use of various motors and pumps, but ultimately all such devices derive their power from the engine. Another way of looking at it is that a motor receives power from an external source, and then converts it into mechanical energy, while an engine creates power from pressure (derived directly from the explosive force of combustion or other chemical reaction, or secondarily from the action of some such force on other substances such as air, water, or steam).

Devices converting heat energy into motion are commonly referred to simply as engines.

History

Antiquity

Simple machines, such as the club and oar (examples of the lever), are prehistoric. More complex engines using human power, animal power, water power, wind power and even steam power date back to antiquity. Human power was focused by the use of simple engines, such as the capstan, windlass or treadmill, and with ropes, pulleys, and block and tackle arrangements; this power was transmitted usually with the forces multiplied and the speed reduced. These were used in cranes and aboard ships in Ancient Greece, as well as in mines, water pumps and siege engines in Ancient Rome. The writers of those times, including Vitruvius, Frontinus and Pliny the Elder, treat these engines as commonplace, so their invention may be more ancient. By the 1st century AD, cattle and horses were used in mills, driving machines similar to those powered by humans in earlier times.

According to Strabo, a water powered mill was built in Kaberia of the kingdom of Mithridates during the 1st century BC. Use of water wheels in mills spread throughout the Roman Empire over the next few centuries. Some were quite complex, with aqueducts, dams, and sluices to maintain and channel the water, along with systems of gears, or toothed-wheels made of wood and metal to regulate the speed of rotation. More sophisticated small devices, such as the Antikythera Mechanism used complex trains of gears and dials to act as calendars or predict astronomical events. In a poem by Ausonius in the 4th century AD, he mentions a stone-cutting saw powered by water. Hero of Alexandria is credited with many such wind and steam powered machines in the 1st century

AD, including the Aeolipile and the vending machine, often these machines were associated with worship, such as animated altars and automated temple doors.

Medieval

Medieval Muslim engineers employed gears in mills and water-raising machines, and used dams as a source of water power to provide additional power to watermills and water-raising machines. In the medieval Islamic world, such advances made it possible to mechanize many industrial tasks previously carried out by manual labour.

In 1206, al-Jazari employed a crank-conrod system for two of his water-raising machines. A rudimentary steam turbine device was described by Taqi al-Din in 1551 and by Giovanni Branca in 1629.

In the 13th century, the solid rocket motor was invented in China. Driven by gunpowder, this, the simplest form of internal combustion engine was unable to deliver sustained power, but was useful for propelling weaponry at high speeds towards enemies in battle and for fireworks. After invention, this innovation spread throughout Europe.

Industrial Revolution

Boulton & Watt engine of 1788

The Watt steam engine was the first type of steam engine to make use of steam at a pressure just above atmospheric to drive the piston helped by a partial vacuum. Improving on the design of the 1712 Newcomen steam engine, the Watt steam engine, developed sporadically from 1763 to 1775, was a great step in the development of the steam engine. Offering a dramatic increase in fuel efficiency, James Watt's design became synonymous with steam engines, due in no small part to his business partner, Matthew Boulton. It enabled rapid development of efficient semi-automated factories on a previously unimaginable scale in places where waterpower was not available. Later development led to steam locomotives and great expansion of railway transportation.

As for internal combustion piston engines, these were tested in France in 1807 by de Rivaz and independently, by the Niépce brothers. They were theoretically advanced by Carnot in 1824. In 1853-57 Eugenio Barsanti and Felice Matteucci invented and patented an engine using the free-piston principle that was possibly the first 4-cycle engine.

The invention of an internal combustion engine which was later commercially successful was made during 1860 by Etienne Lenoir.

The Otto cycle in 1877 was capable of giving a far higher power to weight ratio than steam engines and worked much better for many transportation applications such as cars and aircraft.

Automobiles

The first commercially successful automobile, created by Karl Benz, added to the interest in light and powerful engines. The lightweight petrol internal combustion engine, operating on a four-stroke Otto cycle, has been the most successful for light automobiles, while the more efficient Diesel engine is used for trucks and buses. However, in recent years, turbo Diesel engines have become increasingly popular, especially outside of the United States, even for quite small cars.

Horizontally Opposed Pistons

In 1896, Karl Benz was granted a patent for his design of the first engine with horizontally opposed pistons. His design created an engine in which the corresponding pistons move in horizontal cylinders and reach top dead center simultaneously, thus automatically balancing each other with respect to their individual momentum. Engines of this design are often referred to as flat engines because of their shape and lower profile. They were used in the Volkswagen Beetle, some Porsche and Subaru cars, many BMW and Honda motorcycles, and aircraft engines (for propeller driven aircraft).

Advancement

Continuance of the use of the internal combustion engine for automobiles is partly due to the improvement of engine control systems (onboard computers providing engine management processes, and electronically controlled fuel injection). Forced air induction by turbocharging and supercharging have increased power outputs and engine efficiencies. Similar changes have been applied to smaller diesel engines giving them almost the same power characteristics as petrol engines. This is especially evident with the popularity of smaller diesel engine propelled cars in Europe. Larger diesel engines are still often used in trucks and heavy machinery, although they require special machining not available in most factories. Diesel engines produce lower hydrocarbon and CO_2 emissions, but greater particulate and NOx pollution, than gasoline engines. Diesel engines are also 40% more fuel efficient than comparable gasoline engines.

Increasing Power

The first half of the 20th century saw a trend to increasing engine power, particularly in the American models. Design changes incorporated all known methods of raising engine capacity, including increasing the pressure in the cylinders to improve efficiency, increasing the size of the engine, and increasing the rate at which the engine produces work. The higher forces and pressures created by these changes created engine vibration and size problems that led to stiffer, more compact engines with V and opposed cylinder layouts replacing longer straight-line arrangements.

Combustion Efficiency

The design principles favoured in Europe, because of economic and other restraints such as smaller and twistier roads, leant toward smaller cars and corresponding to the design principles that concentrated on increasing the combustion efficiency of smaller engines. This produced more economical engines with earlier four-cylinder designs rated at 40 horsepower (30 kW) and six-cylinder designs rated as low as 80 horsepower (60 kW), compared with the large volume V-8 American engines with power ratings in the range from 250 to 350 hp, some even over 400 hp (190 to 260 kW).

Engine Configuration

Earlier automobile engine development produced a much larger range of engines than is in common use today. Engines have ranged from 1- to 16-cylinder designs with corresponding differences in overall size, weight, engine displacement, and cylinder bores. Four cylinders and power ratings from 19 to 120 hp (14 to 90 kW) were followed in a majority of the models. Several three-cylinder, two-stroke-cycle models were built while most engines had straight or in-line cylinders. There were several V-type models and horizontally opposed two- and four-cylinder makes too. Overhead camshafts were frequently employed. The smaller engines were commonly air-cooled and located at the rear of the vehicle; compression ratios were relatively low. The 1970s and 1980s saw an increased interest in improved fuel economy, which caused a return to smaller V-6 and four-cylinder layouts, with as many as five valves per cylinder to improve efficiency. The Bugatti Veyron 16.4 operates with a W16 engine, meaning that two V8 cylinder layouts are positioned next to each other to create the W shape sharing the same crankshaft.

The largest internal combustion engine ever built is the Wärtsilä-Sulzer RTA96-C, a 14-cylinder, 2-stroke turbocharged diesel engine that was designed to power the Emma Mærsk, the largest container ship in the world. This engine weighs 2,300 tons, and when running at 102 RPM produces 109,000 bhp (80,080 kW) consuming some 13.7 tons of fuel each hour.

Types

An engine can be put into a category according to two criteria: the form of energy it accepts in order to create motion, and the type of motion it outputs.

Heat Engine

Combustion Engine

Combustion engines are heat engines driven by the heat of a combustion process.

Internal Combustion Engine

The internal combustion engine is an engine in which the combustion of a fuel (generally, fossil fuel) occurs with an oxidizer (usually air) in a combustion chamber. In an internal combustion engine the expansion of the high temperature and high pressure gases, which are produced by the combustion, directly applies force to components of the engine, such as the pistons or turbine blades or a nozzle, and by moving it over a distance, generates useful mechanical energy.

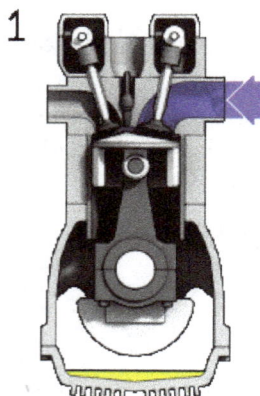

Animation showing the four stages of the four-stroke combustion engine cycle:

1. Induction (Fuel enters)

2. Compression

3. Ignition (Fuel is burnt)

4. Emission (Exhaust out)

External Combustion Engine

An external combustion engine (EC engine) is a heat engine where an internal working fluid is heated by combustion of an external source, through the engine wall or a heat exchanger. The fluid then, by expanding and acting on the mechanism of the engine produces motion and usable work. The fluid is then cooled, compressed and reused (closed cycle), or (less commonly) dumped, and cool fluid pulled in (open cycle air engine).

"Combustion" refers to burning fuel with an oxidizer, to supply the heat. Engines of similar (or even identical) configuration and operation may use a supply of heat from other sources such as nuclear, solar, geothermal or exothermic reactions not involving combustion; but are not then strictly classed as external combustion engines, but as external thermal engines.

The working fluid can be a gas as in a Stirling engine, or steam as in a steam engine or an organic liquid such as n-pentane in an Organic Rankine cycle. The fluid can be of any composition; gas is by far the most common, although even single-phase liquid is sometimes used. In the case of the steam engine, the fluid changes phases between liquid and gas.

Air-breathing Combustion Engines

Air-breathing combustion engines are combustion engines that use the oxygen in atmospheric air to oxidise ('burn') the fuel, rather than carrying an oxidiser, as in a rocket. Theoretically, this should result in a better specific impulse than for rocket engines.

A continuous stream of air flows through the air-breathing engine. This air is compressed, mixed with fuel, ignited and expelled as the exhaust gas.

Examples

Typical air-breathing engines include:

- Reciprocating engine
- Steam engine
- Gas turbine

  airbreathing jet engine

  Turbo-propeller engine
- Pulse detonation engine
- Pulse jet
- Ramjet
- Scramjet
- Liquid air cycle engine/Reaction Engines SABRE.

Environmental Effects

The operation of engines typically has a negative impact upon air quality and ambient sound levels. There has been a growing emphasis on the pollution producing features of automotive power systems. This has created new interest in alternate power sources and internal-combustion engine refinements. Though a few limited-production battery-powered electric vehicles have appeared, they have not proved competitive owing to costs and operating characteristics. In the 21st century the diesel engine has been increasing in popularity with automobile owners. However, the gasoline engine and the Diesel engine, with their new emission-control devices to improve emission performance, have not yet been significantly challenged. A number of manufacturers have introduced hybrid engines, mainly involving a small gasoline engine coupled with an electric motor and with a large battery bank, but these too have yet to make much of an inroad into the market shares of gasoline and Diesel engines.

Air Quality

Exhaust from a spark ignition engine consists of the following: nitrogen 70 to 75% (by volume), water vapor 10 to 12%, carbon dioxide 10 to 13.5%, hydrogen 0.5 to 2%, oxygen 0.2 to 2%, carbon monoxide: 0.1 to 6%, unburnt hydrocarbons and partial oxidation products (e.g. aldehydes) 0.5 to 1%, nitrogen monoxide 0.01 to 0.4%, nitrous oxide <100 ppm, sulfur dioxide 15 to 60 ppm, traces of other compounds such as fuel additives and lubricants, also halogen and metallic compounds, and other particles. Carbon monoxide is highly toxic, and can cause carbon monoxide poisoning, so it is important to avoid any build-up of the gas in a confined space. Catalytic converters can reduce toxic emissions, but not completely eliminate them. Also, resulting greenhouse gas emissions, chiefly carbon dioxide, from the widespread use of engines in the modern industrialized world is contributing to the global greenhouse effect – a primary concern regarding global warming.

Non-combusting Heat Engines

Some engines convert heat from noncombustive processes into mechanical work, for example a nuclear power plant uses the heat from the nuclear reaction to produce steam and drive a steam engine, or a gas turbine in a rocket engine may be driven by decomposing hydrogen peroxide. Apart from the different energy source, the engine is often engineered much the same as an internal or external combustion engine. Another group of noncombustive engines includes thermoacoustic heat engines (sometimes called "TA engines") which are thermoacoustic devices which use high-amplitude sound waves to pump heat from one place to another, or conversely use a heat difference to induce high-amplitude sound waves. In general, thermoacoustic engines can be divided into standing wave and travelling wave devices.

Non-thermal Chemically Powered Motor

Non-thermal motors usually are powered by a chemical reaction, but are not heat engines. Examples include:

- Molecular motor - motors found in living things
- Synthetic molecular motor.

Electric Motor

An electric motor uses electrical energy to produce mechanical energy, usually through the interaction of magnetic fields and current-carrying conductors. The reverse process, producing electrical energy from mechanical energy, is accomplished by a generator or dynamo. Traction motors used on vehicles often perform both tasks. Electric motors can be run as generators and vice versa, although this is not always practical. Electric motors are ubiquitous, being found in applications as diverse as industrial fans, blowers and pumps, machine tools, household appliances, power tools, and disk drives. They may be powered by direct current (for example a battery powered portable device or motor vehicle), or by alternating current from a central electrical distribution grid. The smallest motors may be found in electric wristwatches. Medium-size motors of highly standardized dimensions and characteristics provide convenient mechanical power for industrial uses. The very largest electric motors are used for propulsion of large ships, and for such purposes as pipeline compressors, with ratings in the thousands of kilowatts. Electric motors may be classified by the source of electric power, by their internal construction, and by their application.

The physical principle of production of mechanical force by the interactions of an electric current and a magnetic field was known as early as 1821. Electric motors of increasing efficiency were constructed throughout the 19th century, but commercial exploitation of electric motors on a large scale required efficient electrical generators and electrical distribution networks.

To reduce the electric energy consumption from motors and their associated carbon footprints, various regulatory authorities in many countries have introduced and implemented legislation to encourage the manufacture and use of higher efficiency electric motors. A well-designed motor can convert over 90% of its input energy into useful power for decades. When the efficiency of a motor is raised by even a few percentage points, the savings, in kilowatt hours (and therefore in cost), are enormous. The electrical energy efficiency of a typical industrial induction motor can be improved by: 1) reducing the electrical losses in the stator windings (e.g., by increasing the cross-sectional area of the conductor, improving the winding technique, and using materials with higher electrical conductivities, such as copper), 2) reducing the electrical losses in the rotor coil or casting (e.g., by using materials with higher electrical conductivities, such as copper), 3) reducing magnetic losses by using better quality magnetic steel, 4) improving the aerodynamics of motors to reduce mechanical windage losses, 5) improving bearings to reduce friction losses, and 6) minimizing manufacturing tolerances.

By convention, electric engine refers to a railroad electric locomotive, rather than an electric motor.

Physically Powered Motor

Some motors are powered by potential or kinetic energy, for example some funiculars, gravity plane and ropeway conveyors have used the energy from moving water or rocks, and some clocks have a weight that falls under gravity. Other forms of potential energy include compressed gases (such as pneumatic motors), springs (clockwork motors) and elastic bands.

Historic military siege engines included large catapults, trebuchets, and (to some extent) battering rams were powered by potential energy.

Pneumatic Motor

A pneumatic motor is a machine that converts potential energy in the form of compressed air into mechanical work. Pneumatic motors generally convert the compressed air to mechanical work though either linear or rotary motion. Linear motion can come from either a diaphragm or piston actuator, while rotary motion is supplied by either a vane type air motor or piston air motor. Pneumatic motors have found widespread success in the hand-held tool industry and continual attempts are being made to expand their use to the transportation industry. However, pneumatic motors must overcome efficiency deficiencies before being seen as a viable option in the transportation industry.

Hydraulic Motor

A hydraulic motor is one that derives its power from a pressurized fluid. This type of engine can be used to move heavy loads or produce motion.

Performance

Engine Speed

In the case of engines outputting shaft power, engine speed is measured in revolutions per minute (RPM). Engines may be classified as low-speed, medium-speed or high-speed but these terms are inexact and depend on the type of engine being described. Generally, diesel engines operate at lower speed compared to gasoline engines. Electric motors and turboshafts are capable of very high speeds. In the case of engines producing thrust, it is rather inaccurate to talk of an 'engine speed' since what is moving is not the engine, but the working medium that the engine is accelerating; in this case one talks of an exhaust velocity, which is exactly the I_{sp} outside of a gravitational field and therefore makes one jump straight to a discussion of efficiency.

Thrust

Thrust is the force arising from the interaction between two masses which exert equal but opposite forces on each other due to their speed. The force F can be measured either in newtons (N, SI units) or in pounds-thrust (lb_f, imperial units).

Torque

Torque is the force being exerted on a theoretical lever connected to the output shaft of an engine. This is expressed by the formula:

$$\tau = |\mathbf{r} \times \mathbf{F}| = rF \sin(\mathbf{r}, \mathbf{F})$$

where r is the length of the lever, F is the force applied on it, and r×F is the vector cross product. Torque is measured typically either in newton-metres (N·m, SI units) or in foot-pounds (ft·lb, imperial units).

Power

Power is the amount of work being done, or energy being produced, per unit of time. This is expressed by the formula:

$$P = \frac{\mathrm{d}W}{\mathrm{d}t}$$

With a quick demonstration, it can be shown that:

$$P = \mathbf{F} \cdot \mathbf{v}$$

This formula with linear forces and speeds can be used equally well for both engines outputting thrust and engines exerting torque.

When considering propulsive engines, typically only the raw force of the core mass flow is considered, leading to such engines having their 'power' rated in any of the units discussed above for forces.

If the engine in question outputs its power on a shaft, then:

$$P = \tau\omega.$$

This is the reason why any engine outputting its power on a rotating shaft is usually quoted, along with its rated power, the rotational speed at which that rated power is developed.

Efficiency

Depending on the type of engine employed, different rates of efficiency are attained.

For heat engines, efficiency cannot be greater than the Carnot efficiency.

Sound Levels

In the case of sound levels, engine operation is of greatest impact with respect to mobile sources such as automobiles and trucks. Engine noise is a particularly large component of mobile source noise for vehicles operating at lower speeds, where aerodynamic and tire noise is less significant. Generally speaking, petrol and diesel engines emit less noise than turboshafts of equivalent power output; electric motors very often emit less noise than their fossil fuel-powered equivalents. Thrust-outputting engines, such as turbofans, turbojets and rockets emit the greatest amount of noise because their method of producing thrust is directly related to the production of sound. Various methods have been devised to reduce noise. Petrol and diesel engines are fitted with mufflers (silencers); newer turbofans often have outsized fans (the so-called high-bypass technology) in order to reduce the proportion of noisy, hot exhaust from the integrated turboshaft in the exhaust stream, and hushkits exist for older, low-bypass turbofans. No known methods exist for reducing the noise output of rockets without a corresponding reduction in thrust.

Engines by use

Particularly notable kinds of engines include:

- Aircraft engine
- Automobile engine
- Model engine
- Motorcycle engine
- Marine propulsion engines such as Outboard motor
- Non-road engine is the term used to define engines that are not used by vehicles on roadways.
- Railway locomotive engine
- Spacecraft propulsion engines such as Rocket engine
- Traction engine

Automotive Engine

As of 2013 there were a wide variety of propulsion systems available or potentially available for automobiles and other vehicles. Options included internal combustion engines fueled by petrol, diesel, propane, or natural gas; hybrid vehicles, plug-in hybrids, fuel cell vehicles fueled by hydrogen and all electric cars. Fueled vehicles seemed to have the short term advantage due to the limited range and high cost of batteries. Some options required construction of a network of fueling or charging stations. With no compelling advantage for any particular option car makers pursued parallel development tracks using a variety of options. Reducing the weight of vehicles was one strategy being employed.

Internal combustion engines, like the 1.6 litre (98 cubic inch) petrol engine from 2009 seen here, have been the dominant propulsion system for most of the history of automobiles

Recent Developments

The use of high-technology (such as electronic engine control units) in advanced designs resulting from substantial investments in development research by European and Japanese countries seemed to give an advantage to them over Chinese automakers and parts suppliers who, as of 2013, had low development budgets and lacked capacity to produce parts for high-tech engine and power train designs.

Characteristics

The chief characteristic of an automotive engine (compared to a stationary engine or a marine engine) is a high power-to-weight ratio. This is achieved by using a high rotational speed. However, automotive engines are sometimes modified for marine use, forming a marine automobile engine.

History

In the early years, steam engines and electric motors were tried, but with limited success. In the 20th century, the internal combustion (ic) engine became dominant. In 2015, the ic engine remains the most widely used but a resurgence of electricity seems likely because of increasing concern about ic engine exhaust gas emissions.

References

- Cardwell, Diane; Krauss, Clifford (April 22, 2013). "Trucking Industry Is Set to Expand Its Use of Natural Gas". The New York Times. Retrieved April 23, 2013.

- "Propulsion systems The great powertrain race Carmakers are hedging their bets on powering cars". The Economist (print ed.). April 20, 2013. Retrieved April 19, 2013.

Vehicle Engineering: An Overview

Vehicle engineering includes fields such as automotive engineering, vehicle dynamics, rolling stock and noise, vibration and harshness. The components of vehicle dynamics are automobile layout, electronics stability control, steering and suspension. This section is an overview of the subject matter incorporating all the major aspects of vehicle engineering.

Vehicle Engineering

Vehicle engineering encompasses the fields of automotive engineering, aerospace engineering, rolling stock and marine engineering.

Automotive Engineering

Automotive engineering is the design, manufacture and operation of motorcycles, automobiles and trucks and their respective engineering subsystems. Automotive engineering is one of the most exciting professions you can choose. From the global concerns of sustainable mobility, and teaching cars to drive themselves, to working out how we'll get around on the surface of Mars, automotive engineering is all about the future. The work of an automobile down into three categories:

Design: Designing new products and improving existing ones

Research and Development: Finding solutions to engineering problems

Production: Planning and designing new production processes

Aerospace Engineering

Aerospace engineering is the branch of engineering that deals with the design, development, testing, and production of aircraft and related systems (aeronautical engineering) and of spacecraft, missiles, rocket-propulsion systems, and other equipment operating beyond the earth's atmosphere(astronautical engineering) Further concerned with the science of force and physics that are particular only to performance in Earth's atmosphere and the expanse of space.

Aeronautics

Definition

Aeronautics is the study of the science of flight. Aeronautics is the method of designing an airplane or other flying machine. There are four basic areas that aeronautical engineers must understand in order to be able to design planes. To design a plane, engineers must understand all of these elements.

Design Process

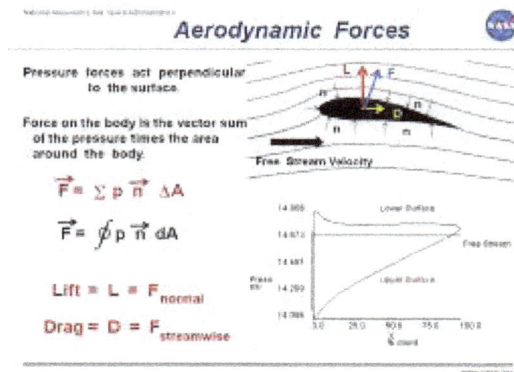

- Aerodynamics: is the study of how air flows around the airplane. By studying the way air flows around the plane the engineers can define the shape of the plane. The wings, the tail, and the main body or fuselage of the plane all affect the way the air will move around the plane.

- Propulsion: is the study of how to design an engine that will provide the thrust that is needed for a plane to take off and fly through the air. The engine provides the power for the airplane. The study of propulsion is what leads the engineers to determine the right kind of engine and the right amount of power that a plane will need

- Materials and Structures: is the study of what materials are to be used on the plane and in the engine and how those materials make the plane strong enough to fly effectively. The choice of materials that are used to make the fuselage wings, tail and engine will affect the strength and stability of the plane. Many airplane materials are now made out of composites, materials that are stronger than most metals and are lightweight.

- Stability and Control: is the study of how to control the speed, direction, altitude and other conditions that affect how a plane flies. The engineers design the controls that are needed in order to fly and instruments are provided for the pilot in the cockpit of the plane. The pilot uses these instruments to control the stability of the plane during flight.

Astronautics

Astronautics is the design and development of spacecraft with an emphasis on spacecraft systems, the design of ground control systems for spacecraft, and the design of orbital mechanics for spacecraft missions.

Rolling Stock

Rolling stock comprises all the vehicles that move on a railway. It usually includes both powered and unpowered vehicles, for example locomotives, railroad cars, coaches, and wagons.

Naval Architecture

Naval architecture also known as Naval engineering is an engineering discipline dealing with the design, construction, maintenance and operation of marine vessels and structures.

Automotive Engineering

Automobile engineering, along with aerospace engineering and marine engineering, is a branch of vehicle engineering, incorporating elements of mechanical, electrical, electronic, software and safety engineering as applied to the design, manufacture and operation of motorcycles, automobiles and trucks and their respective engineering subsystems.

Disciplines

Automobile Engineering

Automobile Engineering is a branch study of engineering which teaches manufacturing, designing, mechanical-mechanisms as well operations of automobiles. It is an introduction to vehicle engineering which deals with motorcycles, cars, buses trucks etc. It include branch study of mechanical, electronic, software and safety elements. Some of the engineering attributes and disciplines that are of importance to the automotive engineer and many of the other aspects are included in it:

Safety engineering: Safety engineering is the assessment of various crash scenarios and their impact on the vehicle occupants. These are tested against very stringent governmental regulations. Some of these requirements include: seat belt and air bag functionality testing, front and side impact testing, and tests of rollover resistance. Assessments are done with various methods and tools, including Computer crash simulation (typically finite element analysis), crash test dummies, and partial system sled and full vehicle crashes.

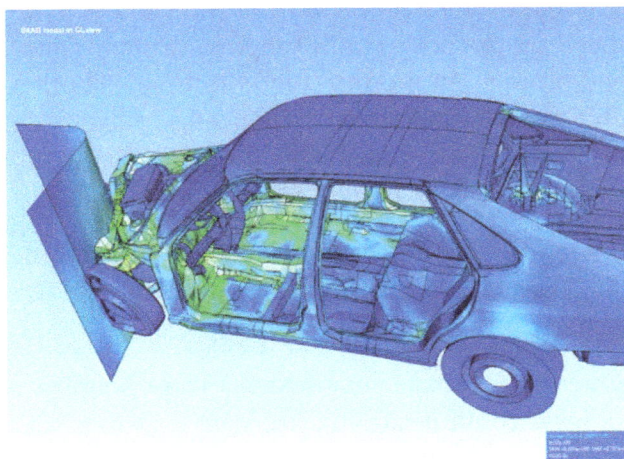

Visualization of how a car deforms in an asymmetrical crash using finite element analysis.

Fuel economy/emissions: Fuel economy is the measured fuel efficiency of the vehicle in miles per gallon or kilometers per litre. Emissions testing includes the measurement of vehicle emissions, including hydrocarbons, nitrogen oxides (NOx), carbon monoxide (CO), carbon dioxide (CO2), and evaporative emissions.

Vehicle dynamics: Vehicle dynamics is the vehicle's response of the following attributes: ride, handling, steering, braking, comfort and traction. Design of the chassis systems of suspension, steering, braking, structure (frame), wheels and tires, and traction control are highly leveraged by the vehicle dynamics engineer to deliver the vehicle dynamics qualities desired.

NVH engineering (noise, vibration, and harshness): NVH is the customer's feedback (both tactile [felt] and audible [heard]) from the vehicle. While sound can be interpreted as a rattle, squeal, or hoot, a tactile response can be seat vibration, or a buzz in the steering wheel. This feedback is generated by components either rubbing, vibrating, or rotating. NVH response can be classified in various ways: powertrain NVH, road noise, wind noise, component noise, and squeak and rattle. Note there are both good and bad NVH qualities. The NVH engineer works to either eliminate bad NVH, or change the "bad NVH" to good (i.e., exhaust tones).

Vehicle Electronics: Automotive electronics is an increasingly important aspect of automotive engineering. Modern vehicles employ dozens of electronic systems. These systems are responsible for operational controls such as the throttle, brake and steering controls; as well as many comfort and convenience systems such as the HVAC, infotainment, and lighting systems. It would not be possible for automobiles to meet modern safety and fuel economy requirements without electronic controls.

Performance: Performance is a measurable and testable value of a vehicles ability to perform in various conditions. Performance can be considered in a wide variety of tasks, but it's generally associated with how quickly a car can accelerate (e.g. standing start 1/4 mile elapsed time, 0–60 mph, etc.), its top speed, how short and quickly a car can come to a complete stop from a set speed (e.g. 70-0 mph), how much g-force a car can generate without losing grip, recorded lap times, cornering speed, brake fade, etc. Performance can also reflect the amount of control in inclement weather (snow, ice, rain).

Shift quality: Shift quality is the driver's perception of the vehicle to an automatic transmission shift event. This is influenced by the powertrain (engine, transmission), and the vehicle (driveline, suspension, engine and powertrain mounts, etc.) Shift feel is both a tactile (felt) and audible (heard) response of the vehicle. Shift quality is experienced as various events: Transmission shifts are felt as an upshift at acceleration (1–2), or a downshift maneuver in passing (4–2). Shift engagements of the vehicle are also evaluated, as in Park to Reverse, etc.

Durability / corrosion engineering: Durability and corrosion engineering is the evaluation testing of a vehicle for its useful life. Tests include mileage accumulation, severe driving conditions, and corrosive salt baths.

Package / ergonomics engineering: Package engineering is a discipline that designs/analyzes the occupant accommodations (seat roominess), ingress/egress to the vehicle, and the driver's field of vision (gauges and windows). The package engineer is also responsible for other areas of the vehicle like the engine compartment, and the component to component placement. Ergonomics is the discipline that assesses the occupant's access to the steering wheel, pedals, and other driver/passenger controls.

Climate control: Climate control is the customer's impression of the cabin environment and level of comfort related to the temperature and humidity. From the windshield defrosting, to the heating and cooling capacity, all vehicle seating positions are evaluated to a certain level of comfort.

Drivability: Drivability is the vehicle's response to general driving conditions. Cold starts and stalls, RPM dips, idle response, launch hesitations and stumbles, and performance levels.

Cost: The cost of a vehicle program is typically split into the effect on the variable cost of the vehicle, and the up-front tooling and fixed costs associated with developing the vehicle. There are also costs associated with warranty reductions, and marketing.

Program timing: To some extent programs are timed with respect to the market, and also to the production schedules of the assembly plants. Any new part in the design must support the development and manufacturing schedule of the model.

Assembly feasibility: It is easy to design a module that is hard to assemble, either resulting in damaged units, or poor tolerances. The skilled product development engineer works with the assembly/manufacturing engineers so that the resulting design is easy and cheap to make and assemble, as well as delivering appropriate functionality and appearance.

Quality management: Quality control is an important factor within the production process, as high quality is needed to meet customer requirements and to avoid expensive recall campaigns. The complexity of components involved in the production process requires a combination of different tools and techniques for quality control. Therefore, the International Automotive Task Force (IATF), a group of the world's leading manufacturers and trade organizations, developed the standard ISO/TS 16949. This standard defines the design, development, production, and when relevant, installation and service requirements. Furthermore, it combines the principles of ISO 9001 with aspects of various regional and national automotive standards such as AVSQ (Italy), EAQF (France), VDA6 (Germany) and QS-9000 (USA). In order to further minimize risks related to product failures and liability claims of automotive electric and electronic systems, the quality discipline functional safety according to ISO/IEC 17025 is applied.

Since the 1950s, the comprehensive business approach total quality management, TQM, helps to continuously improve the production process of automotive products and components. Some of the companies who have implemented TQM include Ford Motor Company, Motorola and Toyota Motor Company.

Job Functions

Development Engineer

A development engineer has the responsibility for coordinating delivery of the engineering attributes of a complete automobile (bus, car, truck, van, SUV, motorcycle etc.) as dictated by the automobile manufacturer, governmental regulations, and the customer who buys the product.

Much like the Systems Engineer, the development engineer is concerned with the interactions of all systems in the complete automobile. While there are multiple components and systems in an automobile that have to function as designed, they must also work in harmony with the complete automobile. As an example, the brake system's main function is to provide braking functionality to the automobile. Along with this, it must also provide an acceptable level of: pedal feel (spongy, stiff), brake system "noise" (squeal, shudder, etc.), and interaction with the ABS (anti-lock braking system)

Another aspect of the development engineer's job is a trade-off process required to deliver all of the automobile attributes at a certain acceptable level. An example of this is the trade-off between engine performance and fuel economy. While some customers are looking for maximum power from their engine, the automobile is still required to deliver an acceptable level of fuel economy. From the engine's perspective, these are opposing requirements. Engine performance is looking for maximum displacement (bigger, more power), while fuel economy is looking for a smaller displacement engine (ex: 1.4 L vs. 5.4 L). The engine size however, is not the only contributing factor to fuel economy and automobile performance. Different values come into play.

Other attributes that involve trade-offs include: automobile weight, aerodynamic drag, transmission gearing, emission control devices, handling/roadholding, ride quality, and tires.

The development engineer is also responsible for organizing automobile level testing, validation, and certification. Components and systems are designed and tested individually by the Product Engineer. The final evaluation is to be conducted at the automobile level to evaluate system to system interactions. As an example, the audio system (radio) needs to be evaluated at the automobile level. Interaction with other electronic components can cause interference. Heat dissipation of the system and ergonomic placement of the controls need to be evaluated. Sound quality in all seating positions needs to be provided at acceptable levels.

Manufacturing Engineer

Manufacturing Engineers are responsible for ensuring proper production of the automotive components or complete vehicles. While the development engineers are responsible for the function of the vehicle, manufacturing engineers are responsible for the safe and effective production of the vehicle. This group of engineers consist of Process Engineers, Logistic Coordinators, Tooling Engineers, Robotics Engineers, and Assembly Planners.

In the automotive industry manufacturers are playing a larger role in the development stages of automotive components to ensure that the products are easy to manufacture. Design for Manufacturability in the automotive world is crucial to make certain whichever design is developed in the Research and Development Stage of automotive design. Once the design is established, the manufacturing engineers take over. They design the machinery and tooling necessary to build the automotive components or vehicle and establish the methods of how to mass-produce the product. It is the manufacturing engineers job to increase the efficiency of the automotive plant and to implement lean manufacturing techniques such as Six Sigma and Kaizen.

Other Automotive Engineering Roles

Other automotive engineers include those listed below:

- Aerodynamics engineers will often give guidance to the styling studio so that the shapes they design are aerodynamic, as well as attractive.

- Body engineers will also let the studio know if it is feasible to make the panels for their designs.

- Change control engineers make sure that all of the design and manufacturing changes that occur are organized, managed and implemented.

- Acoustics engineers are specific types of development engineers who do sound and aerodynamic testing to prevent loud cabin noises while the vehicle is on the road.

The Modern Automotive Product Engineering Process

Studies indicate that a substantial part of the modern vehicle's value comes from intelligent systems, and that these represent most of the current automotive innovation. To facilitate this, the modern automotive engineering process has to handle an increased use of mechatronics. Configuration and performance optimization, system integration, control, component, subsystem and system-level validation of the intelligent systems must become an intrinsic part of the standard vehicle engineering process, just as this is the case for the structural, vibro-acoustic and kinematic design. This requires a vehicle development process that is typically highly simulation-driven.

The V-approach

One way to effectively deal with the inherent multi-physics and the control systems development that is involved when including intelligent systems, is to adopt the V-Model approach to systems development, as has been widely used in the automotive industry for twenty years or more. In this V-approach, system-level requirements are propagated down the V via subsystems to component design, and the system performance is validated at increasing integration levels. Engineering of mechatronic systems requires the application of two interconnected "V-cycles": one focusing on the multi-physics system engineering (like the mechanical and electrical components of an electrically powered steering system, including sensors and actuators); and the other focuses on the controls engineering, the control logic, the software and realization of the control hardware and embedded software.

Predictive Engineering Analytics

An alternative approach is called predictive engineering analytics, and takes the V-approach to the next level. It lets design continue after product delivery. That is important for development of built-in predictive functionality and for creating vehicles that can be optimized while being in use, even based on real use data. This approach is based on the creation of a Digital Twin, a replica of the real product that remains in-sync. Manufacturers try to achieve this by implementing a set of development tactics and tools. Critical is a strong alignment of 1D systems simulation, 3D CAE an physical testing to reach more realism in the simulation process. This is combined with intelligent reporting and data analytics for better insight in the vehicle use. By supporting this with a strong data management structure that spans the entire product lifecycle, they bridge the gap between design, manufacturing and product use.

Aerospace Engineering

Aerospace engineering is the primary field of engineering concerned with the development of aircraft and spacecraft. It has two major and overlapping branches: aeronautical engineering and astronautical engineering.

Aeronautical engineering was the original term for the field. As flight technology advanced to include craft operating in outer space (astronautics), the broader term "aerospace engineering" has largely replaced it in common usage. Aerospace engineering, particularly the astronautics branch, is often colloquially referred to as "rocket science".

Overview

Flight vehicles are subjected to demanding conditions such as those produced by changes in atmospheric pressure and temperature, with structural loads applied upon vehicle components. Consequently, they are usually the products of various technological and engineering disciplines including aerodynamics, propulsion, avionics, materials science, structural analysis and manufacturing. The interaction between these technologies is known as aerospace engineering. Because of the complexity and number of disciplines involved, aerospace engineering is carried out by teams of engineers, each having their own specialized area of expertise.

History

Orville and Wilbur Wright flew the Wright Flyer in 1903 at Kitty Hawk, North Carolina.

The origin of aerospace engineering can be traced back to the aviation pioneers around the late 19th to early 20th centuries, although the work of Sir George Cayley dates from the last decade of the 18th to mid-19th century. One of the most important people in the history of aeronautics, Cayley was a pioneer in aeronautical engineering and is credited as the first person to separate the forces of lift and drag, which are in effect on any flight vehicle. Early knowledge of aeronautical engineering was largely empirical with some concepts and skills imported from other branches of engineering. Scientists understood some key elements of aerospace engineering, like fluid dynamics, in the 18th century. Many years later after the successful flights by the Wright brothers, the 1910s saw the development of aeronautical engineering through the design of World War I military aircraft.

The first definition of aerospace engineering appeared in February 1958. The definition considered the Earth's atmosphere and the outer space as a single realm, thereby encompassing both aircraft (aero) and spacecraft (space) under a newly coined word aerospace. In response to the USSR launching the first satellite, Sputnik into space on October 4, 1957, U.S. aerospace engineers launched the first American satellite on January 31, 1958. The National Aeronautics and Space Administration was founded in 1958 as a response to the Cold War.

Elements

Wernher von Braun, with the F-1 engines of the Saturn V first stage at the US Space and Rocket Center

Soyuz TMA-14M spacecraft engineered for descent by parachute

Some of the elements of aerospace engineering are:

A fighter jet engine undergoing testing. The tunnel behind the engine allows noise and exhaust to escape.

- Radar cross-section – the study of vehicle signature apparent to Radar remote sensing.

- Fluid mechanics – the study of fluid flow around objects. Specifically aerodynamics concerning the flow of air over bodies such as wings or through objects such as wind tunnels.

- Astrodynamics – the study of orbital mechanics including prediction of orbital elements when given a select few variables. While few schools in the United States teach this at the undergraduate level, several have graduate programs covering this topic (usually in conjunction with the Physics department of said college or university).

- Statics and Dynamics (engineering mechanics) – the study of movement, forces, moments in mechanical systems.

- Mathematics – in particular, calculus, differential equations, and linear algebra.

- Electrotechnology – the study of electronics within engineering.

- Propulsion – the energy to move a vehicle through the air (or in outer space) is provided by internal combustion engines, jet engines and turbomachinery, or rockets. A more recent addition to this module is electric propulsion and ion propulsion.

- Control engineering – the study of mathematical modeling of the dynamic behavior of systems and designing them, usually using feedback signals, so that their dynamic behavior is desirable (stable, without large excursions, with minimum error). This applies to the dynamic behavior of aircraft, spacecraft, propulsion systems, and subsystems that exist on aerospace vehicles.

- Aircraft structures – design of the physical configuration of the craft to withstand the forces encountered during flight. Aerospace engineering aims to keep structures lightweight and low-cost, while maintaining structural integrity.

- Materials science – related to structures, aerospace engineering also studies the materials of which the aerospace structures are to be built. New materials with very specific properties are invented, or existing ones are modified to improve their performance.

- Solid mechanics – Closely related to material science is solid mechanics which deals with stress and strain analysis of the components of the vehicle. Nowadays there are several Finite Element programs such as MSC Patran/Nastran which aid engineers in the analytical process.

- Aeroelasticity – the interaction of aerodynamic forces and structural flexibility, potentially causing flutter, divergence, etc.

- Avionics – the design and programming of computer systems on board an aircraft or spacecraft and the simulation of systems.

- Software – the specification, design, development, test, and implementation of computer software for aerospace applications, including flight software, ground control software, test & evaluation software, etc.

- Risk and reliability – the study of risk and reliability assessment techniques and the mathematics involved in the quantitative methods.

- Noise control – the study of the mechanics of sound transfer.

- Aeroacoustics – the study of noise generation via either turbulent fluid motion or aerodynamic forces interacting with surfaces.

- Flight test – designing and executing flight test programs in order to gather and analyze performance and handling qualities data in order to determine if an aircraft meets its design and performance goals and certification requirements.

The basis of most of these elements lies in theoretical physics, such as fluid dynamics for aerodynamics or the equations of motion for flight dynamics. There is also a large empirical component. Historically, this empirical component was derived from testing of scale models and prototypes, either in wind tunnels or in the free atmosphere. More recently, advances in computing have enabled the use of computational fluid dynamics to simulate the behavior of fluid, reducing time and expense spent on wind-tunnel testing. Those studying hydrodynamics or Hydroacoustics often obtained degrees in Aerospace Engineering.

Additionally, aerospace engineering addresses the integration of all components that constitute an aerospace vehicle (subsystems including power, aerospace bearings, communications, thermal control, life support, etc.) and its life cycle (design, temperature, pressure, radiation, velocity, lifetime).

Degree Programs

Aerospace engineering may be studied at the advanced diploma, bachelor's, master's, and Ph.D. levels in aerospace engineering departments at many universities, and in mechanical engineering departments at others. A few departments offer degrees in space-focused astronautical engineering. Some institutions differentiate between aeronautical and astronautical engineering. Graduate degrees are offered in advanced or specialty areas for the aerospace industry.

A background in chemistry, physics, computer science and mathematics is important for students pursuing an aerospace engineering degree.

In Popular Culture

The term "rocket scientist" is sometimes used to describe a person of great intelligence since "rocket science" is seen as a practice requiring great mental ability, especially technical and mathematical ability. The term is used ironically in the expression "It's not rocket science" to indicate that a task is simple. Strictly speaking, the use of "science" in "rocket science" is a misnomer since science is about understanding the origins, nature, and behavior of the universe; engineering is about using scientific and engineering principles to solve problems and develop new technology. However, the media and the public often use "science" and "engineering" as synonyms.

Rolling Stock

The term rolling stock originally referred to any vehicles that move on a railway. It has since expanded to include the wheeled vehicles used by businesses on roadways. It usually includes both powered and unpowered vehicles, for example locomotives, railroad cars, coaches, and wagons.

Variety of rolling stock in rail yard

Overview

Rolling stock is considered to be a liquid asset, or close to it, since the value of the vehicle can be readily estimated and then shipped to the buyer without much cost or delay.

The term contrasts with fixed stock (infrastructure), which is a collective term for the track, signals, stations, other buildings, electric wires, etc., necessary to operate a railway.

Steam and diesel locomotives

Articulated well cars with intermodal containers

Code Names

In Great Britain, types of rolling stock were given code names, often of animals. For example, "Toad" was used as a code name for the Great Western Railway goods brake van, while British Railways wagons used for track maintenance were named after fish, such as "Dogfish" for a ballast hopper. These codes were telegraphese, somewhat analogous to the SMS language of today.

Vehicle Dynamics

Vehicle dynamics refers to the dynamics of vehicles, here assumed to be ground vehicles. Vehicle dynamics is a part of engineering primarily based on classical mechanics.

This article applies primarily to automobiles. For single-track vehicles, specifically the two-wheeled variety.

Components

Components, attributes or aspects of vehicle dynamics include:

- Automobile layout
- Electronic stability control (ESC)
- Steering
- Suspension
- Traction control system (TCS)

Aerodynamic Specific

Some attributes or aspects of vehicle dynamics are purely aerodynamic. These include:

- Automobile drag coefficient
- Automotive aerodynamics
- Center of pressure
- Downforce
- Ground effect in cars

Geometry Specific

Some attributes or aspects of vehicle dynamics are purely geometric. These include:

- Ackermann steering geometry
- Axle track
- Camber angle
- Caster angle
- Ride height
- Roll center
- Scrub radius

- Steering ratio
- Toe
- Wheelbase

Mass Specific

Some attributes or aspects of vehicle dynamics are purely due to mass and its distribution. These include:

- Center of mass
- Moment of inertia
- Roll moment
- Sprung mass
- Unsprung mass
- Weight distribution

Motion Specific

Some attributes or aspects of vehicle dynamics are purely dynamic. These include:

- Body flex
- Body roll
- Bump Steer
- Bundorf analysis
- Directional stability
- Critical speed
- Noise, vibration, and harshness
- Pitch
- Ride quality
- Roll
- Speed wobble
- Understeer, oversteer, lift-off oversteer, and fishtailing
- Weight transfer and load transfer
- Yaw

Tire Specific

Some attributes or aspects of vehicle dynamics can be attributed directly to the tires. These include:

- Camber thrust
- Circle of forces
- Contact patch
- Cornering force
- Ground pressure
- Pacejka's Magic Formula
- Pneumatic trail
- Radial Force Variation
- Relaxation length
- Rolling resistance
- Self aligning torque
- Slip angle
- Slip (vehicle dynamics)
- Steering ratio
- Tire load sensitivity

Roadway Specific

Some attributes or aspects of vehicle dynamics can be attributed directly to the roads on which they travel. These include:

- Banked turn, cross slope, drainage gradient, and cant or superelevation
- Road slipperiness and Split friction
- Surface roughness, International Roughness Index, Profilograph, Texture

Driving Techniques

Driving techniques which relate to, or improve the stability of vehicle dynamics include:

- Cadence braking
- Threshold braking
- Double declutching
- Drifting (motorsport)

- Handbrake turn

- Bootleg turn

- Heel-and-Toe

- Left-foot braking

- Opposite lock

- Scandinavian flick

- Ski (driving stunt)

- Wheelie

Analysis and Simulation

The dynamic behavior of vehicles can be analysed in several different ways. This can be as straightforward as a simple spring mass system, through a three-degree of freedom (DoF) bicycle model, to a large degree of complexity using a multibody system simulation package such as MSC ADAMS or Modelica. As computers have gotten faster, and software user interfaces have improved, commercial packages such as CarSim have become widely used in industry for rapidly evaluating hundreds of test conditions much faster than real time. Vehicle models are often simulated with advanced controller designs provided as software in the loop (SIL) with controller design software such as Simulink, or with physical hardware in the loop (HIL).

Vehicle motions are largely due to the shear forces generated between the tires and road, and therefore the tire model is an essential part of the math model. The tire model must produce realistic shear forces during braking, acceleration, cornering, and combinations, on a range of surface conditions. Many models are in use. Most are semi-empirical, such as the Pacejka Magic Formula model.

Racing car games or simulators are also a form of vehicle dynamics simulation. In early versions many simplifications were necessary in order to get real-time performance with reasonable graphics. However, improvements in computer speed have combined with interest in realistic physics, leading to driving simulators that are used for vehicle engineering using detailed models such as CarSim.

It is important that the models should agree with real world test results, hence many of the following tests are correlated against results from instrumented test vehicles.

Techniques include:

- Linear range constant radius understeer

- Fishhook

- Frequency response

- Lane change

- Moose test

- Sinusoidal steering

- Skidpad

- Swept path analysis.

Components of Vehicle Dynamics

Automobile Layout

In automotive design, the automobile layout describes where on the vehicle the engine and drive wheels are found. Many different combinations of engine location and driven wheels are found in practice, and the location of each is dependent on the application for which the vehicle will be used. Factors influencing the design choice include cost, complexity, reliability, packaging (location and size of the passenger compartment and boot), weight distribution, and the vehicle's intended handling characteristics.

Layouts can roughly be divided into two categories: front- or rear-wheel drive. Four-wheel-drive vehicles may take on the characteristics of either, depending on how power is distributed to the wheels.

Front-wheel-drive Layouts

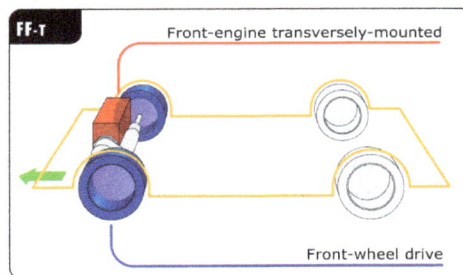

FF layout

Front-wheel-drive layouts are those in which the front wheels of the vehicle are driven. The most popular layout used in cars today is the front-engine, front-wheel drive, with the engine in front of the front axle, driving the front wheels. This layout is typically chosen for its compact packaging; since the engine and driven wheels are on the same side of the vehicle, there is no need for a central tunnel through the passenger compartment to accommodate a prop-shaft between the engine and the driven wheels.

As the steered wheels are also the driven wheels, FF (front-engine, front-wheel-drive layout) cars are generally considered superior to FR (front-engine, rear-wheel-drive layout) cars in conditions such as snow, mud, or wet tarmac. The weight of the engine over the driven wheels also improves grip in such conditions. However, powerful cars rarely use the FF layout because weight transference under acceleration reduces the weight on the front wheels and reduces their traction, limiting the torque which can be utilized. Electronic traction control can avoid wheelspin but largely negates the benefit of extra torque/power.

A transverse engine (also known as "east-west") is commonly used in FF designs, in contrast to FR which uses a longitudinal engine. The FF layout also restricts the size of the engine that can be placed in modern engine compartments, as FF configurations usually have Inline-4 and V6 engines, while longer engines such as Inline-6 and 90° V8 will rarely fit. This is another reason luxury/sports cars avoid the FF layout. Exceptions do exist, such as the Volvo S80 (FWD/4WD) which uses transversely mounted inline 6 and V8 engines, and the Ford Taurus SHO, available with a 60° V8 and front-wheel drive.

Characteristics

Front-wheel drive gives more interior space since the powertrain is a single unit contained in the engine compartment of the vehicle and there is no need to devote interior space for a driveshaft tunnel or rear differential, increasing the volume available for passengers and cargo. There are some exceptions to this as rear engine designs do not take away interior space. It also has fewer components overall and thus lower weight. The direct connection between engine and transaxle reduces the mass and mechanical inertia of the drivetrain compared to a rear-wheel-drive vehicle with a similar engine and transmission, allowing greater fuel economy. In front-wheel-drive cars the mass of the drivetrain is placed over the driven wheels and thus moves the centre of gravity farther forward than a comparable rear-wheel-drive layout, improving traction and directional stability on wet, snowy, or icy surfaces. Front-wheel-drive cars, with a front weight bias, tend to understeer at the limit which, according to Saab engineer Gunnar Larsson, is easier since it makes instinct correct in avoiding terminal oversteer, and less prone to result in fishtailing or a spin.

According to a sales brochure for the 1989 Lotus Elan, the ride and handling engineers at Lotus found that "for a given vehicle weight, power and tyre size, a front-wheel-drive car was always faster over a given section of road." However, this may only apply for cars with moderate power-to-weight ratio. According to road test with two Dodge Daytonas, one FWD and one RWD, the road layout is also important for what configuration is the fastest.

Weight shifting limits the acceleration of a front-wheel-drive vehicle. During heavy acceleration, weight is shifted to the back, improving traction at the rear wheels at the expense of the front driving wheels; consequently, most racing cars are rear-wheel drive for acceleration. However, since front-wheel-drive cars have the weight of the engine over the driving wheels, the problem only applies in extreme conditions in which case the car understeers. On snow, ice, and sand, rear-wheel drive loses its traction advantage to front or all-wheel-drive vehicles which have greater weight over the driven wheels. Rear-wheel-drive cars with rear engine or mid engine configuration retain traction over the driven wheels, although fishtailing remains an issue on hard acceleration while in a turn. Some rear engine cars (e.g., Porsche 911) can suffer from reduced steering ability under heavy acceleration, since the engine is outside the wheelbase and at the opposite end of the car from the wheels doing the steering. A rear-wheel-drive car's centre of gravity is shifted rearward when heavily loaded with passengers or cargo, which may cause unpredictable handling behavior.

On front-wheel-drive cars, the short driveshaft may reduce drivetrain elasticity, improving responsiveness.

Advantages

- Interior space: Since the powertrain is a single unit contained in the engine compartment of the vehicle, there is no need to devote interior space for a driveshaft tunnel or rear differential, increasing the volume available for passengers and cargo.

- Instead, the tunnel may be used to route the exhaust system pipes.

- Weight: Fewer components usually means lower weight.

- Improved fuel efficiency due to less weight.

- Cost: Fewer material components and less installation complexity overall. However, the considerable MSRP differential between a FF and FR car cannot be attributed to layout alone. The difference is more probably explained by production volumes as most rear-wheel cars are usually in the sports/performance/luxury categories (which tend to be more upscale and/or have more powerful engines), while the FF configuration is typically in mass-produced mainstream cars. Few modern "family" cars have rear-wheel drive as of 2009, so a direct cost comparison is not necessarily possible. A contrast could be somewhat drawn between the Audi A4 FrontTrak (which has an FF layout and front-wheel drive) and a rear-wheel-drive BMW 3-Series (which is FR), both which are in the compact executive car classification and use longitudinally mounted engines.

- Improved drivetrain efficiency: the direct connection between engine and transaxle reduce the mass and mechanical inertia of the drivetrain compared to a rear-wheel-drive vehicle with a similar engine and transmission, allowing greater fuel economy.

- Assembly efficiency: the powertrain can often be assembled and installed as a unit, which allows more efficient production.

- Placing the mass of the drivetrain over the driven wheels moves the centre of gravity farther forward than a comparable rear-wheel-drive layout, improving traction and directional stability on wet, snowy, or icy surfaces.

- Predictable handling characteristics: front-wheel-drive cars, with a front weight bias, tend to understeer at the limit, which (according to SAAB engineer Gunnar Larsson) is easier since it makes instinct correct in avoiding terminal oversteer, and less prone to result in fishtailing or a spin.

- A skilled driver can control the movement of the car even while skidding by steering, throttling and pulling the hand brake (given that the hand brake operates the rear wheels as in most cases, with some Citroen and Saab models being notable exceptions).

- It is easier to correct trailing-throttle or trailing-brake oversteer.

- The wheelbase can be extended without building a longer driveshaft (as with rear-wheel-driven cars).

Disadvantages

- Front-engine front-wheel-drive layouts are "nose heavy" with more weight distribution forward, which makes them prone to understeer, especially in high horsepower applications.

  - If a front-engine front-wheel-drive layout is fitted with a four-wheel-drive, plus enthusiast driver aids, such as active front differential, active steering, and ultra-quick electrically adjustable shocks, this somewhat negate the understeer problem and allow the car to perform as well as a front-engine rear-wheel-drive car. These trick differentials, which are found on the Acura TL SH-AWD and Audi S4 3.0 TFSI quattro, and Audi RS5 4.2 FSI quattro, are heavy, complex, and expensive. While these aids do tame front end plow, cars fitted with these systems are still at a disadvantage when track tested against rear-wheel drive vehicles (including those with added four-wheel drive).

- Torque steer is the tendency for some front-wheel-drive cars to pull to the left or right under hard acceleration. It is a result of the offset between the point about which the wheel steers (it is aligned with the points where the wheel is connected to the steering mechanisms) and the centroid of its contact patch. The tractive force acts through the centroid of the contact patch, and the offset of the steering point means that a turning moment about the axis of steering is generated. In an ideal situation, the left and right wheels would generate equal and opposite moments, canceling each other out; however, in reality, this is less likely to happen. Torque steer can be addressed by using a longitudinal layout, equal length drive shafts, half shafts, a multilink suspension or centre-point steering geometry.

- In a vehicle, the weight shifts back during acceleration, giving more traction to the rear wheels. This is one of the main reasons nearly all racing cars are rear-wheel drive. However, since front-wheel-drive cars have the weight of the engine over the driving wheels, the problem only applies in extreme conditions such as attempting to accelerate up a wet hill or attempting to beat another RWD car off the line.

- In some towing situations, front-wheel-drive cars can be at a traction disadvantage since there will be less weight on the driving wheels. Because of this, the weight that the vehicle is rated to safely tow is likely to be less than that of a rear-wheel-drive or four-wheel-drive vehicle of the same size and power.

- Due to geometry and packaging constraints, the CV joints (constant-velocity joints) attached to the wheel hub have a tendency to wear out much earlier than the universal joints typically used in their rear-wheel-drive counterparts (although rear-wheel-drive vehicles with independent rear suspension also employ CV joints and half-shafts). The significantly shorter drive axles on a front-wheel-drive car causes the joint to flex through a much wider degree of motion, compounded by additional stress and angles of steering, while the CV joints of a rear-wheel-drive car regularly see angles and wear of less than half that of front-wheel-drive vehicles.

- Turning circle — FF layouts almost always use a Transverse engine ("east-west") installation, which limits the amount by which the front wheels can turn, thus increasing the turning circle of a front-wheel-drive car compared to a rear-wheel-drive one with the same wheelbase. A notable example is the original Mini. It is widely misconceived that this limitation is due to a limit on the angle at which a CV joint can be operated, but this is easily disproved by considering the turning circle of car models that use a longitudinal FF or F4 layout from Audi and (prior to 1992) Saab

- The FF transverse engine layout (also known as "east-west") restricts the size of the engine that can be placed in modern engine compartments, so it is rarely adopted by powerful luxury and sports cars. FF configurations can usually only accommodate Inline-4 and V6 engines, while longer engines such as Inline-6 and 90° big-bore V8 will rarely fit, though there are exceptions. One way around this problem is using a staggered engine.

- It makes heavier use of the front tyres (i.e., accelerating, braking, and turning), causing more wear in the front than in a rear-wheel-drive layout.

- Under extreme braking (like for instance in a panic stop), the already front heavy layout further reduces traction to the rear wheels. This results in disproportionate gripping forces focused at the front while the rear does not have enough weight to effectively use its brakes. Because the rear tyres' capabilities in braking are not very high, a significant number of cheaper front drive vehicles use drum brakes in the rear even today.

- The steering 'feel' is more numbed than a RWD car. This is due to the extra weight of drive shafts and CV joint components that increase unsprung weight.

Rear-wheel-drive Layouts

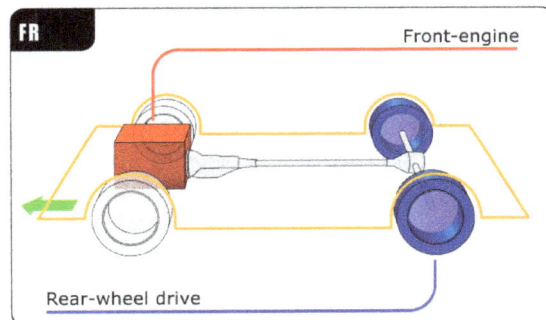

FR layout

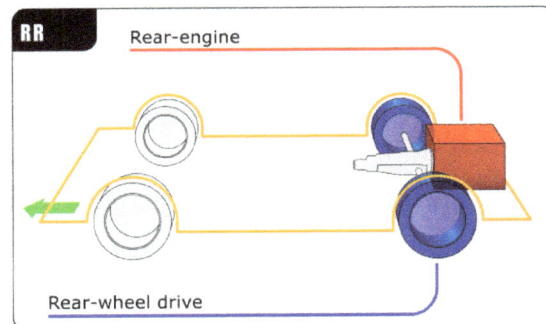

RR layout

Rear-wheel drive (RWD) typically places the engine in the front of the vehicle and the driven wheels are located at the rear, a configuration known as front-engine, rear-wheel drive layout (FR layout). The front mid-engine, rear mid-engine and rear engine layouts are also used. This was the traditional automobile layout for most vehicles up until the 1970s and 1980s. Nearly all motorcycles and bicycles use rear-wheel drive as well, either by driveshaft, chain, or belt, since the front wheel is turned for steering, and it would be very difficult and cumbersome to "bend" the drive mechanism around the turn of the front wheel. A relatively rare exception is with the 'moving bottom bracket' type of recumbent bicycle, where the entire drivetrain, including pedals and chain, pivot with the steering front wheel.

Characteristics

The vast majority of rear-wheel-drive vehicles use a longitudinally mounted engine in the front of the vehicle, driving the rear wheels via a driveshaft linked via a differential between the rear axles. Some FR layout vehicles place the gearbox at the rear, though most attach it to the engine at the front.

The FR layout is often chosen for its simple design and good handling characteristics. Placing the drive wheels at the rear allows ample room for the transmission in the centre of the vehicle and avoids the mechanical complexities associated with transmitting power to the front wheels. For performance-oriented vehicles, the FR layout is more suitable than front-wheel-drive designs, especially with engines that exceed 200 horsepower. This is because weight transfers to the rear of the vehicle during acceleration, which loads the rear wheels and increases their grip.

Another advantage of the FR layout is relatively easy access to the engine compartment, a result of the longitudinal orientation of the drivetrain, as compared to the FF layout (front-engine, front-wheel drive). Powerful engines such as the Inline-6 and 90° big-bore V8 are usually too long to fit in a FF transverse engine ("east-west") layout; the FF configuration can typically accommodate at the maximum an Inline-4 or V6. This is another reason luxury/sports cars almost never use the FF layout.

Advantages

- Even weight distribution — The layout of a rear-wheel-drive car is much closer to an even fore-and-aft weight distribution than a front-wheel-drive car, as more of the engine can lie between the front and rear wheels (in the case of a mid engine layout, the entire engine), and the transmission is moved much farther back.

- Weight transfer during acceleration — During heavy acceleration, weight is placed on the rear, or driving wheels, which improves traction.

- No torque steer (unless it's an all-wheel steer with an offset differential).

- Steering radius — As no complicated drive shaft joints are required at the front wheels, it is possible to turn them further than would be possible using front-wheel drive, resulting in a smaller steering radius for a given wheelbase.

- Better handling at the hands of an expert — the more even weight distribution and weight transfer improve the handling of the car. The front and rear tyres are placed under more even loads, which allows for more grip while cornering.

- Better braking — the more even weight distribution helps prevent lockup from the rear wheels becoming unloaded under heavy braking.

- Towing — Rear-wheel drive puts the wheels which are pulling the load closer to the point where a trailer articulates, helping steering, especially for large loads.

- Serviceability — Drivetrain components on a rear-wheel-drive vehicle are modular and do not involve packing as many parts into as small a space as does front-wheel drive, thus requiring less disassembly or specialized tools in order to service the vehicle.

- Robustness — due to geometry and packaging constraints, the universal joints attached to the wheel hub have a tendency to wear out much later than the CV joints typically used in front-wheel-drive counterparts. The significantly shorter drive axles on a front-wheel-drive car causes the joint to flex through a much wider degree of motion, compounded by additional stress and angles of steering, while the CV joints of a rear-wheel-drive car regularly see angles and wear of less than half that of front-wheel-drive vehicles.

- Can accommodate more powerful engines as a result of the longitudinal orientation of the drivetrain, such as the Inline-6, 90° big-bore V8, V10 and V12 making the FR a common configuration for luxury and sports cars. These engines are usually too long to fit in a FF transverse engine ("east-west") layout; the FF configuration can typically accommodate at the maximum an Inline-4 or V6.

- Road grip feedback — front wheels are not affected by engine and gearbox, thus allowing for better feeling of tyre grip on road surface.

Disadvantages

- Under heavy acceleration (as in racing), oversteer and fishtailing may occur as the rear wheels break free and spin. The corrective action is to let off the throttle (this is what traction control automatically does for RWD vehicles).

- On snow, ice and sand, rear-wheel drive loses its traction advantage to front- or all-wheel-drive vehicles, which have greater weight on the driven wheels. This issue is particularly noticeable on pickup trucks, as the weight of the engine and cab will significantly shift the weight from the rear to the front wheels. Rear-wheel-drive cars with rear engine or mid engine configuration do not suffer from this, although fishtailing remains an issue. To correct this situation, owners of RWD vehicles can load sandbags in the back of the vehicle (either in the bed, or boot) in order to increase the weight over the rear axle, however speeds should be restricted to correctly predicted available grip of the road.

- Some rear engine cars (e.g., Porsche 911) can suffer from reduced steering ability under heavy acceleration, because the engine is outside the wheelbase and at the opposite end of the car from the wheels doing the steering although the engine weight over the rear wheels provides outstanding traction and grip during acceleration.

- Decreased interior space — Though individual designs vary greatly, rear-wheel-drive vehicles may have: Less front leg room as the transmission tunnel takes up a space between the driver and front passenger, less leg room for centre rear passengers (due to the tunnel needed for the drive shaft), and sometimes less boot space (since there is also more hardware that must be placed underneath the boot). Rear engine designs (such as the Porsche 911 and Volkswagen Beetle) do not inherently take away interior space.

  - A rear-wheel drive vehicle with four-wheel drive, compared to a front-wheel drive vehicle with four-wheel drive, will have a less efficient interior packaging since the transmission is often under the front passenger compartment between the two seats, whereas the latter can package all the components under the hood.

- Increased weight — The components of a rear-wheel-drive vehicle's power train are less complex, but they are larger. The driveshaft adds weight. There is extra sheet metal to form the transmission tunnel. There is a rear axle or rear half-shafts, which are typically longer than those in a front-wheel-drive car. A rear-wheel-drive car will weigh slightly more than a comparable front-wheel-drive car (but less than four-wheel drive).

- Rear biased weight distribution when loaded — A rear-wheel-drive car's centre of gravity is shifted rearward when heavily loaded with passengers or cargo, which may cause unpredictable handling behavior at the hands of an inexperienced driver. It needs to be noted that rear engine cars are by their very nature, rear weight biased.

- Higher initial purchase price — Modern rear-wheel-drive vehicles are typically more expensive to purchase than comparable front-wheel-drive vehicles. Part of this can be explained by the added cost of materials and increased labor put in to assembly of FR layouts, as the powertrain is not one compact unit. However, the difference is more probably explained by production volumes as most rear-wheel cars are usually in the sports/performance/luxury categories (which tend to be more upscale and/or have more powerful engines), while the FF configuration is typically in mass-produced mainstream cars.

- The possibility of a slight loss in the mechanical efficiency of the drivetrain (approximately 17% coastdown losses between engine flywheel and road wheels compared to 15% for front-wheel drive — however these losses are highly dependent on the individual transmission). Cars with rear engine or mid engine configuration and a transverse engine layout do not suffer from this.

- The long driveshaft (on front engine cars) adds to drivetrain elasticity. The driveshaft must also be extended for cars with a stretched wheelbase (e.g. limousines, minivans).

Four-wheel-drive Layouts

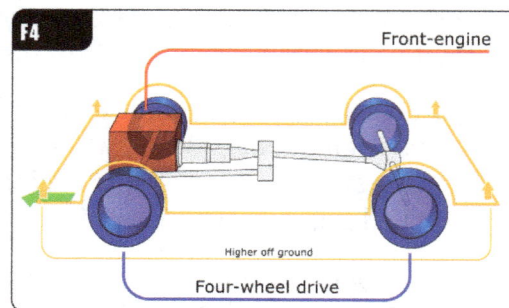

Front-engine, rear-wheel drive derived "F4" layout

Note: in North America, Australia and New Zealand the term "four-wheel drive" usually refers only to drivetrains which are primarily two-wheel drive with a part-time four-wheel-drive capability, as typically found in pickup trucks and other off-road vehicles, while the term "all-wheel drive" is used to refer to full time four-wheel-drive systems found in performance cars and smaller car-based SUVs.

Most 4WD layouts are front-engine and are derivatives of earlier front-engine, two-wheel-drive designs. They fall into two major categories:

- Front-engine, rear-wheel drive derived 4WD systems, standard in most sport utility vehicles and in passenger cars, (usually referred to "front engine, rear-wheel drive/four-wheel drive"), forerunners of today's models include the Jensen FF, AMC Eagle and Mercedes-Benz W124 with the 4Matic system and Suzuki Grand Vitara with/without 4 mode transfer case.

- Transverse and longitudinal engine 4WD systems derived almost exclusively from front-engine, front-drive layouts, fitted to luxury, sporting and heavy duty segments, for example the transverse-engine Mitsubishi 3000GT VR-4 and Toyota RAV4 and the longitudinal-engine Audi Quattro and most of the Subaru line.

Advantages

In terms of handling, traction and performance, 4WD systems generally have most of the advantages of both front-wheel drive and rear-wheel drive. Some unique benefits are:

- Traction is nearly doubled compared to a two-wheel-drive layout. Given sufficient power, this results in unparalleled acceleration and driveability on surfaces with less than ideal grip, and superior engine braking on loose surfaces. The development of 4WD systems for high performance cars was stimulated primarily by rallying.

- Handling characteristics in normal conditions can be configured to emulate FWD or RWD, or some mixture, even to switch between these behaviours according to circumstance. However, at the limit of grip, a well balanced 4WD configuration will not degenerate into either understeer or oversteer, but instead break traction of all 4 wheels at the same time into a four-wheel drift. Combined with modern electronic driving aids, this flexibility allows production car engineers a wide range of freedom in selecting handling characteristics that will allow a 4WD car to be driven more safely at higher speeds by inexpert motorists than 2WD designs.

Disadvantages

- 4WD systems require more machinery and complex transmission components, and so increase the manufacturing cost of the vehicle and complexity of maintenance procedures and repairs compared to 2WD designs

- 4WD systems increase power-train mass, rotational inertia and power transmission losses, resulting in a reduction in performance in ideal dry conditions and increased fuel consumption compared to 2WD designs

- The handbrake cannot be used to induce over-steer for maneuvering purposes, as the drivetrain couples the front and rear axles together. To overcome this limitation, some custom prepared stage rally cars have a special mechanism added to the transmission to disconnect the rear drive if the handbrake is applied while the car is moving.

Unusual 4WD Layouts

- From 1989 onwards, some models of Porsche 911 feature a rear-engine 4WD layout, which is akin to a longitudinal front-engine 4WD layout installed backwards with the engine at the rear of the car

- From 2007 onwards, the Nissan GT-R features a front-engine 4WD longitudinal layout, but with the gearbox at the rear of the vehicle. This provides a more ideal weight balance, and improves directional stability at very high speeds by increasing the vehicle's moment of inertia around the vertical axis. This layout necessitates a second prop-shaft to carry power to the front wheels.

- Some types of farm tractors and construction site machinery use a 4WD layout where the wheels on each side are coupled together, rather than the wheels on each axle, allowing these vehicles to pivot about their centre point. Such vehicles are controlled in a fashion similar to a military tank.

- The Citroën Sahara had a 4WD system using complete Citroën 2CV drivetrains at both ends of the car, such that the engine at the front powered the front wheels and the engine at the back powered the rear wheels.

- A 'through the road' hybrid vehicle uses a conventional piston engine to power two wheels, with electric motor/generators on the other two wheels, giving a form of part-time 4WD.

- The 2005 Jeep Hurricane concept had an all-wheel drive layout that featured two V8 engines powering a single driveshaft, with a gearbox mounted in the centre of the vehicle. The gears connected to two additional driveshafts, one on each side of the vehicle, that delivered power to the wheels via driveshaft joints. This was designed in order to accommodate the vehicle's unique steering system.

- The Ferrari FF features a front-engine 4WD layout in which a separate transmission is used for each pair of driven wheels, rather than the more conventional setup in which a single transmission is used, followed by a centre differential or viscous coupling unit to split power between the front and rear wheels.

History and Current Use

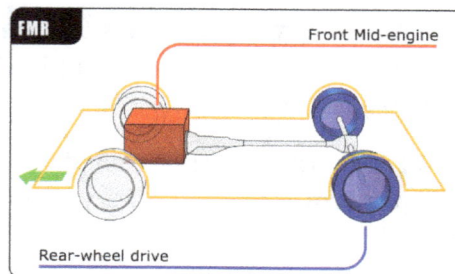

FMR layout, standard in most Front-engine / Rear-wheel-drive cars pre-World War II, where the engine was located behind the front axle.

The first FR car was an 1895 Panhard model, so this layout was known as the "Système Panhard" in the early years. Most American cars used the FR layout until the mid-1980s. The Oil crisis of the

1970s and the success of small FF cars like the Mini, Volkswagen Golf, Toyota Tercel, and Honda Civic led to the widespread adoption of that layout.

After the Arab Oil Embargo of 1973 and the 1979 fuel crises, a majority of American FR vehicles (station wagons, luxury sedans) were phased out for the FF layout — this trend would spawn the SUV/van conversion market. Throughout the 1980s and 1990s, most American companies set as a priority the eventual removal of rear-wheel drive from their mainstream and luxury lineup. Chrysler went 100% FF by 1990 and GM's American production went entirely FF by 1997 except the Firebird, Corvette and Camaro. Ford's full-size cars (the Ford Crown Victoria, Mercury Grand Marquis, and Lincoln Town Car) have always been FR, as was the Lincoln LS. In 2008 Hyundai introduced its own rear-wheel-drive car, the Hyundai Genesis.

In Australia, FR cars have remained popular throughout this period, with the Holden Commodore and Ford Falcon having consistently strong sales. In Europe, front-wheel drive was popularized by small cars like the Mini, Renault 5 and Volkswagen Golf and adopted for virtually all mainstream cars.

Upscale marques like Mercedes-Benz, BMW, and Jaguar remained mostly independent of this trend, and retained a lineup mostly or entirely made up of FR cars. Japanese mainstream marques such as Toyota and Nissan became mostly or entirely FF early on, while reserving for their latterly conceived luxury divisions (Lexus and Infiniti, respectively) a mostly FR lineup. While many auto-makers lost sight of the true sports car, Mazda introduced the highly successful Miata roadster in 1990, a true 2-seater sports car using the traditional FR layout which led to other compaines such as General Motors to produce a FR sports car based on their Kappa platform.

Currently most cars are FF, including virtually all front-engine economy cars, though FR cars are making a return as an alternative to large sport-utility vehicles. In North America, GM returned to production of the FR luxury car with the 2003 Cadillac CTS, and with the removal of the DTS, Cadillac will be entirely FR (with four-wheel drive available as an option on several models) by 2010, and the 2010 Camaro returns as a FR sports car. Chrysler returned its full-size cars to this layout with the Chrysler 300 and related models. Despite Ford's 2011 discontinuation of the rear-wheel drive Panther Platform cars, they are seeking to develop a new FR replacement. Nissan is also bringing back the Silvia to their line-up, Mazda is said to be releasing a new rotary-powered FR car in their RX line-up, and Toyota has produced the FT-86, an affordable RWD car which is the successor to the AE86. Hyundai introduced their affordable RWD car being the 2009 Hyundai Genesis and 2010 Hyundai Genesis Coupe.

In the 21st century, with solutions to the engineering complexities of 4WD being widely under-stood, and consumer demand for increasing performance in production cars, front-engine 4WD layouts are rapidly becoming more common, and most major manufacturers now offer 4WD op-tions on at least some models. Manufacturers with a notable expertise and history in producing 4WD performance cars are Audi and Subaru.

Electronic Stability Control

Electronic stability control (ESC), also referred to as electronic stability program (ESP) or dy-namic stability control (DSC), is a computerized technology that improves a vehicle's stability by detecting and reducing loss of traction (skidding). When ESC detects loss of steering control, it automatically applies the brakes to help "steer" the vehicle where the driver intends to go. Braking is automatically applied to wheels individually, such as the outer front wheel to counter oversteer or the inner rear wheel to counter understeer. Some ESC systems also reduce engine power until control is regained. ESC does not improve a vehicle's cornering performance; instead, it helps to minimize the loss of control. According to Insurance Institute for Highway Safety and the U.S. National Highway Traffic Safety Administration, one-third of fatal accidents could be prevented by the use of the technology.

ESC control light

History

In 1983, a series production Four-wheel electronic anti-skid control is introduced on the Toyota Crown.

In 1987, Mercedes-Benz, BMW and Toyota introduced their first traction control systems. Traction control works by applying individual wheel braking and throttle to keep traction while accelerating but, unlike the ESC, it is not designed to aid in steering.

In 1990, Mitsubishi released the Diamante (Sigma) in Japan. It featured a new electronically controlled active trace & traction control system (the first integration of these two systems in the world) that Mitsubishi developed. Simply named TCL in 1990, the system has now evolved into Mitsubishi's modern Active Skid and Traction Control (ASTC) system. Developed to help the driver maintain the intended line through a corner; an onboard computer monitored several vehicle operating parameters through various sensors. When too much throttle has been used when taking a curve, engine output and braking are automatically regulated to ensure the proper line through a curve and to provide the proper amount of traction under various road surface conditions. While conventional traction control systems at the time featured only a slip control function, Mitsubishi's newly developed TCL system had a preventive (active) safety function which improved the course tracing performance by automatically adjusting the traction force (called "trace control") thereby restraining the development of excessive lateral acceleration while turning. Although not a 'proper' modern stability control system, trace control monitors steering angle, throttle position and individual wheel speeds although there is no yaw input. The TCL system's standard wheel slip control function enables better traction on slippery surfaces or during cornering. In addition to the TCL system's individual effect, it also works together with Diamante's electronic controlled suspension and four-wheel steering that Mitsubishi had equipped to improve total handling and performance.

BMW, working with Robert Bosch GmbH and Continental Automotive Systems, developed a system to reduce engine torque to prevent loss of control and applied it to most of the BMW model line for 1992 excluding the l3 series (E30 and some E36 chassis codes) which could be ordered with the winter package, which came with a limited slip differential, heated seats and mirrors. From 1987 to 1992, Mercedes-Benz and Robert Bosch GmbH co-developed a system called Elektronisches Stabilitätsprogramm (Ger. "Electronic Stability Programme" trademarked as ESP) to control lateral slippage.

Introduction

In 1995, three automobile manufacturers introduced ESC systems. Mercedes-Benz, supplied by Bosch, was the first to implement ESP with their Mercedes-Benz S 600 Coupé.

That same year BMW, supplied by Bosch and ITT Automotive (later acquired by Continental Automotive Systems) introduced the system on the BMW 7 Series (E38) (DSC III).

Toyota's Vehicle Stability Control (VSC) system (also in 2004, a preventive system called Vehicle Dynamics Integrated Management (VDIM) appeared on the Toyota Crown Majesta in 1995.

In 1997, Audi introduced the first series production ESP for all-wheel drive vehicles (Audi A8 and Audi A6 with quattro (four-wheel drive system)).

In 1998, Volvo Cars began to offer their version of ESC called Dynamic Stability and Traction Control (DSTC) on the new Volvo S80.

Meanwhile, others investigated and developed their own systems.

During a moose test (swerving to avoid an obstacle), which became famous in Germany as "the elk test", the Swedish journalist Robert Collin of Teknikens Värld (World of Technology) in October 1997 rolled a Mercedes A-Class (without ESC) at 78 km/h. Because Mercedes-Benz promotes a reputation for safety, they recalled and retrofitted 130,000 A-Class cars with ESC. This produced a significant reduction in crashes and the number of vehicles with ESC rose. Today, virtually all premium brands have made ESC standard on all vehicles, and the number of models with ESC continues to increase. The availability of ESC in small cars like the A-Class ignited a market trend thus ESC became available for all models at least as an option. Consequently, the European Union decided in 2009 to make ESC mandatory. Since November 1, 2011, EU Type Approval is only granted to models equipped with ESC. By November 1, 2014, ESC is required on all newly registered cars in the EU.

General Motors (GM) worked with Delphi Automotive and introduced its version of ESC called "StabiliTrak" in 1997 for select Cadillac models. StabiliTrak was made standard equipment on all GM SUVs and vans sold in the U.S. and Canada by 2007 except for certain commercial and fleet vehicles. While the "StabiliTrak" name is used on most General Motors vehicles for the U.S. market, the "Electronic Stability Control" identity is used for GM overseas brands, such as Opel, Holden and Saab, except in the case of Saab's 9-7X which also uses the "StabiliTrak" name. The same year, Cadillac introduced an integrated vehicle handling and software control system, called Integrated Chassis Control System (ICCS), on the Cadillac Eldorado. It involves an omnibus computer integration of engine, traction control, Stabilitrak electronic stability control, steering, and adaptive continuously variable road sensing suspension CVRSS, with the intent of improving responsiveness to driver input, performance, and overall safety. Similar to Toyota/Lexus Vehicle Dynamics Integrated Management VDIM.

Ford's version of ESC, called AdvanceTrac, was launched in the year 2000. Ford later added Roll Stability Control to AdvanceTrac which was first introduced in Volvo XC90 in 2003 when Volvo Cars was fully owned by Ford and it is now being implemented in many Ford vehicles.

Ford and Toyota announced that all their North American vehicles would be equipped with ESC standard by the end of 2009 (it was standard on Toyota SUVs as of 2004, and after the 2011 model-year, All Lexus, Toyota, and Scion vehicles have ESC; the last one to get it was the 2011 model-year Scion tC). However, as recent as November 2010, Ford still sells models in North America without ESC. General Motors had made a similar announcement for the end of 2010. The NHTSA requires all new passenger vehicles sold in the US to be equipped with ESC as of the 2012 model year, and estimates it will prevent 5,300–9,600 annual fatalities. A similar requirement has been proposed for new truck tractors and certain buses, but it hasn't yet been finalized.

Operation

During normal driving, ESC works in the background and continuously monitors steering and vehicle direction. It compares the driver's intended direction (determined through the measured steering wheel angle) to the vehicle's actual direction (determined through measured lateral acceleration, vehicle rotation (yaw), and individual road wheel speeds).

ESC intervenes only when it detects a probable loss of steering control, i.e. when the vehicle is not going where the driver is steering. This may happen, for example, when skidding during emergency evasive swerves, understeer or oversteer during poorly judged turns on slippery roads, or hydroplaning. ESC may also intervene in an unwanted way during high-performance driving, because steering input may not always be directly indicative of the intended direction of travel (i.e. controlled drifting). ESC estimates the direction of the skid, and then applies the brakes to individual wheels asymmetrically in order to create torque about the vehicle's vertical axis, opposing the skid and bringing the vehicle back in line with the driver's commanded direction. Additionally, the system may reduce engine power or operate the transmission to slow the vehicle down.

ESC can work on any surface, from dry pavement to frozen lakes. It reacts to and corrects skidding much faster and more effectively than the typical human driver, often before the driver is even aware of any imminent loss of control. In fact, this led to some concern that ESC could allow drivers to become overconfident in their vehicle's handling and/or their own driving skills. For this reason, ESC systems typically inform the driver when they intervene, so that the driver knows that the vehicle's handling limits have been approached. Most activate a dashboard indicator light and/or alert tone; some intentionally allow the vehicle's corrected course to deviate very slightly from the driver-commanded direction, even if it is possible to more precisely match it.

Indeed, all ESC manufacturers emphasize that the system is not a performance enhancement nor a replacement for safe driving practices, but rather a safety technology to assist the driver in recovering from dangerous situations. ESC does not increase traction, so it does not enable faster cornering (although it can facilitate better-controlled cornering). More generally, ESC works within inherent limits of the vehicle's handling and available traction between the tyres and road. A reckless maneuver can still exceed these limits, resulting in loss of control. For example, in a severe hydroplaning scenario, the wheels that ESC would use to correct a skid may not even initially be in contact with the road, reducing its effectiveness.

In July 2004, on the Crown Majesta, Toyota offered a Vehicle Dynamics Integrated Management (VDIM) system that incorporated formerly independent systems, including ESC. This worked not only after the skid was detected but also to prevent the skid from occurring in the first place. Using electric variable gear ratio steering power steering, this more advanced system could also alter steering gear ratios and steering torque levels to assist the driver in evasive manoeuvres.

Due to the fact that stability control can sometimes be incompatible with high-performance driving (i.e. when the driver intentionally loses traction as in drifting), many vehicles have an over-ride control which allows the system to be partially or fully shut off. In simpler systems, a single button may disable all features, while more complicated setups may have a multi-position switch or may never be truly turned fully off.

Effectiveness

Numerous studies around the world confirm that ESC is highly effective in helping the driver maintain control of the car, thereby saving lives and reducing the severity of crashes. In the fall of 2004 in the U.S., the National Highway and Traffic Safety Administration confirmed the international studies, releasing results of a field study in the U.S. of ESC effectiveness. The NHTSA in United States concluded that ESC reduces crashes by 35%. Additionally, Sport utility vehicles (SUVs) with stability control are involved in 67% fewer accidents than SUVs without the system. The United States Insurance Institute for Highway Safety (IIHS) issued its own study in June 2006 showing that up to 10,000 fatal US crashes could be avoided annually if all vehicles were equipped with ESC. The IIHS study concluded that ESC reduces the likelihood of all fatal crashes by 43%, fatal single-vehicle crashes by 56%, and fatal single-vehicle rollovers by 77–80%.

ESC is described as the most important advance in auto safety by many experts, including Nicole Nason, Administrator of the NHTSA, Jim Guest and David Champion of Consumers Union of the Fédération Internationale de l'Automobile (FIA), E-Safety Aware, Csaba Csere, editor of Car and Driver, and Jim Gill, long time ESC proponent of Continental Automotive Systems. The European New Car Assessment Program (EuroNCAP) "strongly recommends" that people buy cars fitted with stability control.

The IIHS requires that a vehicle must have ESC as an available option in order for it to qualify for their Top Safety Pick award for occupant protection and accident avoidance.

Components and Design

ESC incorporates yaw rate control into the anti-lock braking system (ABS). Yaw is a rotation around the vertical axis; i.e. spinning left or right. Anti-lock brakes enable ESC to brake individual wheels. Many ESC systems also incorporate a traction control system (TCS or ASR), which senses drive-wheel slip under acceleration and individually brakes the slipping wheel or wheels and/or reduces excess engine power until control is regained. However, ESC achieves a different purpose than ABS or Traction Control.

The ESC system uses several sensors to determine what the driver wants (input). Other sensors indicate the actual state of the vehicle (response). The control algorithm compares driver input to vehicle response and decides, when necessary, to apply brakes and/or reduce throttle by the amounts calculated through the state space (set of equations used to model the dynamics of the vehicle). The ESC controller can also receive data from and issue commands to other controllers on the vehicle such as an all wheel drive system or an active suspension system to improve vehicle stability and controllability.

The sensors used for ESC have to send data at all times in order to detect possible defects as soon as possible. They have to be resistant to possible forms of interference (rain, holes in the road, etc.). The most important sensors are:

*   Steering wheel angle sensor: determines the driver's intended rotation; i.e. where the driver wants to steer. This kind of sensor is often based on AMR-elements.

*   Yaw rate sensor: measures the rotation rate of the car; i.e. how much the car is actually turning. The data from the yaw sensor is compared with the data from the steering wheel angle sensor to determine regulating action.

*   Lateral acceleration sensor: often an accelerometer

*   Wheel speed sensor: measures the wheel speed.

Other sensors can include:

*   Longitudinal acceleration sensor: similar to the lateral acceleration sensor in design, but can offer additional information about road pitch and also provide another source of vehicle acceleration and speed.

*   Roll rate sensor: similar to the yaw rate sensor in design but improves the fidelity of the controller's vehicle model and correct for errors when estimating vehicle behavior from the other sensors alone.

ESC uses a hydraulic modulator to assure that each wheel receives the correct brake force. A similar modulator is used in ABS. ABS needs to reduce pressure during braking, only. ESC additionally needs to increase pressure in certain situations and an active vacuum brake booster unit may be utilized in addition to the hydraulic pump to meet these demanding pressure gradients.

The brain of the ESC system is the electronic control unit (ECU). The various control techniques are embedded in it. Often, the same ECU is used for diverse systems at the same time (ABS, Traction control system, climate control, etc.). The input signals are sent through the input-circuit to the digital controller. The desired vehicle state is determined based upon the steering wheel angle, its gradient and the wheel speed. Simultaneously, the yaw sensor measures the actual state. The controller computes the needed brake or acceleration force for each wheel and directs via the driver circuits the valves of the hydraulic modulator. Via a Controller Area Network interface the ECU is connected with other systems (ABS, etc.) in order to avoid giving contradictory commands.

Many ESC systems have an "off" override switch so the driver can disable ESC, which may be desirable when badly stuck in mud or snow, or driving on a beach, or if using a smaller-sized spare tire which would interfere with the sensors. Some systems also offer an additional mode with raised thresholds so that a driver can utilize the limits of adhesion with less electronic intervention. However, ESC defaults to "On" when the ignition is restarted. Some ESC systems that lack an "off switch", such as on many recent Toyota and Lexus vehicles, can be temporarily disabled through an undocumented series of brake pedal and handbrake operations. Furthermore, unplugging a wheel speed sensor is another method of disabling most ESC systems. The ESC implementation on newer Ford vehicles cannot be completely disabled even through the use of the "off switch". The ESC will automatically reactivate at highway speeds, and below that if it detects a skid with the brake pedal depressed.

Availability and Cost

ESC is built on top of an anti-lock brake (ABS) system, and all ESC-equipped vehicles are fitted with traction control. The ESC components include a yaw rate sensor, a lateral acceleration sensor, a steering wheel sensor, and an upgraded integrated control unit. In the US, Federal regulations require that ESC be installed as a standard feature on all passenger cars and light trucks as of the 2012 model year. According to National Highway Traffic Safety Administration (NHTSA) research, ABS in 2005 cost an estimated US$368; ESC cost a further US$111. The retail price of ESC varies; as a stand-alone option it retails for as little as $250 USD. ESC was once rarely offered as a sole option, and was generally not available for aftermarket installation. Instead, it was frequently bundled with other features or more expensive trims, so the cost of a package that included ESC was several thousand dollars. Nonetheless, ESC is considered highly cost-effective and it might pay for itself in reduced insurance premiums.

Availability of ESC in passenger vehicles varies between manufacturers and countries. In 2007, ESC was available in roughly 50% of new North American models compared to about 75% in Sweden. However, consumer awareness affects buying patterns so that roughly 45% of vehicles sold in North America and the UK were purchased with ESC, contrasting with 78–96% in other European countries such as Germany, Denmark, and Sweden. While few vehicles had ESC prior to 2004, increased awareness will increase the number of vehicles with ESC on the used car market.

ESC is available on cars, SUVs and pickup trucks from all major auto makers. Luxury cars, sports cars, SUVs, and crossovers are usually equipped with ESC. Midsize cars were also gradually catching on, though the 2008 model years of the Nissan Altima and Ford Fusion only offered ESC on their V6 engine-equipped cars; however, some midsize cars, such as the Honda Accord had it as standard equipment by then. While ESC includes traction control, there are vehicles such as the 2008 Chevrolet Malibu LS and 2008 Mazda6 that have traction control but not ESC. ESC is rare among subcompact cars as of 2008. The 2009 Toyota Corolla in the United States (but not Canada) has stability control as a $250 option on all trims below that of the XRS which has it as standard. In Canada, for the 2010 Mazda3, ESC is as an option on the midrange GS trim as part of the moonroof package, and is standard on the top-of-the-line GT version. The 2009 Ford Focus has ESC as an option for the S and SE models, and standard on the SEL and SES models

In the UK, even mass-market superminis such as the Ford Fiesta Mk.6 and VW Polo Mk.5 come with ESC as standard.

ESC is also available on some motor homes. Elaborate ESC and ESP systems (including Roll Stability Control (RSC)) are available for many commercial vehicles, including transport trucks, trailers, and buses from manufacturers such as Bendix Corporation, WABCO Daimler, Scania AB, and Prevost, and light passenger vehicles.

The ChooseESC! campaign, run by the EU's eSafetyAware! project, provides a global perspective on ESC. One ChooseESC! publication shows the availability of ESC in EU member countries.

In the US, the Insurance Institute for Highway Safety (IIHS) website shows availability of ESC in individual US models and the National Highway Traffic Safety Administration (NHTSA website) lists US models with ESC.

In Australia, the National Roads and Motorists' Association NRMA shows the availability of ESC in Australian models.

Future

The market for ESC is growing quickly, especially in European countries such as Sweden, Denmark, and Germany. For example, in 2003 in Sweden the purchase rate on new cars with ESC was 15%. The Swedish road safety administration issued a strong ESC recommendation and in September 2004, 16 months later, the purchase rate was 58%. A stronger ESC recommendation was then given and in December 2004, the purchase rate on new cars had reached 69% and by 2008 it had grown to 96%. ESC advocates around the world are promoting increased ESC use through legislation and public awareness campaigns and by 2012, most new vehicles should be equipped with ESC.

Just as ESC is founded on the Anti-lock braking system (ABS), ESC is the foundation for new advances such as Roll Stability Control (RSC) or Active rollover protection that works in the vertical plane much like ESC works in the horizontal plane. When RSC detects impending rollover (usually on transport trucks or SUVs), RSC applies brakes, reduces throttle, induces understeer, and/or slows down the vehicle.

The computing power of ESC facilitates the networking of active and passive safety systems, addressing other causes of crashes. For example, sensors may detect when a vehicle is following too closely and slow down the vehicle, straighten up seat backs, and tighten seat belts, avoiding and/or preparing for a crash.

Regulation

While Sweden used public awareness campaigns to promote ESC use, others implemented or proposed legislation.

The Canadian province of Quebec was the first jurisdiction to implement an ESC law, making it compulsory for carriers of dangerous goods (without data recorders) in 2005.

The United States was next, requiring ESC for all passenger vehicles under 10,000 pounds (4536 kg), phasing in the regulation starting with 55% of 2009 models (effective 1 September 2008), 75% of 2010 models, 95% of 2011 models, and all 2012 models.

Canada requires all new passenger vehicles to have ESC from 1 September 2011.

The Australian Government announced on 23 June 2009 that ESC would be compulsory from 1

November 2011 for all new passenger vehicles sold in Australia, and for all new vehicles from November 2013. The New Zealand government followed suit in February 2014 making it compulsory on all new vehicles from 1 July 2015 with a staggered roll-out to all used-import passenger vehicles by 1 January 2020.

The European Parliament has also called for the accelerated introduction of ESC. The European Commission has confirmed a proposal for the mandatory introduction of ESC on all new cars and commercial vehicle models sold in the EU from 2012, with all new cars being equipped by 2014.

The United Nations Economic Commission for Europe has passed a Global Technical Regulation to harmonize ESC standards. Global Technical Regulation No. 8 ELECTRONIC STABILITY CONTROL SYSTEMS was sponsored by the United States of America, and is based on Federal Motor Vehicle Safety Standard FMVSS126.

Product Names

Electronic stability control (ESC) is the generic term recognised by the European Automobile Manufacturers Association (ACEA), the North American Society of Automotive Engineers (SAE), the Japan Automobile Manufacturers Association, and other worldwide authorities. However, vehicle manufacturers may use a variety of different trade names for ESC:

- Acura: Vehicle Stability Assist (VSA) (formerly CSL 4-Drive TCS)
- Alfa Romeo: Vehicle Dynamic Control (VDC)
- Audi: Electronic Stability Program (ESP)
- Bentley: Electronic Stability Program (ESP)
- BMW: Co engineering partner and inventor with Robert BOSCH GmbH and Continental (TEVES) Dynamic Stability Control (DSC) (including Dynamic Traction Control)
- Bugatti: Electronic Stability Program (ESP)
- Buick: StabiliTrak
- Cadillac: StabiliTrak" and "StabiliTrak3.0 with Active Front Steering (AFS)
- Chery: Electronic Stability Program
- Chevrolet: StabiliTrak; Active Handling (Corvette & Camaro only)
- Chrysler: Electronic Stability Program (ESP)
- Citroën: Electronic Stability Program (ESP)
- Daihatsu: Vehicle Stability Control (VSC)
- Dodge: Electronic Stability Program (ESP)
- Daimler: Electronic Stability Program (ESP)
- Fiat: Electronic Stability Program (ESP) and Vehicle Dynamic Control (VDC)

- Ferrari: Controllo Stabilità (CST)

- Ford: AdvanceTrac with Roll Stability Control (RSC) and Interactive Vehicle Dynamics (IVD) and Electronic Stability Program (ESP); Dynamic Stability Control (DSC) (Australia only)

- General Motors: StabiliTrak

- Honda: Vehicle Stability Assist (VSA) (formerly CSL 4-Drive TCS)

- Holden: Electronic Stability Program (ESP)

- Hyundai: Electronic Stability Program (ESP), Electronic Stability Control (ESC) and Vehicle Stability Assist (VSA)

- Infiniti: Vehicle Dynamic Control (VDC)

- Isuzu: Electronic Vehicle Stability Control (EVSC)

- Jaguar: Dynamic Stability Control (DSC)

- Jeep: Electronic Stability Program (ESP)

- Kia: Electronic Stability Control (ESC) and Electronic Stability Program (ESP)

- Lamborghini: Electronic Stability Program (ESP)

- Land Rover: Dynamic Stability Control (DSC)

- Lexus: Vehicle Dynamics Integrated Management (VDIM) with Vehicle Stability Control (VSC)

- Lincoln: AdvanceTrac

- Maserati: Maserati Stability Program (MSP)

- Mazda: Dynamic Stability Control (DSC) (including Dynamic Traction Control)

- Mercedes-Benz (co-inventor) with Robert BOSCH GmbH: Electronic Stability Program (ESP)

- Mercury: AdvanceTrac

- MINI: Dynamic Stability Control

- Mitsubishi: Active Skid and Traction Control MULTIMODE and Active Stability Control (ASC)

- Nissan: Vehicle Dynamic Control (VDC)

- Oldsmobile: Precision Control System (PCS)

- Opel: Electronic Stability Program (ESP) and Trailer Stability Program (TSP)

- Peugeot: Electronic Stability Program (ESP)

- Pontiac: StabiliTrak

- Porsche: Porsche Stability Management (PSM)
- Proton: Electronic Stability Control (ESC) or Vehicle Dynamics Control (VDC)
- Renault: Electronic Stability Program (ESP)
- Rover Group: Dynamic Stability Control (DSC)
- Saab: Electronic Stability Program (ESP)
- Saturn: StabiliTrak
- Scania: Electronic Stability Program (ESP)
- SEAT: Electronic Stability Program (ESP)
- Škoda: Electronic Stability Program (ESP) and Electronic Stability Control (ESC)
- Smart: Electronic Stability Program (ESP)
- Subaru: Vehicle Dynamics Control (VDC)
- Suzuki: Electronic Stability Program (ESP)
- Tata: Corner Stability Control (CSC)
- Toyota: Either Vehicle Stability Control (VSC) or Vehicle Dynamics Integrated Management (VDIM)
- Tesla: Electronic Stability Control
- Vauxhall: Electronic Stability Program (ESP)
- Volvo: Dynamic Stability and Traction Control (DSTC)
- Volkswagen: Electronic Stability Program (ESP)

System Manufacturers

ESC system manufacturers include:

- Fujitsu Ten Ltd.
- Robert Bosch GmbH
- Aisin Advics
- Bendix Corporation
- Continental Automotive Systems
- BeijingWest Industries
- Hitachi
- ITT Automotive, since 1982 part of Continental AG
- Johnson Electric

- Mando Corporation

- Nissin Kogyo

- Teves, now part of Continental AG

- TRW

- WABCO

- Hyundai Mobis

- Knorr-Bremse

Steering

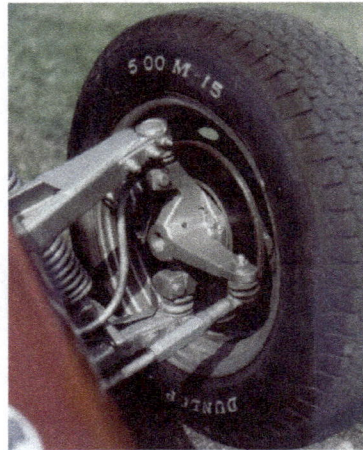

Part of car steering mechanism: tie rod, steering arm, king pin axis (using ball joints)

Steering is the collection of components, linkages, etc. which allows any vehicle (car, motorcycle, bicycle) to follow the desired course. An exception is the case of rail transport by which rail tracks combined together with railroad switches (and also known as 'points' in British English) provide the steering function. The primary purpose of the steering system is to allow the driver to guide the vehicle.

Bell-crank steering linkage

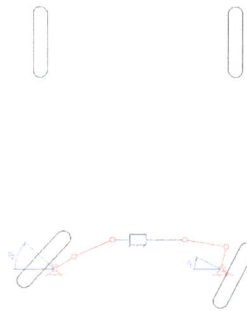

Rack-and-pinion steering linkage

Introduction

The most conventional steering arrangement is to turn the front wheels using a hand–operated steering wheel which is positioned in front of the driver, via the steering column, which may contain universal joints (which may also be part of the collapsible steering column design), to allow it to deviate somewhat from a straight line. Other arrangements are sometimes found on different types of vehicles, for example, a tiller or rear–wheel steering. Tracked vehicles such as bulldozers and tanks usually employ differential steering — that is, the tracks are made to move at different speeds or even in opposite directions, using clutches and brakes, to bring about a change of course or direction.

Wheeled Vehicle Steering

Basic Geometry

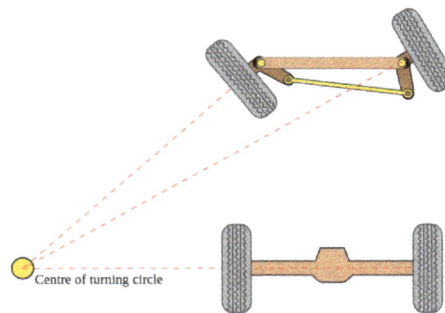

Centre of turning circle

Ackermann steering geometry

Ackerman steering linkage

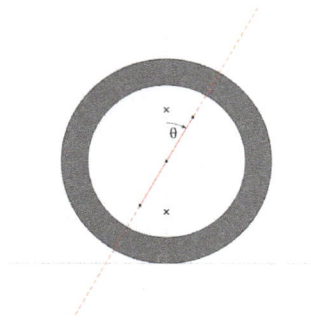

Caster angle θ indicates kingpin pivot line and gray area indicates vehicle's tire with the wheel moving from right to left. A positive caster angle aids in directional stability, as the wheel tends to trail, but a large angle makes steering more difficult.

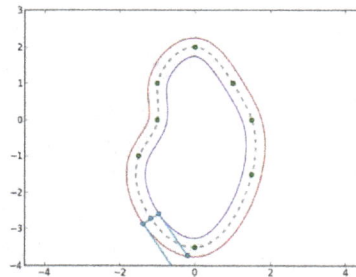

Curves described by the rear wheels of a conventional automobile. While the vehicle moves with a constant speed its inner and outer rear wheels do not.

The basic aim of steering is to ensure that the wheels are pointing in the desired directions. This is typically achieved by a series of linkages, rods, pivots and gears. One of the fundamental concepts is that of caster angle – each wheel is steered with a pivot point ahead of the wheel; this makes the steering tend to be self-centering towards the direction of travel.

The steering linkages connecting the steering box and the wheels usually conform to a variation of Ackermann steering geometry, to account for the fact that in a turn, the inner wheel is actually travelling a path of smaller radius than the outer wheel, so that the degree of toe suitable for driving in a straight path is not suitable for turns. The angle the wheels make with the vertical plane also influences steering dynamics as do the tires.

Rack and Pinion, Recirculating Ball, Worm and Sector

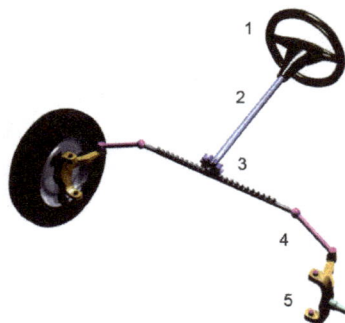

Rack and pinion steering mechanism: 1 steering wheel; 2 steering column; 3 rack and pinion; 4 tie rod; 5 kingpin

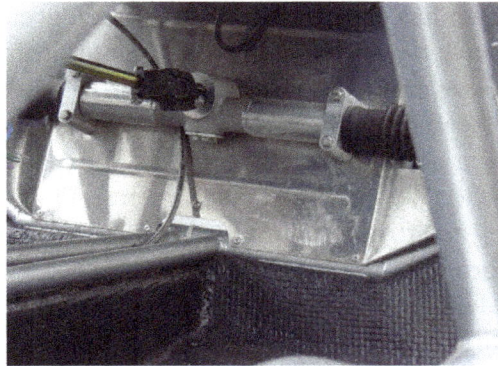

Rack and pinion unit mounted in the cockpit of an Ariel Atom sports car chassis. For most high volume production, this is usually mounted on the other side of this panel

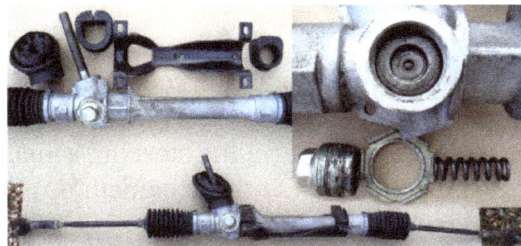

Steering box of a motor vehicle, the traditional (non-assisted), you may notice that the system allows you to adjust the braking and steering systems.

Many modern cars use rack and pinion steering mechanisms, where the steering wheel turns the pinion gear; the pinion moves the rack, which is a linear gear that meshes with the pinion, converting circular motion into linear motion along the transverse axis of the car (side to side motion). This motion applies steering torque to the swivel pin ball joints that replaced previously used kingpins of the stub axle of the steered wheels via tie rods and a short lever arm called the steering arm.

The rack and pinion design has the advantages of a large degree of feedback and direct steering "feel". A disadvantage is that it is not adjustable, so that when it does wear and develop lash, the only cure is replacement.

BMW began to use rack and pinion steering systems in the 1930s, and many other European manufacturers adopted the technology. American automakers adopted rack and pinion steering beginning with the 1974 Ford Pinto.

Older designs use two main principles: the worm and sector design and the screw and nut. Both types were enhanced by reducing the friction; for screw and nut it is the recirculating ball mechanism, which is still found on trucks and utility vehicles. The steering column turns a large screw which meshes with nut by recirculating balls. The nut moves a sector of a gear, causing it to rotate about its axis as the screw is turned; an arm attached to the axis of the sector moves the Pitman arm, which is connected to the steering linkage and thus steers the wheels. The recirculating ball version of this apparatus reduces the considerable friction by placing large ball bearings between the screw and the nut; at either end of the apparatus the balls exit from between the two pieces into a channel internal to the box which connects them with the other end of the apparatus, thus they are "recirculated".

The recirculating ball mechanism has the advantage of a much greater mechanical advantage, so that it was found on larger, heavier vehicles while the rack and pinion was originally limited to smaller and lighter ones; due to the almost universal adoption of power steering, however, this is no longer an important advantage, leading to the increasing use of rack and pinion on newer cars. The recirculating ball design also has a perceptible lash, or "dead spot" on center, where a minute turn of the steering wheel in either direction does not move the steering apparatus; this is easily adjustable via a screw on the end of the steering box to account for wear, but it cannot be entirely eliminated because it will create excessive internal forces at other positions and the mechanism will wear very rapidly. This design is still in use in trucks and other large vehicles, where rapidity of steering and direct feel are less important than robustness, maintainability, and mechanical advantage.

The worm and sector was an older design, used for example in Willys and Chrysler vehicles, and the Ford Falcon (1960's). To reduce friction the sector is replaced by a roller or rotating pins on the rocker shaft arm.

Other systems for steering exist, but are uncommon on road vehicles. Children's toys and go-karts often use a very direct linkage in the form of a bellcrank (also commonly known as a Pitman arm) attached directly between the steering column and the steering arms, and the use of cable-operated steering linkages (e.g. the capstan and bowstring mechanism) is also found on some home-built vehicles such as soapbox cars and recumbent tricycles.

Power Steering

Power steering helps the driver of a vehicle to steer by directing some of its power to assist in swiveling the steered road wheels about their steering axes. As vehicles have become heavier and switched to front wheel drive, particularly using negative offset geometry, along with increases in tire width and diameter, the effort needed to turn the wheels about their steering axis has increased, often to the point where major physical exertion would be needed were it not for power assistance. To alleviate this auto makers have developed power steering systems, or more correctly power-assisted steering, since on road-going vehicles there has to be a mechanical linkage as a fail-safe. There are two types of power steering systems: hydraulic and electric/electronic. A hydraulic-electric hybrid system is also possible.

A hydraulic power steering (HPS) uses hydraulic pressure supplied by an engine-driven pump to assist the motion of turning the steering wheel. Electric power steering (EPS) is more efficient than hydraulic power steering, since the electric power steering motor only needs to provide assistance when the steering wheel is turned, whereas the hydraulic pump must run constantly. In EPS, the amount of assistance is easily tunable to the vehicle type, road speed, and even driver preference. An added benefit is the elimination of environmental hazard posed by leakage and disposal of hydraulic power steering fluid. In addition, electrical assistance is not lost when the engine fails or stalls, whereas hydraulic assistance stops working if the engine stops, making the steering doubly heavy as the driver must now turn not only the very heavy steering—without any help—but also the power-assistance system itself.

Speed Sensitive Steering

An outgrowth of power steering is speed sensitive steering, where the steering is heavily assisted at low speed and lightly assisted at high speed. Auto makers perceive that motorists might need to make large steering inputs while manoeuvering for parking, but not while traveling at high speed. The first vehicle with this feature was the Citroën SM with its Diravi layout, although rather than altering the amount of assistance as in modern power steering systems, it altered the pressure on a centring cam which made the steering wheel try to "spring" back to the straight-ahead position. Modern speed-sensitive power steering systems reduce the mechanical or electrical assistance as the vehicle speed increases, giving a more direct feel. This feature is gradually becoming more common.

Four-wheel Steering

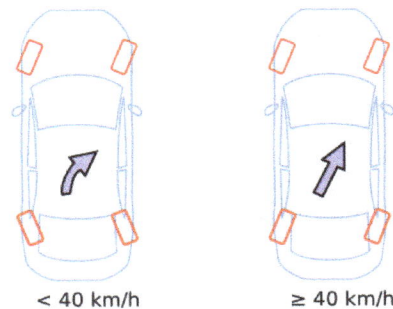

< 40 km/h   ≥ 40 km/h

Speed-dependent four-wheel steering.

Early example of four-wheel steering. 1910 photograph of 80 hp Caldwell Vale tractor in action.

Four-wheel steering is a system employed by some vehicles to improve steering response, increase vehicle stability while maneuvering at high speed, or to decrease turning radius at low speed.

1937 Mercedes-Benz Type G 5 with four-wheel steering.

Sierra Denali with Quadrasteer, rear steering angle.

Hamm DV70 tandem roller using crab steering to cover maximum road surface (2010).

Agricultural slurry applicator using crab steering to minimise soil compaction (2009).

Active Four-wheel Steering

In an active four-wheel steering system, all four wheels turn at the same time when the driver steers. In most active four-wheel steering systems, the rear wheels are steered by a computer and actuators. The rear wheels generally cannot turn as far as the front wheels. There can be controls to switch off the rear steer and options to steer only the rear wheels independently of the front wheels. At low speed (e.g. parking) the rear wheels turn opposite to the front wheels, reducing the turning radius by twenty-five percent, sometimes critical for large trucks or tractors and vehicles with trailers, while at higher speeds both front and rear wheels turn alike (electronically controlled), so that the vehicle may change position with less yaw, enhancing straight-line stability. The "snaking effect" experienced during motorway drives while towing a travel trailer is thus largely nullified.

Four-wheel steering found its most widespread use in monster trucks, where maneuverability in small arenas is critical, and it is also popular in large farm vehicles and trucks. Some of the modern European Intercity buses also utilize four-wheel steering to assist maneuverability in bus terminals, and also to improve road stability. The first rally vehicle to use the technology was the Peugeot 405 Turbo 16. Its debut was at the 1988 Pikes Peak International Hill Climb, where it set a record breaking time of 10:47.77. The car would go on to victory in the 1989 and 1990 Paris-Dakar Rally, again driven by Ari Vatanen.

Previously, Honda had four-wheel steering as an option in their 1987–2001 Prelude and Honda Ascot Innova models (1992–1996). Mazda also offered four-wheel steering on the 626 and MX6 in 1988. General Motors offered Delphi's Quadrasteer in their consumer Silverado/Sierra and Suburban/Yukon. However, only 16,500 vehicles were sold with this system from its introduction in 2002 through 2004. Due to this low demand, GM discontinued the technology at the end of the 2005 model year. Nissan/Infiniti offer several versions of their HICAS system as standard or as an option in much of their line-up. A new "Active Drive" system is introduced on the 2008 version of the Renault Laguna line. It was designed as one of several measures to increase security and stability. The Active Drive should lower the effects of under steer and decrease the chances of spinning by diverting part of the G-forces generated in a turn from the front to the rear tires. At low speeds the turning circle can be tightened so parking and maneuvering is easier.

Production Cars with Active Four Wheel Steering

- Audi Q7 (all-wheel steering, on second generation from 2015)
- Acura RLX (P-AWS)
- Acura TLX (P-AWS), front drive models
- BMW 850CSi (only Euro spec models)
- BMW 7-Series (2009 onward, part of sport package)
- BMW 6-series (2011 onwards, Integral Active Steering option)
- BMW 5-series (2011 onwards, Integral Active Steering option)
- Chevrolet Silverado (2002–2005) (high and low speed)
- Efini MS-9 (high and low speed)
- Ferrari GTC4Lusso
- GMC Sierra (2002–2005) (high and low speed)
- GMC Sierra Denali (2002–2004) (high and low speed)
- Honda Prelude (high and low speed, mechanical from 1987 to 1991, computerized from 1992–2001)
- Honda Accord (1991) (high and low speed, mechanical)
- Honda Ascot Innova (1992) (high and low speed, computerized from 1992–1996)
- Infiniti FX50 AWD (option on Sports package) (2008–Present) (high and low speed, fully electronic)
- Infiniti G35 Sedan (option on Sport models) (2007–Present) (high speed only?)

- Infiniti G35 Coupe (option on Sport models) (2006–Present) (high speed only)
- Infiniti J30t (touring package) (1993–1994)
- Infiniti M35 (option on Sport models) (2006–Present) (high speed only?)
- Infiniti M45 (option on Sport models) (2006–Present) (high speed only?)
- Infiniti Q45t (1989–1994) (high speed only?)
- Lexus GS (2013 onwards, if equipped with optional Lexus Dynamic Handling)
- Mazda 929 (1992–1995)(computerised, high and low speed)(all models)
- Mazda 626 (1988) (high and low speed)
- Mazda MX-6 (1989–1997) (high and low speed)
- Mazda RX-7 (optional, computerized, high and low speed)
- Mazda Xedos 9/Mazda Eunos 800 (1996–2003) (Optional, computerized, high and low speed)
- Mercedes-Benz Vito (London Taxi variant)
- Mitsubishi Galant (high speed only)
- Mitsubishi GTO (also sold as the Mitsubishi 3000GT and the Dodge Stealth) (Mechanical) (high speed only)
- Nissan Cefiro (A31) (high speed only)
- Nissan 180SX (HICAS option)
- Nissan 240SX/Silvia (option on SE models) (high speed only)
- Nissan 300ZX (all Twin-Turbo Z32 models) (high speed only)
- Nissan Laurel (later versions) (high speed only)
- Nissan Fuga/Infiniti M (high speed only)
- Nissan Silvia (option on all S13 models) (high speed only)
- Nissan Skyline GTS, GTS-R, GTS-X (1986) (high speed only)
- Nissan Skyline GT-R (high and low speed)
- Porsche 911 GT3 (Model 991) (high and low speed)
- Porsche 911 GT3 RS (Model 991) (high and low speed)
- Porsche 911 Turbo (Model 991/991.2) (high and low speed)
- Porsche 911 Turbo S (Model 991/991.2) (high and low speed)
- Porsche 918 Spyder (high and low speed)
- Renault Espace (part of "Multi-sense" system, optional from 5th generation)

- Renault Laguna (only in GT version of 3rd generation which was launched October 2007, GT launched on April 2008)

- Renault Mégane (GT versions of 4th generation)

- Renault Talisman

- Saab 9-3 Aero (2003-2008) (low speed only)

- Subaru Alcyone SVX JDM (1991–1996) (Japanese version: "L-CDX" only) (high speed only)

- Toyota Aristo (1997) (high and low speed?)

- Toyota Camry / Vista JDM 1988–1999 (Optional)

- Toyota Carina ED / Toyota Corona EXiV (world's first dual-mode switchable 2WS to 4WS)

- Toyota Celica (option on 5th and 6th generation, 1990–1993 ST183 and 1994–1997 ST203) (Dual-mode, high and low speed)

- Toyota Soarer (UZZ32)

Crab Steering

Crab steering is a special type of active four-wheel steering. It operates by steering all wheels in the same direction and at the same angle. Crab steering is used when the vehicle needs to proceed in a straight line but under an angle (i.e. when moving loads with a reach truck, or during filming with a camera dolly), or when the rear wheels may not follow the front wheel tracks (i.e. to reduce soil compaction when using rolling farm equipment).

Passive Rear Wheel Steering

Many modern vehicles have passive rear steering. On many vehicles, when cornering, the rear wheels tend to steer slightly to the outside of a turn, which can reduce stability. The passive steering system uses the lateral forces generated in a turn (through suspension geometry) and the bushings to correct this tendency and steer the wheels slightly to the inside of the corner. This improves the stability of the car, through the turn. This effect is called compliance understeer and it, or its opposite, is present on all suspensions. Typical methods of achieving compliance understeer are to use a Watt's link on a live rear axle, or the use of toe control bushings on a twist beam suspension. On an independent rear suspension it is normally achieved by changing the rates of the rubber bushings in the suspension. Some suspensions typically have compliance oversteer due to geometry, such as Hotchkiss live axles or a semi-trailing arm IRS, but may be mitigated by revisions to the pivot points of the leaf spring or trailing arm.

Passive rear wheel steering is not a new concept, as it has been in use for many years, although not always recognised as such.

Articulated Steering

Articulated steering is a system by which a four-wheel drive vehicle is split into front and rear halves which are connected by a vertical hinge. The front and rear halves are connected with one or more hydraulic cylinders that change the angle between the halves, including the front and rear axles and wheels, thus steering the vehicle. This system does not use steering arms, king pins, tie rods, etc. as does four-wheel steering. If the vertical hinge is placed equidistant between the two axles, it also eliminates the need for a central differential, as both front and rear axles will follow the same path, and thus rotate at the same speed. Long road trains, articulated buses, and internal transport trolley trains use articulated steering to achieve smaller turning circles, comparable to those of shorter conventional vehicles. Articulated haulers have very good off-road performance.

Front loader with articulated steering (2007).

Rear Wheel Steering

A few types of vehicle use only rear wheel steering, notably fork lift trucks, camera dollies, early pay loaders, Buckminster Fuller's Dymaxion car, and the ThrustSSC.

Rear wheel steering tends to be unstable because in turns the steering geometry changes hence decreasing the turn radius (over steer), rather than increase it (under steer). Rear wheel steering is meant for slower vehicles that need high-maneuverability in tight spaces, e.g. fork lifts.

Steer-by-wire

1971 Lunar Roving Vehicle (LRV) with joystick steering controls.

2012 Honda EV-STER "Twin Lever Steering" concept.

The aim of steer-by-wire technology is to completely do away with as many mechanical components (steering shaft, column, gear reduction mechanism, etc.) as possible. Completely replacing conventional steering system with steer-by-wire holds several advantages, such as:

- The absence of steering column simplifies the car interior design.

- The absence of steering shaft, column and gear reduction mechanism allows much better space utilization in the engine compartment.

- The steering mechanism can be designed and installed as a modular unit.

- Without mechanical connection between the steering wheel and the road wheel, it is less likely that the impact of a frontal crash will force the steering wheel to intrude into the driver's survival space.

- Steering system characteristics can easily and infinitely be adjusted to optimize the steering response and feel.

As of 2007 there are no production cars available that rely solely on steer-by-wire technology due to safety, reliability and economic concerns, but this technology has been demonstrated in numerous concept cars and the similar fly-by-wire technology is in use in both military and civilian aviation applications. Removing the mechanical steering linkage in road going vehicles would require new legislation in most countries.

Safety

For safety reasons all modern cars feature a collapsible steering column (energy absorbing steering column) which will collapse in the event of a heavy frontal impact to avoid excessive injuries to the driver. Airbags are also generally fitted as standard. Non-collapsible steering columns fitted to older vehicles very often impaled drivers in frontal crashes, particularly when the steering box or rack was mounted in front of the front axle line, at the front of the crumple zone. This was particularly a problem on vehicles that had a rigid separate chassis frame, with no crumple zone. Most modern vehicle steering boxes/racks are mounted behind the front axle on the front bulkhead, at the rear of the front crumple zone.

Collapsible steering columns were invented by Bela Barenyi and were introduced in the 1959 Mercedes-Benz W111 Fintail, along with crumple zones. This safety feature first appeared on cars built by General Motors after an extensive and very public lobbying campaign enacted by Ralph Nader. Ford started to install collapsible steering columns in 1968.

Audi used a retractable steering wheel and seat belt tensioning system called procon-ten, but it has since been discontinued in favor of airbags and pyrotechnic seat belt pre-tensioners.

Cycles

Steering is crucial to the stability of bicycles and motorcycles. Steering monocycles and unicycles is especially complicated.

Watercraft Steering

Ships and boats are usually steered with a rudder. Depending on the size of the vessel, rudders can be manually actuated, or operated using a servomechanism, or a trim tab/servo tab system. Boats using outboard motors steer by rotating the entire drive unit. Boats with inboard motors sometimes steer by rotating the propeller pod only (i.e. Volvo Penta IPS drive). Modern ships with diesel-electric drive use azimuth thrusters. Boats driven by oars (i.e. rowing boats, including gondolas) or paddles (i.e. canoes, kayaks, rafts) are steered by generating a higher propulsion force on the side of the boat opposite of the direction of turn. Jet skis are steered by weight-shift induced roll and water jet thrust vectoring. Water skis and surfboards are steered by weight-shift induced roll only.

Aircraft and Hovercraft Steering

Airplanes are normally steered by the use of ailerons to bank the aircraft into a turn - the rudder is used to minimise adverse yaw, rather than as a means to directly cause the turn. Missiles, airships and hovercraft are usually steered by rudder and/or thrust vectoring. Jet packs and flying platforms are steered by thrust vectoring only. Helicopters are steered by cyclic control, changing the thrust vector of the main rotor(s), and by anti-torque control, usually provided by a tail rotor.

Other Types of Steering

Tunnel boring machines are steered by hydraulic tilting of the cutter head. Rail track vehicles (i.e. trains, trams) are steered by curved guide tracks, including switches, and articulated undercarriages. Land yachts on wheels and kite buggies are steered similarly to cars. Ice yachts and bobsleighs are steered by rotating the front runners out of the direction of travel. Snowmobiles steer the same way by rotating the front skis. Tracked vehicles (i.e. tanks) steer by increasing the drive force on the side opposite of the direction of turn. Horse-drawn sleighs and dog sleds are steered by changing the direction of pull. Zero-turn lawn mowers use independent hydraulic wheel drive to turn on the spot.

Suspension (Vehicle)

The front suspension components of a Ford Model T.

The rear suspension on a truck: a leaf spring.

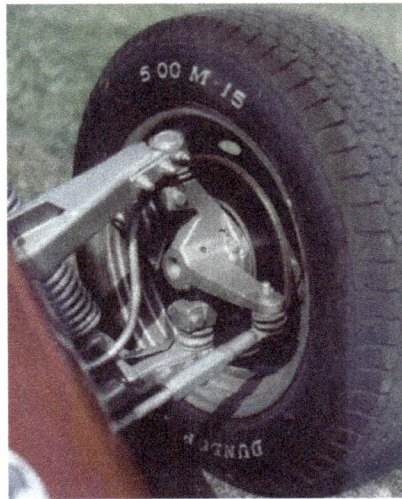

Part of car front suspension and steering mechanism: tie rod, steering arm, king pin axis (using ball joints).

Van Diemen RF01 Racing Car Suspension.

Suspension is the system of tires, tire air, springs, shock absorbers and linkages that connects a vehicle to its wheels and allows relative motion between the two. Suspension systems serve a dual purpose — contributing to the vehicle's roadholding/handling and braking for good active safety and driving pleasure, and keeping vehicle occupants comfortable and a ride quality reasonably well isolated from road noise, bumps, vibrations,etc. These goals are generally at odds, so the tuning of suspensions involves finding the right compromise. It is important for the suspension to keep the road wheel in contact with the road surface as much as possible, because all the road or ground forces acting on the vehicle do so through the contact patches of the tires. The suspension also protects the vehicle itself and any cargo or luggage from damage and wear. The design of front and rear suspension of a car may be different.

History

Strap suspension 1605

Strap suspension 2008
note the transverse limiting straps

An early form of suspension on ox-drawn carts had the platform swing on iron chains attached to the wheeled frame of the carriage. This system remained the basis for all suspension systems until the turn of the 19th century, although the iron chains were replaced with the use of leather straps by the 17th century. No modern automobiles use the 'strap suspension' system.

Automobiles were initially developed as self-propelled versions of horse-drawn vehicles. However, horse-drawn vehicles had been designed for relatively slow speeds, and their suspension was not well suited to the higher speeds permitted by the internal combustion engine.

The first workable spring-suspension required advanced metallurgical knowledge and skill, and only became possible with the advent of industrialisation. Obadiah Elliott registered the first patent for a spring-suspension vehicle; - each wheel had two durable steel leaf springs on each side and the body of the carriage was fixed directly to the springs attached to the axles. Within a decade, most British horse carriages were equipped with springs; wooden springs in the case of light one-horse vehicles to avoid taxation, and steel springs in larger vehicles. These were often made of low-carbon steel and usually took the form of multiple layer leaf springs.

Leaf springs have been around since the early Egyptians. Ancient military engineers used leaf springs in the form of bows to power their siege engines, with little success at first. The use of leaf springs in catapults was later refined and made to work years later. Springs were not only made of metal, a sturdy tree branch could be used as a spring, such as with a bow. Horse-drawn carriages and the Ford Model T used this system, and it is still used today in larger vehicles, mainly mounted in the rear suspension.

This was the first modern suspension system and, along with advances in the construction of roads, heralded the single greatest improvement in road transport until the advent of the automobile. The British steel springs were not well suited for use on America's rough roads of the time, so the Abbot Downing Company of Concord, New Hampshire re-introduced leather strap suspension, which gave a swinging motion instead of the jolting up and down of a spring suspension.

Henri Fournier on his uniquely damped and racewinning 'Mors Machine', photo taken 1902

In 1901 Mors of Paris first fitted an automobile with shock absorbers. With the advantage of a damped suspension system on his 'Mors Machine', Henri Fournier won the prestigious Paris-to-Berlin race on 20 June 1901. Fournier's superior time was 11 hrs 46 min 10 sec, while the best competitor was Léonce Girardot in a Panhard with a time of 12 hrs 15 min 40 sec.

Coil springs first appeared on a production vehicle in 1906 in the Brush Runabout made by the Brush Motor Company. Today, coil springs are used in most cars.

In 1920, Leyland Motors used torsion bars in a suspension system.

In 1922, independent front suspension was pioneered on the Lancia Lambda and became more common in mass market cars from 1932. Today most cars have independent suspension on all four wheels.

In 2002, a new passive suspension component was invented by Malcolm C. Smith, the inerter. This has the ability to increase the effective inertia of a wheel suspension using a geared flywheel, but without adding significant mass. It was initially employed in Formula 1 in secrecy but has since spread to other motorsport.

Difference between Rear Suspension and Front Suspension

Obviously any four wheel vehicle needs suspension for both the front wheels and the rear suspension, but in two wheel drive vehicles these can be very different configuration. For front-wheel drive cars, rear suspension has few constraints and a variety of beam axles and independent suspensions are used. For rear-wheel drive cars, rear suspension has many constraints and the development of the superior but more expensive independent suspension layout has been difficult. Four-wheel drive often have suspensions that are similar for both the front and rear wheels.

History

Henry Ford's Model T used a torque tube to restrain this force, for his differential was attached to the chassis by a lateral leaf spring and two narrow rods. The torque tube surrounded the true driveshaft and exerted the force to its ball joint at the extreme rear of the transmission, which was attached to the engine. A similar method was used by the late-1930s Buick and by Hudson's bath-tub car of 1948, which used helical springs which could not take fore-and-aft thrust.

The Hotchkiss drive, invented by Albert Hotchkiss, was the most popular rear suspension system used in American cars from the 1930s to the 1970s. The system uses longitudinal leaf springs attached both forward and behind the differential of the live axle. These springs transmit the torque to the frame. Although scorned by many European car makers of the time, it was accepted by American car makers because it was inexpensive to manufacture. Also, the dynamic defects of this design were suppressed by the enormous weight of US passenger vehicles before implementation the Corporate Average Fuel Economy standard.

Another Frenchman invented the De Dion tube, which is sometimes called "semi-independent". Like a true independent rear suspension, this employs two universal joints or their equivalent from the center of the differential to each wheel. But the wheels cannot entirely rise and fall independently of each other; they are tied by a yoke that goes around the differential, below and behind it. This method has had little use in the United States, though it does not evidence the bump steer that a more expensive, true independent suspension does. Its use around 1900 was probably due to the poor quality of tires, which wore out quickly. By removing a good deal of unsprung weight, as independent rear suspensions do, it made them last longer.

Rear wheel drive vehicles today frequently use a fairly complex fully independent, multi-link suspension to locate the rear wheels securely while providing decent ride quality.

Spring, Wheel, and Roll Rates

Citroën BX Hydropneumatic suspension - maximum to minimum demonstration

Spring Rate

The spring rate (or suspension rate) is a component in setting the vehicle's ride height or its location in the suspension stroke. When a spring is compressed or stretched, the force it exerts is proportional to its change in length. The spring rate or spring constant of a spring is the change in the force it exerts, divided by the change in deflection of the spring. Vehicles which carry heavy loads will often have heavier springs to compensate for the additional weight that would otherwise collapse a vehicle to the bottom of its travel (stroke). Heavier springs are also used in performance applications where the loading conditions experienced are more extreme.

Springs that are too hard or too soft cause the suspension to become ineffective because they fail to properly isolate the vehicle from the road. Vehicles that commonly experience suspension loads heavier than normal have heavy or hard springs with a spring rate close to the upper limit for that vehicle's weight. This allows the vehicle to perform properly under a heavy load when control is limited by the inertia of the load. Riding in an empty truck used for carrying loads can be uncomfortable for passengers because of its high spring rate relative to the weight of the vehicle. A race car would also be described as having heavy springs and would also be uncomfortably bumpy. However, even though we say they both have heavy springs, the actual spring rates for a 2,000 lb (910 kg) race car and a 10,000 lb (4,500 kg) truck are very different. A luxury car, taxi, or passenger bus would be described as having soft springs. Vehicles with worn out or damaged springs ride lower to the ground which reduces the overall amount of compression available to the suspension and increases the amount of body lean. Performance vehicles can sometimes have spring rate requirements other than vehicle weight and load.

Mathematics of the Spring Rate

Spring rate is a ratio used to measure how resistant a spring is to being compressed or expanded during the spring's deflection. The magnitude of the spring force increases as deflection increases according to Hooke's Law. Briefly, this can be stated as

$$F = -kx$$

where

F is the force the spring exerts

k is the spring rate of the spring.

x is the deflection of the spring from its equilibrium position (i.e., when no force is applied on the spring)

Negative sign indicates direction of applied force and force exerted by spring are opposite. Spring rate is confined to a narrow interval by the weight of the vehicle, load the vehicle will carry, and to a lesser extent by suspension geometry and performance desires.

Spring rates typically have units of N/mm (or lbf/in). An example of a linear spring rate is 500 lbf/in. For every inch the spring is compressed, it exerts 500 lbf. A non-linear spring rate is one for which the relation between the spring's compression and the force exerted cannot be fitted adequately to a linear model. For example, the first inch exerts 500 lbf force, the second inch exerts an additional 550 lbf (for a total of 1050 lbf), the third inch exerts another 600 lbf (for a total of 1650 lbf). In contrast a 500 lbf/in linear spring compressed to 3 inches will only exert 1500 lbf.

The spring rate of a coil spring may be calculated by a simple algebraic equation or it may be measured in a spring testing machine. The spring constant k can be calculated as follows:

$$k = \frac{d^4 G}{8ND^3}$$

where d is the wire diameter, G is the spring's shear modulus (e.g., about 12,000,000 lbf/in² or 80 GPa for steel), N is the number of wraps and D is the diameter of the coil.

Wheel Rate

Wheel rate is the effective spring rate when measured at the wheel. This is as opposed to simply measuring the spring rate alone.

Wheel rate is usually equal to or considerably less than the spring rate. Commonly, springs are mounted on control arms, swing arms or some other pivoting suspension member. Consider the example above where the spring rate was calculated to be 500 lbs/inch (87.5 N/mm), if you were to move the wheel 1 in (2.5 cm) (without moving the car), the spring more than likely compresses a smaller amount. Let's assume the spring moved 0.75 in (19 mm), the lever arm ratio would be 0.75:1. The wheel rate is calculated by taking the square of the ratio (0.5625) times the spring rate, thus obtaining 281.25 lbs/inch (49.25 N/mm). Squaring the ratio is because the ratio has two effects on the wheel rate. The ratio applies to both the force and distance traveled.

Wheel rate on independent suspension is fairly straightforward. However, special consideration must be taken with some non-independent suspension designs. Take the case of the straight axle. When viewed from the front or rear, the wheel rate can be measured by the means above. Yet because the wheels are not independent, when viewed from the side under acceleration or braking the pivot point is at infinity (because both wheels have moved) and the spring is directly inline with the wheel contact patch. The result is often that the effective wheel rate under cornering is different from what it is under acceleration and braking. This variation in wheel rate may be minimized by locating the spring as close to the wheel as possible.

Wheel rates are usually summed and compared with the sprung mass of a vehicle to create a "ride rate" and corresponding suspension natural frequency in ride (also referred to as "heave"). This can be useful in creating a metric for suspension stiffness and travel requirements for a vehicle.

Roll Rate

Roll rate is analogous to a vehicle's ride rate, but for actions that include lateral accelerations, causing a vehicle's sprung mass to roll about its roll axis. It is expressed as torque per degree of roll of the vehicle sprung mass. It is influenced by factors including but not limited to vehicle sprung mass, track width, CG height, spring and damper rates, roll center heights of front and rear, anti-roll bar stiffness and tire pressure/construction. The roll rate of a vehicle can, and usually does, differ front to rear, which allows for the tuning ability of a vehicle for transient and steady state handling. The roll rate of a vehicle does not change the total amount of weight transfer on the vehicle, but shifts the speed at which and percentage of weight transferred on a particular axle to another axle through the vehicle chassis. Generally, the higher the roll rate on an axle of a vehicle, the faster and higher percentage the weight transfer on that axle.

Roll Couple Percentage

Roll couple percentage is a simplified method of describing lateral load transfer distribution front to rear, and subsequently handling balance. It is the effective wheel rate, in roll, of each axle of the vehicle as a ratio of the vehicle's total roll rate. It is commonly adjusted through the use of anti-roll bars, but can also be changed through the use of different springs.

Weight Transfer

Weight transfer during cornering, acceleration or braking is usually calculated per individual wheel and compared with the static weights for the same wheels.

The total amount of weight transfer is only affected by four factors: the distance between wheel centers (wheelbase in the case of braking, or track width in the case of cornering) the height of the center of gravity, the mass of the vehicle, and the amount of acceleration experienced.

The speed at which weight transfer occurs as well as through which components it transfers is complex and is determined by many factors including but not limited to roll center height, spring and damper rates, anti-roll bar stiffness and the kinematic design of the suspension links. In most conventional applications, when weight is transferred through intentionally compliant elements such as springs, dampers and anti-roll bars, the weight transfer is said to be "elastic", while the weight which is transferred through more rigid suspension links such as A-arms and toe links is said to be "geometric".

Unsprung Weight Transfer

Unsprung weight transfer is calculated based on the weight of the vehicle's components that are not supported by the springs. This includes tires, wheels, brakes, spindles, half the control arm's weight and other components. These components are then (for calculation purposes) assumed to be connected to a vehicle with zero sprung weight. They are then put through the same dynamic loads. The weight transfer for cornering in the front would be equal to the total unsprung front weight times the G-Force times the front unsprung center of gravity height divided by the front track width. The same is true for the rear.

Sprung Weight Transfer

Sprung weight transfer is the weight transferred by only the weight of the vehicle resting on the springs, not the total vehicle weight. Calculating this requires knowing the vehicle's sprung weight (total weight less the unsprung weight), the front and rear roll center heights and the sprung center of gravity height (used to calculate the roll moment arm length). Calculating the front and rear sprung weight transfer will also require knowing the roll couple percentage.

The roll axis is the line through the front and rear roll centers that the vehicle rolls around during cornering. The distance from this axis to the sprung center of gravity height is the roll moment arm length. The total sprung weight transfer is equal to the G-force times the sprung weight times the roll moment arm length divided by the effective track width. The front sprung weight transfer is calculated by multiplying the roll couple percentage times the total sprung weight transfer. The rear is the total minus the front transfer.

Jacking Forces

Jacking forces are the sum of the vertical force components experienced by the suspension links.

The resultant force acts to lift the sprung mass if the roll center is above ground, or compress it if underground. Generally, the higher the roll center, the more jacking force is experienced.

Other Properties

Travel

Travel is the measure of distance from the bottom of the suspension stroke (such as when the vehicle is on a jack and the wheel hangs freely) to the top of the suspension stroke (such as when the vehicle's wheel can no longer travel in an upward direction toward the vehicle). Bottoming or lifting a wheel can cause serious control problems or directly cause damage. "Bottoming" can be caused by the suspension, tires, fenders, etc. running out of space to move or the body or other components of the car hitting the road. The control problems caused by lifting a wheel are less severe if the wheel lifts when the spring reaches its unloaded shape than they are if travel is limited by contact of suspension members. Many off-road vehicles, such as desert racers, use straps called "limiting straps" to limit the suspensions downward travel to a point within safe limits for the linkages and shock absorbers. This is necessary, since these trucks are intended to travel over very rough terrain at high speeds, and even become airborne at times. Without something to limit the travel, the suspension bushings would take all the force when the suspension reaches "full droop", and it can even cause the coil springs to come out of their "buckets" if they are held in by compression forces only. A limiting strap is a simple strap, often nylon of a predetermined length, that stops the downward movement at a preset point before the theoretical maximum travel is reached. The opposite of this is the "bump-stop", which protects the suspension and vehicle (as well as the occupants) from violent "bottoming" of the suspension, caused when an obstruction (or hard landing) causes the suspension to run out of upward travel without fully absorbing the energy of the stroke. Without bump-stops, a vehicle that "bottoms out" will experience a very hard shock when the suspension contacts the bottom of the frame or body, which is transferred to the occupants and every connector and weld on the vehicle. Factory vehicles often come with plain rubber "nubs" to absorb the worst of the forces, and insulate the shock. A desert race vehicle, which must routinely absorb far higher impact forces, may be provided with pneumatic or hydro-pneumatic bump-stops. These are essentially miniature shock absorbers (dampeners) that are fixed to the vehicle in a location such that the suspension will contact the end of the piston when it nears the upward travel limit. These absorb the impact far more effectively than a solid rubber bump-stop will, essential because a rubber bump-stop is considered a "last-ditch" emergency insulator for the occasional accidental bottoming of the suspension; it is entirely insufficient to absorb repeated and heavy bottomings such as a high-speed off road vehicle encounters.

Damping

Damping is the control of motion or oscillation, as seen with the use of hydraulic gates and valves in a vehicle's shock absorber. This may also vary, intentionally or unintentionally. Like spring rate, the optimal damping for comfort may be less than for control.

Damping controls the travel speed and resistance of the vehicle's suspension. An undamped car will oscillate up and down. With proper damping levels, the car will settle back to a normal state in a minimal amount of time. Most damping in modern vehicles can be controlled by increasing or decreasing the resistance to fluid flow in the shock absorber.

Camber Control

Camber changes due to wheel travel, body roll and suspension system deflection or compliance. In general, a tire wears and brakes best at -1 to -2° of camber from vertical. Depending on the tire and the road surface, it may hold the road best at a slightly different angle. Small changes in camber, front and rear, can be used to tune handling. Some race cars are tuned with -2 to -7° camber depending on the type of handling desired and the tire construction. Often, too much camber will result in the decrease of braking performance due to a reduced contact patch size through excessive camber variation in the suspension geometry. The amount of camber change in bump is determined by the instantaneous front view swing arm (FVSA) length of the suspension geometry, or in other words, the tendency of the tire to camber inward when compressed in bump.

Roll Center Height

Roll center height is a product of suspension instant center heights and is a useful metric in analyzing weight transfer effects, body roll and front to rear roll stiffness distribution. Conventionally, roll stiffness distribution is tuned adjusting antiroll bars rather than roll center height (as both tend to have a similar effect on the sprung mass), but the height of the roll center is significant when considering the amount of jacking forces experienced.

Instant Center

Due to the fact that the wheel and tire's motion is constrained by the suspension links on the vehicle, the motion of the wheel package in the front view will scribe an imaginary arc in space with an "instantaneous center" of rotation at any given point along its path. The instant center for any wheel package can be found by following imaginary lines drawn through the suspension links to their intersection point.

A component of the tire's force vector points from the contact patch of the tire through instant center. The larger this component is, the less suspension motion will occur. Theoretically, if the resultant of the vertical load on the tire and the lateral force generated by it points directly into the instant center, the suspension links will not move. In this case, all weight transfer at that end of the vehicle will be geometric in nature. This is key information used in finding the force-based roll center as well.

In this respect the instant centers are more important to the handling of the vehicle than the kinematic roll center alone, in that the ratio of geometric to elastic weight transfer is determined by the forces at the tires and their directions in relation to the position of their respective instant centers.

Variations in Suspension Design

Anti-dive and anti-squat

Anti-dive and anti-squat are percentages that indicate the degree to which the front dives under braking and the rear squats under acceleration. They can be thought of as the counterparts for braking and acceleration, as jacking forces are to cornering. The main reason for the difference is due to the different design goals between front and rear suspension, whereas suspension is usually symmetrical between the left and right of the vehicle.

The method of determining the anti-dive or anti-squat depends on whether the suspension linkages react to the torque of braking and accelerating. For example, with inboard brakes and half-shaft driven rear wheels, the suspension linkages do not, but with outboard brakes and a swing-axle driveline, they do.

To determine the percentage of front suspension braking anti-dive for outboard brakes, it is first necessary to determine the tangent of the angle between a line drawn, in side view, through the front tire patch and the front suspension instant center, and the horizontal. In addition, the percentage of braking effort at the front wheels must be known. Then, multiply the tangent by the front wheel braking effort percentage and divide by the ratio of the center of gravity height to the wheelbase. A value of 50% would mean that half of the weight transfer to the front wheels, during braking, is being transmitted through the front suspension linkage and half is being transmitted through the front suspension springs.

For inboard brakes, the same procedure is followed but using the wheel center instead of contact patch center.

Forward acceleration anti-squat is calculated in a similar manner and with the same relationship between percentage and weight transfer. Anti-squat values of 100% and more are commonly used in drag racing, but values of 50% or less are more common in cars that have to undergo severe braking. Higher values of anti-squat commonly cause wheel hop during braking. It is important to note that, while the value of 100% means that all of the weight transfer is being carried through the suspension linkage. However, this does not mean that the suspension is incapable of carrying additional loads (aerodynamic, cornering, etc.) during an episode of braking or forward acceleration. In other words, no "binding" of the suspension is to be implied.

Flexibility and Vibration Modes of the Suspension Elements

In some modern cars, the flexibility is mainly in the rubber bushings, which are subject to decay over time. For high-stress suspensions, such as off-road vehicles, polyurethane bushings are available, which offer more longevity under greater stresses. However, due to weight and cost considerations, structures are not made more rigid than necessary. Some vehicles exhibit detrimental vibrations involving the flexing of structural parts, such as when accelerating while turning sharply. Flexibility of structures such as frames and suspension links can also contribute to springing, especially to damping out high frequency vibrations. The flexibility of wire wheels contributed to their popularity in times when cars had less advanced suspensions.

Load Levelling

Automobiles can be heavily laden with luggage, passengers, and trailers. This loading will cause a vehicle's tail to sink downwards. Maintaining a steady chassis level is essential to achieving the proper handling the vehicle was designed for. Oncoming drivers can be blinded by the headlight beam. Self-levelling suspension counteracts this by inflating cylinders in the suspension to lift the chassis higher.

Isolation from High Frequency Shock

For most purposes, the weight of the suspension components is unimportant, but at high frequencies, caused by road surface roughness, the parts isolated by rubber bushings act as a multistage filter to suppress noise and vibration better than can be done with only the tires and springs. (The springs work mainly in the vertical direction.)

Contribution to Unsprung Weight and Total Weight

These are usually small, except that the suspension is related to whether the brakes and differential(s) are sprung.

This is the main functional advantage of aluminum wheels over steel wheels. Aluminum suspension parts have been used in production cars, and carbon fiber suspension parts are common in racing cars.

Space Occupied

Designs differ as to how much space they take up and where it is located. It is generally accepted that MacPherson struts are the most compact arrangement for front-engined vehicles, where space between the wheels is required to place the engine.

Inboard brakes (which reduce unsprung weight) are probably avoided more due to space considerations than to cost.

Force Distribution

The suspension attachment must match the frame design in geometry, strength and rigidity.

Air Resistance (Drag)

Certain modern vehicles have height adjustable suspension in order to improve aerodynamics and fuel efficiency. Modern formula cars that have exposed wheels and suspension typically use streamlined tubing rather than simple round tubing for their suspension arms to reduce aerodynamic drag. Also typical is the use of rocker arm, push rod, or pull rod type suspensions that, among other things, place the spring/damper unit inboard and out of the air stream to further reduce air resistance.

Cost

Production methods improve, but cost is always a factor. The continued use of the solid rear axle, with unsprung differential, especially on heavy vehicles, seems to be the most obvious example.

Springs and Dampers

Most conventional suspensions use passive springs to absorb impacts and dampers (or shock absorbers) to control spring motions.

Some notable exceptions are the hydropneumatic systems, which can be treated as an integrated unit of gas spring and damping components, used by the French manufacturer Citroën and the hydrolastic, hydragas and rubber cone systems used by the British Motor Corporation, most notably on the Mini. A number of different types of each have been used:

Passive Suspensions

Traditional springs and dampers are referred to as passive suspensions — most vehicles are suspended in this manner.

Springs

Pneumatic spring on a semitrailer

The majority of land vehicles are suspended by steel springs, of these types:

- Leaf spring – AKA Hotchkiss, Cart, or semi-elliptical spring
- Torsion beam suspension
- Coil spring

Automakers are aware of the inherent limitations of steel springs, that they tend to produce undesirable oscillations, and have developed other types of suspension materials and mechanisms in attempts to improve performance:

- Rubber bushing
- Gas under pressure - air spring
- Gas and hydraulic fluid under pressure - hydropneumatic suspension and oleo strut

Dampers or Shock Absorbers

The shock absorbers damp out the (otherwise simple harmonic) motions of a vehicle up and down on its springs. They also must damp out much of the wheel bounce when the unsprung weight of a wheel, hub, axle and sometimes brakes and differential bounces up and down on the springiness of a tire. Some have suggested that the regular bumps found on dirt roads (nicknamed "corduroy", but properly corrugations or washboarding) are caused by this wheel bounce, though some evidence exists that it is unrelated to suspension at all.

Semi-active and Active Suspensions

If the suspension is externally controlled then it is a semi-active or active suspension — the suspension is reacting to signals from an electronic controller.

For example, a hydropneumatic Citroën will "know" how far off the ground the car is supposed to be and constantly reset to achieve that level, regardless of load. It will not instantly compensate for body roll due to cornering however. Citroën's system adds about 1% to the cost of the car versus passive steel springs.

Semi-active suspensions include devices such as air springs and switchable shock absorbers, various self-levelling solutions, as well as systems like hydropneumatic, hydrolastic, and hydragas suspensions. Mitsubishi developed the world's first production semi-active electronically controlled suspension system in passenger cars; the system was first incorporated in the 1987 Galant model. Delphi currently sells shock absorbers filled with a magneto-rheological fluid, whose viscosity can be changed electromagnetically, thereby giving variable control without switching valves, which is faster and thus more effective.

Fully active suspension systems use electronic monitoring of vehicle conditions, coupled with the means to impact vehicle suspension and behavior in real time to directly control the motion of the car. Lotus Cars developed several prototypes, from 1982 onwards, and introduced them to F1, where they have been fairly effective, but have now been banned. Nissan introduced a low bandwidth active suspension in circa 1990 as an option that added an extra 20% to the price of luxury models. Citroën has also developed several active suspension models. A recently publicised fully active system from Bose Corporation uses linear electric motors (i.e., solenoids) in place of hydraulic or pneumatic actuators that have generally been used up until recently. Mercedes introduced an active suspension system called Active Body Control in its top-of-the-line Mercedes-Benz CL-Class in 1999.

Several electromagnetic suspensions have also been developed for vehicles. Examples include the electromagnetic suspension of Bose, and the electromagnetic suspension developed by prof. Laurentiu Encica. In addition, the new Michelin wheel with embedded suspension working on an electromotor is also similar.

With the help of control system, various semi-active/active suspensions realize an improved design compromise among different vibrations modes of the vehicle, namely bounce, roll, pitch and warp modes. However, the applications of these advanced suspensions are constrained by the cost, packaging, weight, reliability, and/or the other challenges.

Interconnected Suspensions

Interconnected suspension, unlike semi-active/active suspensions, could easily decouple different vehicle vibration modes in a passive manner. The interconnections can be realized by various means, such as mechanical, hydraulic and pneumatic. Anti-roll bars are one of the typical examples of mechanical interconnections, while it has been stated that fluidic interconnections offer greater potential and flexibility in improving both the stiffness and damping properties.

Considering the considerable commercial potentials of hydro-pneumatic technology (Corolla, 1996), interconnected hydropneumatic suspensions have also been explored in some recent studies, and their potential benefits in enhancing vehicle ride and handling have been demonstrated. The control system can also be used for further improving performance of interconnected suspensions. Apart from academic research, an Australian company, Kinetic, is having some success (WRC: 3 Championships, Dakar Rally: 2 Championships, Lexus GX470 2004 4x4 of the year with KDSS, 2005 PACE award) with various passive or semi-active systems, which generally decouple at least two vehicle modes (roll, warp (articulation), pitch and/or heave (bounce)) to simultaneously control each mode's stiffness and damping, by using interconnected shock absorbers, and other methods. In 1999, Kinetic was bought out by Tenneco. Later developments by a Catalan company, Creuat has devised a simpler system design based on single-acting cylinders. After some projects on competition Creuat is active in providing retrofit systems for some vehicle models.

Historically, the first mass production car with front to rear mechanical interconnected suspension was the 1948 Citroën 2CV. The suspension of the 2CV was extremely soft — the longitudinal link was making pitch softer instead of making roll stiffer. It relied on extreme antidive and antisquat geometries to compensate for that. This redunded into a softer axle crossing stiffness that anti-roll bars would have otherwise compromised. The leading arm / trailing arm swinging arm, fore-aft linked suspension system together with inboard front brakes had a much smaller unsprung weight than existing coil spring or leaf designs. The interconnection transmitted some of the force deflecting a front wheel up over a bump, to push the rear wheel down on the same side. When the rear wheel met that bump a moment later, it did the same in reverse, keeping the car level front to rear. The 2CV had a design brief to be able to be driven at speed over a ploughed field. It originally featured friction dampers and tuned mass dampers. Later models had tuned mass dampers at the front with telescopic dampers/shock absorbers front and rear.

The British Motor Corporation was also an early adopter of interconnected suspension. A system dubbed Hydrolastic was introduced in 1962 on the Morris 1100 and went on to be used on a variety of BMC models. Hydrolastic was developed by suspension engineer Alex Moulton and used rubber cones as the springing medium (these were first used on the 1959 Mini) with the suspension units on each side connected to each other by a fluid filled pipe. The fluid transmitted the force of road bumps from one wheel to the other (on the same principle as the Citroen 2CV's mechanical system described above) and because each suspension unit contained valves to restrict the flow of fluid also served as a shock absorber. Moulton went on to develop a replacement for Hydrolastic for BMC's successor, British Leyland. This system, called Hydragas worked on the same principle but instead of rubber spring units it used metal spheres divided internally by a rubber diaphragm. The top half contained pressurised gas and the lower half the same fluid as used on the Hydrolastic system. The fluid transmitted suspension forces between the units on each side whilst the gas acted as the springing medium via the diaphragm. This is the same principle as the Citroen hydropneumatic system and provides a similar ride quality but is self-contained and doesn't require an engine-driven pump to provide hydraulic pressure. The downside is that Hydragas is, unlike the Citroen system, not height adjustable or self-levelling. Hydragas was introduced in 1973 on the Austin Allegro and was used on several models, the last car to use it being the MG F in 2002.

Some of the last post-war Packard models also featured interconnected suspension.

Suspension Geometry

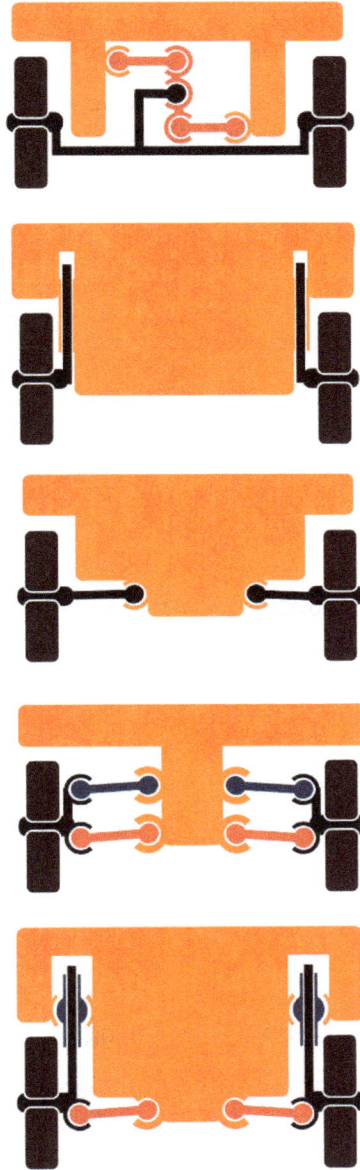

Common types seen from behind; in order:

- Live axle with Watt bar
- Suspension like on a bike fork
- Swing axle
- Double wishbone suspension
- MacPherson

This diagram is not exhaustive; notably excluding elements such as trailing arm links and those that are flexible.

Suspension systems can be broadly classified into two subgroups: dependent and independent. These terms refer to the ability of opposite wheels to move independently of each other.

A dependent suspension normally has a beam (a simple 'cart' axle) or (driven) live axle that holds wheels parallel to each other and perpendicular to the axle. When the camber of one wheel changes, the camber of the opposite wheel changes in the same way (by convention on one side this is a positive change in camber and on the other side this a negative change). De Dion suspensions are also in this category as they rigidly connect the wheels together.

An independent suspension allows wheels to rise and fall on their own without affecting the opposite wheel. Suspensions with other devices, such as sway bars that link the wheels in some way are still classed as independent.

A third type is a semi-dependent suspension. In this case, the motion of one wheel does affect the position of the other but they are not rigidly attached to each other. A twist-beam rear suspension is such a system.

Dependent Suspensions

Dependent systems may be differentiated by the system of linkages used to locate them, both longitudinally and transversely. Often both functions are combined in a set of linkages.

Examples of location linkages include:
- Satchell link
- Panhard rod
- Watt's linkage
- WOBLink
- Mumford linkage
- Leaf springs used for location (transverse or longitudinal)
    - Fully elliptical springs usually need supplementary location links and are no longer in common use
    - Longitudinal semi-elliptical springs used to be common and still are used in heavy-duty trucks and aircraft. They have the advantage that the spring rate can easily be made progressive (non-linear).
    - A single transverse leaf spring for both front wheels and/or both back wheels, supporting solid axles, was used by Ford Motor Company, before and soon after World War II, even on expensive models. It had the advantages of simplicity and low unsprung weight (compared to other solid axle designs).

In a front engine, rear-drive vehicle, dependent rear suspension is either "live axle" or deDion axle, depending on whether or not the differential is carried on the axle. Live axle is simpler but the un-sprung weight contributes to wheel bounce.

Because it assures constant camber, dependent (and semi-independent) suspension is most common on vehicles that need to carry large loads as a proportion of the vehicle weight, that have relatively soft springs and that do not (for cost and simplicity reasons) use active suspensions. The use of dependent front suspension has become limited to heavier commercial vehicles.

Independent Suspensions

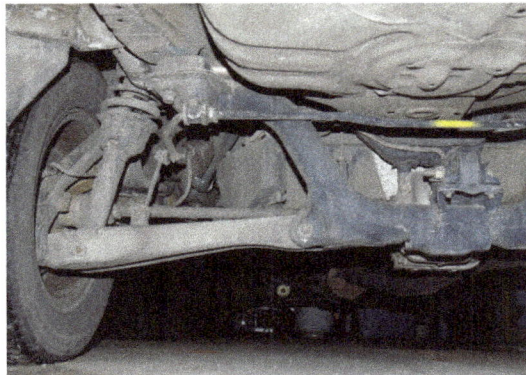

A rear independent suspension on an AWD car.

The variety of independent systems is greater and includes:

- Swing axle
- Sliding pillar
- MacPherson strut/Chapman strut
- Upper and lower A-arm (double wishbone)
- Multi-link suspension
- Semi-trailing arm suspension
- Swinging arm
- Leaf springs
    - Transverse leaf springs when used as a suspension link, or four quarter elliptics on one end of a car are similar to wishbones in geometry, but are more compliant. Examples are the front of the original Fiat 500, the Panhard Dyna Z and the early examples of Peugeot 403 and the back of the AC Ace and AC Aceca.

Because the wheels are not constrained to remain perpendicular to a flat road surface in turning, braking and varying load conditions, control of the wheel camber is an important issue. Swinging arm was common in small cars that were sprung softly and could carry large loads, because the camber is independent of load. Some active and semi-active suspensions maintain the ride height, and therefore the camber, independent of load. In sports cars, optimal camber change when turning is more important.

Wishbone and multi-link allow the engineer more control over the geometry, to arrive at the best compromise, than swing axle, MacPherson strut or swinging arm do; however the cost and space requirements may be greater. Semi-trailing arm is in between, being a variable compromise between the geometries of swinging arm and swing axle.

Semi-independent Suspension

In a semi-independent suspensions, the wheels of an axle are able to move relative to one another as in an independent suspension but the position of one wheel has an effect on the position and attitude of the other wheel. This effect is achieved via the twisting or deflecting of suspension parts under load. The most common type of semi-independent suspension is the twist beam.

- Twist beam

Tilting Suspension System

Tilting Suspension System (Also known as Leaning Suspension System) is not actually a different type or different geometry of construction, moreover it is a technology addition to the conventional suspension system.

This kind of suspension system mainly consist of independent suspension (e.g- MacPherson strut, A-arm (double wishbone)). With addition with these suspension system there is a further tilting or leaning mechanism which connects the suspension system with the vehicle body (chassis).

This kind of suspension system improves stability, traction, turning radius of vehicle and comfort of riders as well. While turning right or left passengers or objects on a vehicle feel G-force or inertial force outward the radius of curvature that is why Two Wheeler riders lean towards the center of curvature while turning which improves stability and decrease the chances of toppling. But for vehicle more than two wheels and with conventional suspension system could not do the same till now so the passengers feel the outward inertial force which reduce the stability of passengers and comfort as well. This kind of tilting suspension system is the solution of the problem. If the road do not have super-elevation or banking it will not affect the comfort with this suspension system, the vehicle tilt and decrease the height of center of gravity with increase the stability. This is also used in fun vehicle.

Some trains also use tilting suspension (Tilting Train) with increase the speed at cornering.

Rocker Bogie Mechanism

The rocker-bogie system is the suspension arrangement in which there are some trailing arm fitted with some idler wheels, due to the articulation between the driving section and the followers this suspension is very flexible. this kind of suspension is appropriate for extremely rough terrain.

This kind of suspension was used in the Curiosity rover.

Tracked Vehicles

Some vehicles such as trains run on long rail tracks fixed to the ground, and some such as tractors, snow vehicles and tanks run on continuous tracks that are part of the vehicle. Though either sort helps to smooth the path and reduce ground pressure, many of the same considerations apply.

Armoured Fighting Vehicle Suspension

This Grant I tank's suspension has road wheels mounted on wheel trucks, or bogies.

Military AFVs, including tanks, have specialized suspension requirements. They can weigh more than seventy tons and are required to move as quickly as possible over very rough or soft ground. Their suspension components must be protected from land mines and antitank weapons. Tracked AFVs can have as many as nine road wheels on each side. Many wheeled AFVs have six or eight large wheels. Some have a Central Tire Inflation System to reduce ground loading on poor surfaces. Some wheels are too big and confined to turn, so skid steering is used with some wheeled, as well as with tracked, vehicles.

The earliest tanks of World War I had fixed suspension with no designed movement whatsoever. This unsatisfactory situation was improved with leaf spring or coil spring suspensions adopted from agricultural, automotive or railway machinery, but even these had very limited travel.

Speeds increased due to more powerful engines, and the quality of ride had to be improved. In the 1930s, the Christie suspension was developed, which allowed the use of coil springs inside a vehicle's armored hull, by changing the direction of force deforming the spring, using a bell crank. The T-34's suspension was directly descended from Christie designs. Horstmann suspension was a variation which used a combination of bell crank and exterior coil springs, in use from the 1930s to the 1990s. The bogie, but nonetheless independent, suspension of the M3 Lee/Grant and the M4 Sherman was similar to the Hortsmann type, with the suspension contained within the track oval.

By World War II the other common type was torsion bar suspension, getting spring force from twisting bars inside the hull — this sometimes had less travel than the Christie-type, but was significantly more compact, allowing more space inside the hull, with consequent possibility to install larger turret rings and thus a heavier main armament. The torsion-bar suspension, sometimes including shock absorbers, has been the dominant heavy armored vehicle suspension since World War II. Torsion bars may take space under or near the floor, which may interfere with making the tank low to reduce exposure.

As with cars, wheel travel and spring rate affect the bumpiness of the ride and the speed at which rough terrain can be negotiated. It may be significant that a smooth ride, which is often associated with comfort, increases the accuracy when firing on the move (analogously to battle ships with reduced stability, due to reduced metacentric height). It also reduces shock on optics and other equipment. The unsprung weight and track link weight may limit speed on roads and affect the life of the track and other components.

Most German WW II half tracks and their tanks introduced during the war such as the Panther tank had overlapping and sometimes interleaved road wheels to distribute the load more evenly on the track and therefore on the ground. This apparently made a significant contribution to speed, range and track life, as well as providing a continuous band of protection. It has not been used since the end of that war, probably due to the maintenance requirements of more complicated mechanical parts working in mud, sand, rocks, snow and ice, as well as to cost. Rocks and frozen mud often got stuck between the overlapping wheels, which could prevent them from turning or cause damage to the road wheels. If one of the interior road wheels were damaged, it would require other road wheels to be removed in order to access the damaged road wheel, making the process more complicated and time-consuming.

Traction Control System

A traction control system (TCS), also known as ASR (from German Antriebsschlupfregelung, anti slip regulation), is typically (but not necessarily) a secondary function of the electronic stability control (ESC) on production motor vehicles, designed to prevent loss of traction of driven road wheels. TCS is activated when throttle input and engine torque are mismatched to road surface conditions.

Intervention consists of one or more of the following:

- Brake force applied to one or more wheels

- Reduction or suppression of spark sequence to one or more cylinders

- Reduction of fuel supply to one or more cylinders

- Closing the throttle, if the vehicle is fitted with drive by wire throttle

- In turbocharged vehicles, a boost control solenoid is actuated to reduce boost and therefore engine power.

Typically, traction control systems share the electrohydraulic brake actuator (which does not use the conventional master cylinder and servo) and wheel speed sensors with ABS.

History

The predecessor of modern electronic traction control systems can be found in high-torque, high-power rear-wheel drive cars as a limited slip differential. A limited slip differential is a purely mechanical system that transfers a relatively small amount of power to the non-slipping wheel, while still allowing some wheel spin to occur.

In 1971, Buick introduced MaxTrac, which used an early computer system to detect rear wheel spin and modulate engine power to those wheels to provide the most traction. A Buick exclusive item at the time, it was an option on all full-size models, including the Riviera, Estate Wagon, Electra 225, Centurion, and LeSabre.

Cadillac introduced the Traction Monitoring System (TMS) in 1979 on the redesigned Eldorado.

Overview

The basic idea behind the need for a traction control system is the loss of road grip that compromises steering control and stability of vehicles because of the difference in traction of the drive wheels. Difference in slip may occur due to turning of a vehicle or varying road conditions for different wheels. When a car turns, its outer and inner wheels rotate at different speeds; this is conventionally controlled by using a differential. A further enhancement of the differential is to employ an active differential that can vary the amount of power being delivered to outer and inner wheels as needed. For example, if outward slip is sensed while turning, the active differential may deliver more power to the outer wheel in order to minimize the yaw (essentially the degree to which the front and rear wheels of a car are out of line.) Active differential, in turn, is controlled by an assembly of electromechanical sensors collaborating with a traction control unit.

Operation

When the traction control computer (often incorporated into another control unit, such as the ABS module) detects one or more driven wheels spinning significantly faster than another, it invokes the ABS electronic control unit to apply brake friction to wheels spinning with lessened traction. Braking action on slipping wheel(s) will cause power transfer to wheel axle(s) with traction due to the mechanical action within the differential. All-wheel drive (AWD) vehicles often have an electronically controlled coupling system in the transfer case or transaxle engaged (active part-time AWD), or locked-up tighter (in a true full-time set up driving all wheels with some power all the time) to supply non-slipping wheels with torque.

This often occurs in conjunction with the powertrain computer reducing available engine torque by electronically limiting throttle application and/or fuel delivery, regarding ignition spark, completely shutting down engine cylinders, and a number of other methods, depending on the vehicle and how much technology is used to control the engine and transmission. There are instances when traction control is undesirable, such as trying to get a vehicle unstuck in snow or mud. Allowing one wheel to spin can propel a vehicle forward enough to get it unstuck, whereas both wheels applying a limited amount of power will not produce the same effect. Many vehicles have a traction control shut-off switch for such circumstances.

Components of Traction Control

Generally, the main hardware for traction control and ABS are mostly the same. In many vehicles traction control is provided as an additional option to ABS.

- Each wheel is equipped with a sensor which senses changes in its speed due to loss of traction.

- The sensed speed from the individual wheels is passed on to an electronic control unit (ECU).

- The ECU processes the information from the wheels and initiates braking to the affected wheels via a cable connected to an automatic traction control (ATC) valve.

In all vehicles, traction control is automatically started when the sensors detect loss of traction at any of the wheels.

Use of Traction Control

- In road cars: Traction control has traditionally been a safety feature in premium high-performance cars, which otherwise need sensitive throttle input to prevent spinning driven wheels when accelerating, especially in wet, icy or snowy conditions. In recent years, traction control systems have become widely available in non-performance cars, minivans, and light trucks and in some small hatchbacks.

- In race cars: Traction control is used as a performance enhancement, allowing maximum traction under acceleration without wheel spin. When accelerating out of a turn, it keeps the tires at optimal slip ratio.

- In motorcycles: Traction control for production motorcycles was first available with the BMW K1 in 1988. By 2009, traction control was an option for several models offered by BMW and Ducati, and the model year 2010 Kawasaki Concours 14 (1400GTR).

- In off-road vehicles: Traction control is used instead of, or in addition to, the mechanical limited slip or locking differential. It is often implemented with an electronic limited slip differential, as well as other computerized controls of the engine and transmission. The spinning wheel is slowed down with short applications of brakes, diverting more torque to the non-spinning wheel; this is the system adopted by Range Rover in 1993, for example. ABS brake traction control has several advantages over limited-slip and locking differentials, such as steering control of a vehicle is easier, so the system can be continuously enabled. It also creates less stress on powertrain and driveline components, and increases durability as there are fewer moving parts to fail.

When programmed or calibrated for off-road use, traction control systems like Ford's four-wheel electronic traction control (ETC) which is included with AdvanceTrac, and Porsche's four-wheel automatic brake differential (ABD), can send 100 percent of torque to any one wheel or wheels, via an aggressive brake strategy or "brake locking", allowing vehicles like the Expedition and Cayenne to keep moving, even with two wheels (one front, one rear) completely off the ground.

Controversy in Motorsports

Very effective yet small units are available that allow the driver to remove the traction control system after an event if desired. In Formula One, an effort to ban traction control led to a change of rules for 2008: every car must have a standard (but custom mappable) ECU, issued by FIA, which is relatively basic and does not have traction control capabilities. NASCAR suspended a Whelen Modified Tour driver, crew chief, and car owner for one race and disqualified the team after crossing the finish line first in a September 20, 2008 race at Martinsville Speedway after finding questionable wiring in the ignition system, which can often be used to implement traction control.

Traction Control in Cornering

Traction control is not just used for improving acceleration under slippery conditions. It can also help a driver to corner more safely. If too much throttle is applied during cornering, the drive wheels will lose traction and slide sideways. This occurs as understeer in front wheel drive vehicles and oversteer in rear wheel drive vehicles. Traction control can prevent this from happening by limiting power to the wheels. It cannot increase the limits of grip available and is used only to decrease the effect of driver error or compensate for a driver's inability to react quickly enough to wheel slip.

Automobile manufacturers state in vehicle manuals that traction control systems should not encourage dangerous driving or encourage driving in conditions beyond the driver's control.

Automotive Aerodynamics

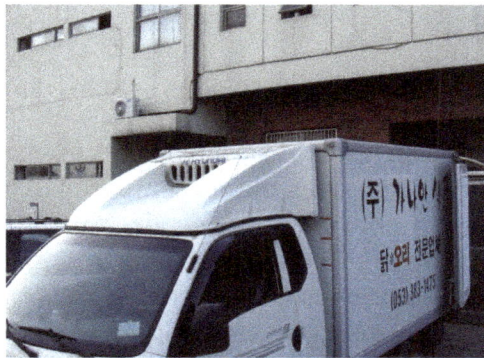

A truck with added bodywork on top of the cab to reduce drag.

Automotive aerodynamics is the study of the aerodynamics of road vehicles. Its main goals are reducing drag and wind noise, minimizing noise emission, and preventing undesired lift forces and other causes of aerodynamic instability at high speeds. Air is also considered a fluid in this case. For some classes of racing vehicles, it may also be important to produce downforce to improve traction and thus cornering abilities.

History

The frictional force of aerodynamic drag increases significantly with vehicle speed. As early as the 1920s engineers began to consider automobile shape in reducing aerodynamic drag at higher speeds. By the 1950s German and British automotive engineers were systematically analyzing the effects of automotive drag for the higher performance vehicles. By the late 1960s scientists also became aware of the significant increase in sound levels emitted by automobiles at high speed. These effects were understood to increase the intensity of sound levels for adjacent land uses at a non-linear rate. Soon highway engineers began to design roadways to consider the speed effects of aerodynamic drag produced sound levels, and automobile manufacturers considered the same factors in vehicle design.

Features of Aerodynamic Vehicles

An aerodynamic automobile will integrate the wheel arcs and lights into the overall shape to reduce drag. It will be streamlined; for example, it does not have sharp edges crossing the wind stream above the windshield and will feature a sort of tail called a fastback or Kammback or liftback. Note that the Aptera 2e, the Loremo, and the Volkswagen 1-litre car try to reduce the area of their back. It will have a flat and smooth floor to support the Venturi effect and produce desirable downwards aerodynamic forces. The air that rams into the engine bay, is used for cooling, combustion, and for passengers, then reaccelerated by a nozzle and then ejected under the floor. For mid and rear engines air is decelerated and pressurized in a diffuser, loses some pressure as it passes the engine bay, and fills the slipstream. These cars need a seal between the low pressure region around the wheels and the high pressure around the gearbox. They all have a closed engine bay floor. The suspension is either streamlined (Aptera) or retracted. Door handles, the antenna, and roof rails can have a streamlined shape. The side mirror can only have a round fairing as a nose. Air flow through the wheel-bays is said to increase drag (German source) though race cars need it for brake cooling and many cars emit the air from the radiator into the wheel bay.

Comparison with Aircraft Aerodynamics

Automotive aerodynamics differs from aircraft aerodynamics in several ways. First, the characteristic shape of a road vehicle is much less streamlined compared to an aircraft. Second, the vehicle operates very close to the ground, rather than in free air. Third, the operating speeds are lower (and aerodynamic drag varies as the square of speed). Fourth, a ground vehicle has fewer degrees of freedom than an aircraft, and its motion is less affected by aerodynamic forces. Fifth, passenger and commercial ground vehicles have very specific design constraints such as their intended purpose, high safety standards (requiring, for example, more 'dead' structural space to act as crumple zones), and certain regulations.

Methods of Studying Aerodynamics

One of the side effects of automotive aerodynamics is seed dispersal.

Automotive aerodynamics is studied using both computer modelling and wind tunnel testing. For the most accurate results from a wind tunnel test, the tunnel is sometimes equipped with a rolling road. This is a movable floor for the working section, which moves at the same speed as the air flow. This prevents a boundary layer from forming on the floor of the working section and affecting the results. An example of such a rolling road wind tunnel is Wind Shear's Full Scale, Rolling Road, Automotive Wind Tunnel in Concord, North Carolina and Auto Research Center in Indianapolis, Indiana USA.

Drag Coefficient

Drag coefficient (C_d) is a commonly published rating of a car's aerodynamic smoothness, related to the shape of the car. Multiplying C_d by the car's frontal area gives an index of total drag. The result is called drag area, and is listed below for several cars. The width and height of curvy cars lead to gross overestimation of frontal area. These numbers use the manufacturer's frontal area specifications from the Mayfield Company unless noted.

Some examples:

Drag area (C_d x Ft²)	Year	Automobile
3.0 sq ft (0.28 m²)	2012	Volkswagen XL1
3.95	1996	GM EV1
5.10	1999	Honda Insight
5.40	1989	Opel Calibra
5.54	1980	Ferrari 308 GTB
5.61	1993	Mazda RX-7
5.61	1993	McLaren F1
5.63	1991	Opel Calibra
5.64	1990	Bugatti EB110
5.71	1990	Honda CR-X
5.74	2002	Acura NSX
5.76	1968	Toyota 2000GT
5.88	1990	Nissan 240SX
5.86	2001	Audi A2 1.2 TDI 3L
5.91	1986	Citroën AX
5.92	1994	Porsche 911 Speedster
5.95	1994	McLaren F1
6.00	2011	Lamborghini Aventador S
6.00	1992	Subaru SVX
6.06	2003	Opel Astra Coupe Turbo
6.08	2008	Nissan GT-R
6.13	1991	Acura NSX
6.15	1989	Suzuki Swift GT
6.17	1995	Lamborghini Diablo
6.19	1969	Porsche 914
6.2	2012	Tesla Model S
6.24	2004	Toyota Prius
6.27	1986	Porsche 911 Carrera
6.27	1992	Chevrolet Corvette
6.35	1999	Lotus Elise
6.77	1995	BMW M3
6.79	1993	Corolla DX
6.81	1989	Subaru Legacy
6.96	1988	Porsche 944 S
7.02	1992	BMW 325I
7.10	1978	Saab 900
7.13	2007	SSC Ultimate Aero
7.31	2015	Mazda3
7.48	1993	Chevrolet Camaro Z28
7.57	1992	Toyota Camry
8.70	1990	Volvo 740 Turbo

8.71	1991	Buick LeSabre Limited
9.54	1992	Chevrolet Caprice Wagon
10.7	1992	Chevrolet S-10 Blazer
11.63	1991	Jeep Cherokee
13.10	1990	Range Rover Classic
13.76	1994	Toyota T100 SR5 4x4
14.52	1994	Toyota Land Cruiser
17.43	1992	Land Rover Discovery
18.03	1992	Land Rover Defender 90
18.06	1993	Hummer H1
20.24	1993	Land Rover Defender 110
26.32	2006	Hummer H2

Downforce

Downforce describes the downward pressure created by the aerodynamic characteristics of a car that allows it to travel faster through a corner by holding the car to the track or road surface. Some elements to increase vehicle downforce will also increase drag. It is very important to produce a good downward aerodynamic force because it affects the car's speed and traction.

Noise, Vibration, and Harshness

Noise, vibration, and harshness (NVH), also known as noise and vibration (N&V), is the study and modification of the noise and vibration characteristics of vehicles, particularly cars and trucks. While noise and vibration can be readily measured, harshness is a subjective quality, and is measured either via "jury" evaluations, or with analytical tools that can provide results reflecting human subjective impressions. These latter tools belong to the field known as "psychoacoustics."

Interior NVH deals with noise and vibration experienced by the occupants of the cabin, while exterior NVH is largely concerned with the noise radiated by the vehicle, and includes drive-by noise testing.

NVH is mostly engineering, but often objective measurements fail to predict or correlate well with the subjective impression on human observers. For example, although the ear's response at moderate noise levels is approximated by A-weighting, two different noises with the same A-weighted level are not necessarily equally disturbing. The field of psychoacoustics is partly concerned with this correlation.

In some cases the NVH engineer is asked to change the sound quality, by adding or subtracting particular harmonics, rather than making the vehicle quieter.

Sources of NVH

The sources of noise in a vehicle are many, including the engine, driveline, tire contact patch and road surface, brakes, and wind. Noise from cooling fans, or the HVAC, alternator, and other engine accessories is also fairly common. Many problems are generated as either vibration or noise, transmitted via a variety of paths, and then radiated acoustically into the cabin. These are classified as "structure-borne" noise. Others are generated acoustically and propagated by airborne paths. Structure-borne noise is attenuated by isolation, while airborne noise is reduced by absorption or through the use of barrier materials. Vibrations are sensed at the steering wheel, the seat, armrests, or the floor and pedals. Some problems are sensed visually - such as the vibration of the rear-view mirror or header rail on open-topped cars

Tonal Versus Broadband

NVH can be tonal such as engine noise, or broadband, such as road noise or wind noise, normally. Some resonant systems respond at characteristic frequencies, but in response to random excitation. Therefore, although they look like tonal problems on any one spectrum, their amplitude varies considerably. Other problems are self resonant, such as whistles from antennas.

Tonal noises often have harmonics. Here is the noise spectrum of Michael Schumacher's Ferrari at 16680 rpm, showing the various harmonics. The x axis is given in terms of multiples of engine speed. The y axis is logarithmic, and uncalibrated.

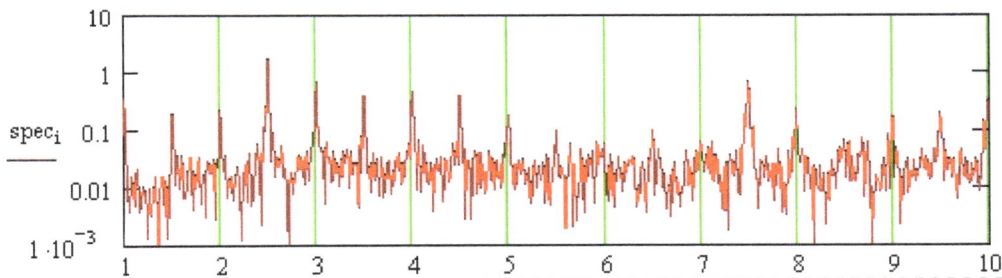

Instrumentation

Typical instrumentation used to measure NVH include microphones, accelerometers and force gauges, or load cells. Many NVH facilities will have semi-anechoic chambers, and rolling road dynamometers. Typically signals are recorded direct to hard disk via an analog-to-digital converter. In the past magnetic or DAT tape recorders were used. The integrity of the signal chain is very important, typically each of the instruments used are fully calibrated in a lab once per year, and any given setup is calibrated as a whole once per day.

Laser scanning vibrometry is an essential tool for effective NVH optimization. The vibrational characteristics of a sample is acquired full field under operational or excited conditions. The results represent the actual vibrations. No added mass is influencing the measurement, as the sensor is light itself.

Investigative Techniques

Techniques used to help identify NVH include part substitution, modal analysis, rig squeak and rattle tests (complete vehicle or component/system tests), lead cladding, acoustic intensity, transfer path analysis, and partial coherence. Most NVH work is done in the frequency domain, using fast Fourier transforms to convert the time domain signals into the frequency domain. Wavelet analysis, order analysis, statistical energy analysis, and subjective evaluation of signals modified in real time are also used.

Computer-based Modeling

NVH needs good representative prototypes of the production vehicle for testing. These are needed early in the design process as the solutions often need substantial modification to the design, forcing in engineering changes which are much cheaper when made early. These early prototypes are very expensive, so there has been great interest in computer aided predictive techniques for NVH. Sometimes these work. Back-of-envelope calculations are very useful.

One example is the modelling works for structure borne noise and vibration analysis. When the phenomenon being considered occurs below, say, 25–30 Hz, for example the idle shaking of the powertrain, a multi-body model can be used. In contrast, when the phenomenon being considered occurs at relatively high frequency, for example above 1 kHz, a Statistical Energy Analysis (SEA) model may be a better approach.

For the mid-frequency band, various methodologies exist, such as vibro-acoustic finite element analysis, and boundary element analysis. The structure can be coupled to the interior cavity and form a fully coupled equation system. Also other techniques exist that can mix measured data with finite element or boundary element data.

Typical Solutions

There are three principal means of improving NVH:

1. reducing the source strength, as in making a noise source quieter with a muffler, or improving the balance of a rotating mechanism;

2. interrupting the noise or vibration path, with barriers (for noise) or isolators (for vibration); or

3. absorption of the noise or vibration energy, as for example with foam noise absorbers, or tuned vibration dampers.

Deciding which of these (or what combination) to use in solving a particular problem is one of the challenges facing the NVH engineer.

Specific methods for improving NVH include the use of tuned mass dampers, subframes, balancing, modifying the stiffness or mass of structures, retuning exhausts and intakes, modifying the characteristics of elastomeric isolators, adding sound deadening or absorbing materials, or using active noise control. In some circumstances, substantial changes in vehicle architecture may be the only way to cure some problems cost effectively.

References

- Tuncer Cebeci, Jian P. Shao, Fassi Kafyeke, Eric Laurendeau, Computational Fluid Dynamics for Engineers: From Panel to Navier-Stokes, Springer, 2005, ISBN 3-540-24451-4

- Wang, Xu (2010). Vehicle noise and vibration refinement. Cambridge, UK: Woodhead Publishing Ltd. ISBN 978-1-84569-497-5. Retrieved 5 December 2016.

- Campillo-Davo and Rassili (eds.). NVH Analysis Techniques for Design and Optimization of Hybrid and Electric Vehicles. ISBN 978-3-8440-4356-3

- Stewart, Ben (2004-09-13). "Comparison Test: Front-Wheel Drive Vs. Rear-Wheel Drive". Popular Mechanics. US. Retrieved 2016-05-13.

- A Study on Total Quality Management and Lean Manufacturing: Through Lean Thinking Approach World Applied Sciences Journal 12 (9): 1585–1596, 2011, Retrieved 11/16/2012

- Quiroga, Tony (August 2010). "2011 Audi RS5 vs. 2010 BMW M3, 2011 Cadillac CTS-V Comparison Tests". Car and Driver. Retrieved 2012-12-09.

- "Electronic Stability Control (ESC) | National Highway Traffic Safety Administration(NHTSA) | U.S. Department of Transportation". Nhtsa.dot.gov. Archived from the original on January 11, 2010. Retrieved 2011-11-13.

- "Jianbo Lu, Dave Messih, and Albert Salib, "Roll Rate Based Stability Control – The Roll Stability Control System," Proceedings of the 20th Enhanced Safety of Vehicles Conference, 2007" (PDF). Retrieved 2011-11-13.

- "Tseng, H.E.; Ashrafi, B.; Madau, D.; Allen Brown, T.; Recker, D.; The development of vehicle stability control at Ford". Ieeexplore.ieee.org. 2002-08-06. doi:10.1109/3516.789681. Retrieved 2011-11-13.

- "The Effectiveness of Electronic Stability Control in Reducing Real-World Crashes: A Literature Review – Traffic Injury Prevention". Informaworld.com. 2007-12-11. Retrieved 2011-11-13.

- Sullivan, Bob. "Red Tape – Who's driving? Toyota woes raise car tech fears". Redtape.msnbc.com. Archived from the original on March 16, 2011. Retrieved 2011-11-13.

Technologies of Automotive Engines

Some of the technologies of automotive engines that have been explained in the section are antifreeze, engine coolant temperature sensor, block heater, cold air intake, choke valve and radiator. Antifreeze helps in lowering the freezing point of any water based liquid. It is used to achieve freezing-point depression for environments that are cold.

Antifreeze

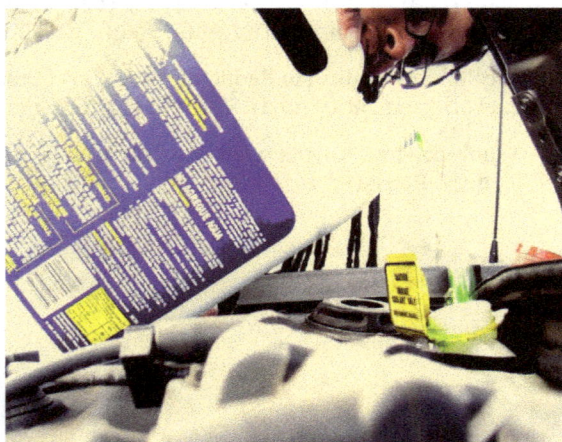

"Topping up" the antifreeze solution in a car's cooling system is a routine maintenance item for most modern cars.

An antifreeze is an additive which lowers the freezing point of a water-based liquid. An antifreeze mixture is used to achieve freezing-point depression for cold environments and also achieves boiling-point elevation ("anti-boil") to allow higher coolant temperature. Freezing and boiling points are colligative properties of a solution, which depend on the concentration of the dissolved substance.

Because water has good properties as a coolant, water plus antifreeze is used in internal combustion engines and other heat transfer applications, such as HVAC chillers and solar water heaters. The purpose of antifreeze is to prevent a rigid enclosure from bursting due to expansion when water freezes. Commercially, both the additive (pure concentrate) and the mixture (diluted solution) are called antifreeze, depending on the context. Careful selection of an antifreeze can enable a wide temperature range in which the mixture remains in the liquid phase, which is critical to efficient heat transfer and the proper functioning of heat exchangers.

Salts are frequently used for de-icing, but salt solutions are not used for cooling systems because they can cause severe corrosion to metals. Instead, non-corrosive antifreezes are commonly used for critical de-icing, such as for aircraft wings.

Automotive and Internal Combustion Engine Use

Fluorescent green-dyed antifreeze is visible in the radiator header tank when car radiator cap is removed

Most automotive engines are "water"-cooled to remove waste heat, although the "water" is actually antifreeze/water mixture and not plain water. The term engine coolant is widely used in the automotive industry, which covers its primary function of convective heat transfer for internal combustion engines. When used in an automotive context, corrosion inhibitors are added to help protect vehicles' radiators, which often contain a range of electrochemically incompatible metals (aluminum, cast iron, copper, brass, solder, et cetera). Water pump seal lubricant is also added.

Antifreeze was developed to overcome the shortcomings of water as a heat transfer fluid. In some engines freeze plugs (engine block expansion plugs) are placed in areas of the engine block where coolant flows in order to protect the engine from freeze damage if the ambient temperature drops below the freezing point of the antifreeze/water mixture. These should not be confused with core plugs, whose purpose is to allow removal of sand used in the casting process of engine blocks (core plugs will be pushed out if the coolant freezes, though, assuming that they adjoin the coolant passages, which is not always the case).

On the other hand, if the engine coolant gets too hot, it might boil while inside the engine, causing voids (pockets of steam), leading to localized hot spots and the catastrophic failure of the engine. If plain water were to be used as an engine coolant, it would promote galvanic corrosion. Proper engine coolant and a pressurized coolant system can help obviate the problems which make plain water incompatible with automotive engines. With proper antifreeze, a wide temperature range can be tolerated by the engine coolant, such as −34 °F (−37 °C) to +265 °F (129 °C) for 50% (by volume) propylene glycol diluted with water and a 15 psi pressurized coolant system.

Early engine coolant antifreeze was methanol (methyl alcohol). Methanol was widely used in windshield fluids, however, in Europe, due to new REACH legislation, the use of methanol in windshield fluids is limited to 5% and in the near future will be further reduced to 3%. As radiator caps were vented, not sealed, the methanol was lost to evaporation, requiring frequent replenishment to avoid freezing of the coolant. Methanol also accelerates corrosion of the metals, especially aluminum, used in the engine and cooling systems. Ethylene glycol was developed, and soon[when?] replaced methanol as an engine cooling system antifreeze. It has a very low volatility compared to methanol and to water. Before the 1950s, coolant systems were unpressurized and the engine was often cooler than modern automotive engines. By pressurizing the coolant system with a radiator cap, the boiling point of the fluid is increased, permitting higher engine temperatures and better fuel efficiency. Pressurized systems do not appreciably change the freezing point.

Other Uses

The most common water-based antifreeze solutions used in electronics cooling are mixtures of water and either ethylene glycol (EGW) or propylene glycol (PGW). The use of ethylene glycol has a longer history, especially in the automotive industry. However, EGW solutions formulated for the automotive industry often have silicate based rust inhibitors that can coat and/or clog heat exchanger surfaces. Ethylene glycol is listed as a toxic chemical requiring care in handling and disposal.

Ethylene glycol has desirable thermal properties, including a high boiling point, low freezing point, stability over a wide range of temperatures, and high specific heat and thermal conductivity. It also has a low viscosity and, therefore, reduced pumping requirements. Although EGW has more desirable physical properties than PGW, the latter coolant is used in applications where toxicity might be a concern. PGW is generally recognized as safe for use in food or food processing applications, and can also be used in enclosed spaces.

Similar mixtures are commonly used in HVAC and industrial heating or cooling systems as a high-capacity heat transfer medium. Many formulations have corrosion inhibitors, and it is expected that these chemicals will be replenished (manually or under automatic control) to keep expensive piping and equipment from corroding.

Primary Agents

Most antifreeze is made by mixing distilled water with some kind of alcohol.

Methanol

Methanol (also known as methyl alcohol, carbinol, wood alcohol, wood naphtha or wood spirits) is a chemical compound with chemical formula CH_3OH. It is the simplest alcohol, and is a light, volatile, colorless, flammable, poisonous liquid with a distinctive odor that is somewhat milder and sweeter than ethanol (ethyl alcohol). At room temperature, it is a polar solvent and is used as an antifreeze, solvent, fuel, and as a denaturant for ethyl alcohol. It is not popular for machinery, but may be found in automotive windshield washer fluid, de-icers, and gasoline additives.

Ethylene Glycol

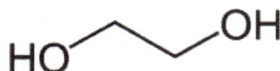

Ethylene glycol

Ethylene glycol solutions became available in 1926 and were marketed as "permanent antifreeze" since the higher boiling points provided advantages for summertime use as well as during cold weather. They are used today for a variety of applications, including automobiles, but gradually being replaced by propylene glycol due to its lower toxicity.

When ethylene glycol is used in a system, it may become oxidized to five organic acids (formic, oxalic, glycolic, glyoxalic and acetic acid). Inhibited ethylene glycol antifreeze mixes are available, with additives that buffer the pH and reserve alkalinity of the solution to prevent oxidation of ethylene glycol and formation of these acids. Nitrites, silicates, theodin, borates and azoles may also be used to prevent corrosive attack on metal.

Poisoning

Ethylene glycol is poisonous to humans and other animals, and should be handled carefully and disposed of properly. Its sweet taste can lead to accidental ingestion or allow its deliberate use as a murder weapon. Ethylene glycol is difficult to detect in the body, and causes symptoms—including intoxication, severe diarrhea, and vomiting—that can be confused with other illnesses or diseases. Its metabolism produces calcium oxalate, which crystallizes in the brain, heart, lungs, and kidneys, damaging them; depending on the level of exposure, accumulation of the poison in the body can last weeks or months before causing death, but death by acute kidney failure can result within 72 hours if the individual does not receive appropriate medical treatment for the poisoning. Some ethylene glycol antifreeze mixtures contain an embittering agent, such as denatonium, to discourage accidental or deliberate consumption.

Propylene Glycol

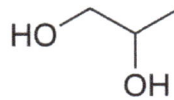

Propylene glycol

Propylene glycol, on the other hand, is considerably less toxic than ethylene glycol and may be labeled as "non-toxic antifreeze". It is used as antifreeze where ethylene glycol would be inappropriate, such as in food-processing systems or in water pipes in homes where incidental ingestion may be possible. As confirmation of its relative non-toxicity, the FDA allows propylene glycol to be added to a large number of processed foods, including ice cream, frozen custard, salad dressings and baked goods, it is also often a main ingredient of E-cigarette liquid (E-liquid or E-juice).

Propylene glycol oxidizes when exposed to air and heat, forming lactic acid. If not properly inhibited, this fluid can be very corrosive, so pH buffering agents such as dipotassium phosphate, Protodin and potassium bicarbonate are often added to propylene glycol, to prevent acidic corrosion of metal components. Pre-inhibited propylene glycol solutions like Dowfrost (manufactured by Dow Chemicals, US) and Tonofrost (manufactured by Chemtex Speciality Ltd, India) can also be used instead of pure propylene glycol to prevent corrosion.

Besides cooling system corrosion, biological fouling also occurs. Once bacterial slime starts to grow, the corrosion rate of the system increases. Maintenance of systems using glycol solution includes regular monitoring of freeze protection, pH, specific gravity, inhibitor level, color, and biological contamination. Propylene glycol should be replaced when it turns a reddish color.

Glycerol

Once used for automotive antifreeze, glycerol has the advantage of being non-toxic, withstands relatively high temperatures, and is noncorrosive.

Like ethylene glycol and propylene glycol, glycerol is a non-ionic kosmotrope that forms strong hydrogen bonds with water molecules, competing with water-water hydrogen bonds. This disrupts the crystal lattice formation of ice unless the temperature is significantly lowered. The minimum freezing point temperature is at about −36 °F / −37.8 °C corresponding to 60–70% glycerol in water.

Glycerol was historically used as an antifreeze for automotive applications before being replaced by ethylene glycol, which has a lower freezing point. While the minimum freezing point of a glycerol-water mixture is higher than an ethylene glycol-water mixture, glycerol is not toxic and is being re-examined for use in automotive applications. Glycerol is mandated for use as an antifreeze in many sprinkler systems.

In the laboratory, glycerol is a common component of solvents for enzymatic reagents stored at temperatures below 0 °C due to the depression of the freezing temperature of solutions with high concentrations of glycerol. It is also used as a cryoprotectant where the glycerol is dissolved in water to reduce damage by ice crystals to laboratory organisms that are stored in frozen solutions, such as bacteria, nematodes, and mammalian embryos.

Measuring the Freeze Point

Once antifreeze has been mixed with water and put into use, it periodically needs to be maintained. If engine coolant leaks, boils, or if the cooling system needs to be drained and refilled, the antifreeze's freeze protection will need to be considered. In other cases a vehicle may need to be operated in a colder environment, requiring more antifreeze and less water. Three methods are commonly employed to determine the freeze point of the solution:

1. Specific gravity—(using a hydrometer or some sort of floating indicator),

2. Refractometer—which measures the refractive index of the antifreeze solution and translates it into freeze point, and

3. Test strips—specialized, disposable indicators made for this purpose.

Although ethylene glycol hydrometers are widely available and mass-marketed for antifreeze testing, they give false readings at high temperatures because specific gravity changes with temperature. Propylene glycol solutions cannot be tested using specific gravity because of ambiguous results (40% and 100% solutions have the same specific gravity).

Corrosion Inhibitors

Most commercial antifreeze formulations include corrosion inhibiting compounds, and a colored dye (commonly a fluorescent green, red, orange, yellow, or blue) to aid in identification. A 1:1 dilution with water is usually used, resulting in a freezing point of about −34 °F (−37 °C), depending on the formulation. In warmer or colder areas, weaker or stronger dilutions are used, respectively, but a range of 40%/60% to 60%/40% is frequently specified to ensure corrosion protection, and 70%/30% for maximum freeze prevention down to −84 °F (−64 °C).

Maintenance

In the absence of leaks, antifreeze chemicals such as ethylene glycol or propylene glycol may retain their basic properties indefinitely. By contrast, corrosion inhibitors are gradually used up, and must be replenished from time to time. Larger systems (such as HVAC systems) are often monitored by specialist firms which take responsibility for adding corrosion inhibitors and regulating coolant composition. For simplicity, most automotive manufacturers recommend periodic complete replacement of engine coolant, to simultaneously renew corrosion inhibitors and remove accumulated contaminants.

Traditional Inhibitors

Traditionally, there were two major corrosion inhibitors used in vehicles: silicates and phosphates. American made vehicles traditionally used both silicates and phosphates. European makes contain silicates and other inhibitors, but no phosphates. Japanese makes traditionally use phosphates and other inhibitors, but no silicates.

Organic Acid Technology

Certain cars are built with organic acid technology (OAT) antifreeze (e.g., DEX-COOL), or with a hybrid organic acid technology (HOAT) formulation (e.g., Zerex G-05), both of which are claimed to have an extended service life of five years or 240,000 km (150,000 mi).

DEX-COOL specifically has caused controversy. Litigation has linked it with intake manifold gasket failures in General Motors' (GM's) 3.1L and 3.4L engines, and with other failures in 3.8L and 4.3L engines. One of the anti-corrosion components presented as sodium or Potassium 2-ethylhexanoate and ethylhexanoic acid is incompatible with nylon 6,6 and silicone rubber, and is a known plasticizer. Class action lawsuits were registered in several states, and in Canada, to address some of these claims. The first of these to reach a decision was in Missouri, where a settlement was announced early in December 2007. Late in March 2008, GM agreed to compensate complainants in the remaining 49 states. GM (Motors Liquidation Company) filed for bankruptcy in 2009, which tied up the outstanding claims until a court determines who gets paid.

According to the DEX-COOL manufacturer, "mixing a 'green' [non-OAT] coolant with DEX-COOL reduces the batch's change interval to 2 years or 30,000 miles, but will otherwise cause no damage to the engine". DEX-COOL antifreeze uses two inhibitors: sebacate and 2-EHA (2-ethylhexanoic acid), the latter which works well with the hard water found in the United States, but is a plasticizer that can cause gaskets to leak.

According to internal GM documents, the ultimate culprit appears to be operating vehicles for long periods of time with low coolant levels. The low coolant is caused by pressure caps that fail in the open position. (The new caps and recovery bottles were introduced at the same time as DEX-COOL). This exposes hot engine components to air and vapors, causing corrosion and contamination of the coolant with iron oxide particles, which in turn can aggravate the pressure cap problem as contamination holds the caps open permanently.

Honda and Toyota's new extended life coolant use OAT with sebacate, but without the 2-EHA. Some added phosphates provide protection while the OAT builds up. Honda specifically excludes 2-EHA from their formulas.

Typically, OAT antifreeze contains an orange dye to differentiate it from the conventional glycol-based coolants (green or yellow). Some of the newer OAT coolants claim to be compatible with all types of OAT and glycol-based coolants; these are typically green or yellow in color.

Hybrid Organic Acid Technology

HCAT coolants typically mix an OAT with a traditional inhibitor, such as silicates or phosphates.

G05 is a low-silicate, phosphate free formula that includes the benzoate inhibitor.

Additives

All automotive antifreeze formulations, including the newer organic acid (OAT antifreeze) formulations, are environmentally hazardous because of the blend of additives (around 5%), including lubricants, buffers and corrosion inhibitors. Because the additives in antifreeze are proprietary, the safety data sheets (SDS) provided by the manufacturer list only those compounds which are considered to be significant safety hazards when used in accordance with the manufacturer's recommendations. Common additives include sodium silicate, disodium phosphate, sodium molybdate, sodium borate, denatonium benzoate and dextrin (hydroxyethyl starch). Disodium fluorescein dyes are added to antifreeze to help trace the source of leaks, and as an identifier since some different formulations are incompatible.

Automotive antifreeze has a characteristic odor due to the additive tolytriazole, a corrosion inhibitor. The unpleasant odor in industrial use tolytriazole comes from impurities in the product that are formed from the toluidine isomers (ortho-, meta- and para-toluidine) and meta-diamino toluene which are side-products in the manufacture of tolytriazole. These side-products are highly reactive and produce volatile aromatic amines which are responsible for the unpleasant odor.

Engine Coolant Temperature Sensor

The coolant temperature sensor is used to measure the temperature of the engine coolant of an internal combustion engine. The readings from this sensor are then fed back to the Engine control unit (ECU), which uses this data to adjust the fuel injection and ignition timing. On some vehicles the sensor may also be used to switch on the electric cooling fan. The data may also be used to provide readings for a coolant temperature gauge on the dashboard.

Types of Sensor

As the sensor's temperature changes, its resistance changes accordingly. Depending on the type of sensor, the resistance will either increase or decrease.

- In negative temperature coefficient (NTC) sensors, the internal resistance will decrease as the temperature rises (and vice versa).

- In a positive temperature coefficient (PTC) sensor, the opposite is true. Its resistance will increase with rising temperature.

Most automotive coolant temperature sensors are NTC sensors.

Operation

The ECU sends out a regulated reference voltage (typically 5 volts) to the coolant temperature sensor. The voltage drop across the sensor will change according to the temperature because its resistance changes. The ECU is then able to calculate the temperature of the engine, and then (with inputs from other engine sensors) uses lookup tables to carry out adjustments to the engine actuators, i.e. change the fuel injection or ignition timing.

This is necessary because in order to run smoothly, a cold engine requires different timing and fuel mixture than an engine at operating temperature.

Block Heater

A North American block heater cord and plug

A block heater warms an engine to increase the chances that the engine will start as well as warm up the vehicle faster than it normally would in extremely cold weather.

The most common type is an electric heating element in the cylinder block, connected through a power cord often routed through the vehicle's grille. The block heater may replace one of the engine's core plugs. In this fashion, the heater element is immersed in the engine's coolant, which then keeps most of the engine warm. This type of heater does not come with a pump. They may also be installed in line with one of the radiator or heater hoses. Some heaters pump and circulate the engine coolant while heating, others only heat the still coolant in the reservoir.

Block heaters that run directly on the vehicle's own gasoline or diesel fuel supply are also available; these do not require an external power source. The coolant is heated and circulated, usually by thermosiphon, through the engine and the vehicle's heater core.

Heaters are also available for engine oil so that warm oil can immediately circulate throughout the engine during start up. The easier starting results from warmer, less viscous engine oil and less condensation of fuel on cold metal surfaces inside the engine; thus an engine block heater reduces a vehicle's emission of unburned hydrocarbons and carbon monoxide; also heat is available more instantly for the passenger compartment and glass defogging.

Block heaters or coolant heaters are also found on permanently installed systems using diesel engines to allow standby generator sets to take up load quickly in an emergency.

Usage

The car's block heater is plugged into an electrical outlet

Block heaters are frequently used in regions with cold winters such as the northern United States, Canada, Russia and Scandinavia. In colder climates, block heaters are often standard equipment in new vehicles. In extremely cold climates, electrical outlets are sometimes found in public or private parking lots, especially in multi-storey car parks. Some parking lots cycle the power on for 20 minutes and off for 20 minutes, to reduce electricity costs.

Research by the Agricultural Engineering Department of the University of Saskatchewan has shown that operating a block heater for longer than four hours prior to starting a vehicle is a waste of energy. It was found that coolant temperature increased by almost 20 °C (36 °F) degrees in that period, regardless of the initial temperature (4 tests were run at ambient temperatures ranging from −11 to −29 °C or 12 to −20 °F; continued use of the heater for a further one or two, or more, hours achieved a mere 2 or 3 more degrees Celsius as conditions stabilized. Engine oil temperature was found to increase over these periods by just 5 °C (9.0 °F).

There are alternatives to a block heater that offer some of the same benefits. These include heaters attached to the engine's oil pan, usually with magnets. Dipstick heaters can be installed in place of the engine's oil dipstick. Heated blankets are available for the entire engine area, as well. A timer can be used with any of these heaters, so that it does not have to be left on all the time. This can help lower the electrical costs of using a block heater.

Some cars, such as the second generation Toyota Prius, pump hot coolant from the cooling system into a 3-litre insulated thermos-style reservoir at shutdown, where it stays warm for up to 3 days.

History

Andrew Freeman of Grand Forks, North Dakota, invented the head bolt heater around 1940 and received a patent for it on November 8, 1949. In 1951, Freeman received another patent on an improved head bolt heater. These early heaters replaced one of the engine's cylinder head bolts with a hollow, threaded shank containing a resistive heating element. Before the block heater was introduced, people used a variety of methods to warm engines before starting them, such as pouring hot water on the engine block or draining the engine's oil for storage inside overnight. Some even shoveled embers underneath their vehicle's engine to obtain the same effect.

During the dawn of aviation in pre-war Northern Canada, aviators flew with flight engineers who were responsible for preparing the radial engines for shutdown and startup to reduce the effects of subzero temperatures. The flight engineer was responsible for draining the oil into buckets at night, and preheating the engine and buckets of oil using a blanket wrapped around the engine and a device known as a blow pot – essentially a kerosene jet-heater used for several hours prior to flight.

Boost Controller

A boost controller is a device to control the boost level produced in the intake manifold of a turbocharged or supercharged engine by affecting the air pressure delivered to the pneumatic and mechanical wastegate actuator.

A boost controller can be a simple manual control which can be easily fabricated, or it may be included as part of the engine management computer in a factory turbocharged car, or an aftermarket electronic boost controller.

Principles of Operation

Without a boost controller, air pressure is fed from the charge air (compressed side) of the turbocharger directly to the wastegate actuator via a vacuum hose. This air pressure can come from anywhere on the intake after the turbo, including after the throttle body, though that is less common. This air pressure pushes against the force of a spring located in the wastegate actuator to allow the wastegate to open and re-direct exhaust gas so that it does not reach the turbine wheel. In this simple configuration, the spring's springrate and preload determine how much boost pressure the system will achieve. Springs are classified by the boost pressure they typically achieve, such as a "7 psi spring" that will allow the turbocharger to reach equilibrium at approximately 7 psi (0.48 bar).

One primary problem of this system is the wastegate will start to open well before the actual desired boost pressure is achieved. This negatively affects the threshold of boost onset and also increases turbocharger lag. For instance, a spring rated at 7 psi may allow the wastegate to begin to (but not fully) open at as little as 3.5 psi (0.24 bar).

Achieving moderate boost levels consistently is also troublesome with this configuration. At partial throttle, full boost may still be reached, making the vehicle difficult to control with precision. Electronic systems can allow the throttle to control the level of boost, so that only at full throttle will maximum boost levels be achieved and intermediate levels of boost can be held consistently at partial throttle levels.

Also to be noted is the way in which boost control is achieved, depending on the type of wastegate used. Typically manual "bleed type" boost controllers are only used on swing type (single port) wastegate actuators. To increase boost, pressure is taken away from the actuator control line, therefore increasing boost. Dual port swing type wastegate actuators and external wastegates generally require electronic boost control although adjustable boost control can also be achieved on both of these with an air pressure regulator, this is not the same as a bleed type boost controller. To increase boost with an external or dual port wastegate, pressure is added to the top control port to increase boost. When boost control is not fitted, this control port is open to the atmosphere.

Manual Boost Control

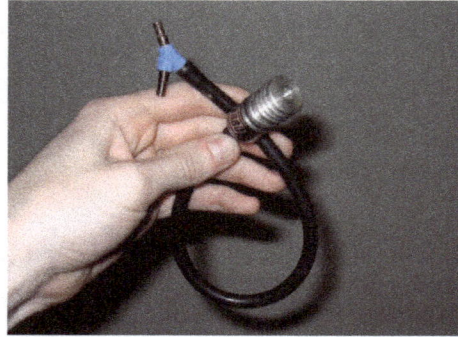

A simple manual boost controller. A small screw is located in the top of the aluminum body to adjust bleed rate. This model is placed in the engine bay, however the vacuum line could be extended to allow it to reach into the passenger compartment.

A bleed-type manual boost controller simple mechanical and pneumatic control to allow some pressure from the wastegate actuator to escape or bleed out to the atmosphere or back into the intake system. This can be as simple as a T-fitting on the boost control line near the actuator with a small bleeder screw. The screw can be turned out to varying degrees to allow air to bleed out of the system, relieving pressure on the wastegate actuator, thus increasing boost levels. These devices are popular due to their negligible cost compared to other devices that may offer the same power increase.

A ball & spring type boost controller uses the force of a spring acting against the boost pressure to control boost. This is installed with one boost signal line coming from the intake somewhere after the turbocharger, and one boost signal line going to the wastegate. A knob changes the force on the spring which in turn dictates how much pressure is on the ball. The tighter the spring, the more boost that is needed to unseat the ball, and allow the boost pressure to reach the wastegate actuator. There is a bleed hole on the boost controller after the ball, to allow the pressurized air that would be trapped between the wastegate actuator and the ball after it is seated again. These type of Manual boost controllers are very popular since they do not provide a boost leak, allowing faster spool times and better control than a "bleed type" boost controller.

There are several different designs of ball-and-spring controllers on the market that range greatly in terms of cost and quality. Common body materials are brass and aluminum vary from inline to 90 degree designs. Another design aspect is the ball valve seat which is critical for performance stability.

Generally a manual boost controller will not be located within the cabin of the vehicle as the lengthy vacuum piping run between the turbo/wastegate & controller can introduce response issues into the system. It is possible to use two manual boost controllers at different settings with a solenoid to switch between them for two different boost pressure settings. Some factory turbocharged cars have a switch to regulate boost pressure, such as a setting designed for fuel economy and a setting for performance.

Manual boost controllers cannot be used to set a specific boost level at a given throttle position (& therefore be used to optimise driveability & control issues), although a ball-spring type boost controller does allow the boost threshold to be as low as is possible on a given engine configuration, and also keeps turbo spool as fast as is possible as the wastegate remains completely shut until the desired boost pressure is reached, ensuring 100% of the exhaust gases are diverted through the turbocharger exhaust turbine. They can be used in conjunction with some electronic systems.

Electronic Boost Control

A 3-port pneumatic solenoid. This solenoid allows interrupt or blocking of the boost pressure rather than just bleed type control.

Electronic boost control adds an air control solenoid and/or a stepper motor controlled by an electronic control unit. The same general principle of a manual controller is present, which is to control the air pressure presented to the wastegate actuator. Further control and intelligent algorithms can be introduced, refining and increasing control over actual boost pressure delivered to the engine.

At the component level, boost pressure can either be bled out of the control lines or blocked outright. Either can achieve the goal of reducing pressure pushing against the wastegate. In a bleed-type system air is allowed to pass out of the control lines, reducing the load on the wastegate actuator. On a blocking configuration, air traveling from the charge air supply to the wastegate actuator is blocked while simultaneously bleeding any pressure that has previously built up at the wastegate actuator.

Control Details

Control for the solenoids and stepper motors can be either closed loop or open loop. Closed loop systems rely on feedback from a manifold pressure sensor to meet a predetermined boost pressure. Open loop systems have a predetermined control output where control output is merely based on other inputs such as throttle angle and/or engine RPM. Open loop specifically leaves out a desired boost level, while closed loop attempts to target a specific level of boost pressure. Since open loop systems do not modify control levels based on MAP sensor, differing boost pressure levels may be reached based on outside variables such as weather conditions or engine coolant temperature. For this reason, systems that do not feature closed loop operation are not as widespread.

▲ 4-port pneumatic solenoid installed to control a dual port wastegate controlled by a single PWM PID controller

Boost controllers often use pulse width modulation (PWM) techniques to bleed off boost pressure on its way to the reference port on the wastegate actuator diaphragm in order to (on occasion) under report boost pressure in such a way that the wastegate permits a turbocharger to build more boost pressure in the intake than it normally could. In effect, a boost-control solenoid valve lies to the wastegate under the engine control unit´s (ECU) control. The boost control solenoid contains a needle valve that can open and close very fast. By varying the pulse width to the solenoid, the solenoid valve can be commanded to be open a certain percentage of the time. This effectively alters the flow rate of air pressure through the valve, changing the rate at which air bleeds out of the T in the manifold pressure reference line to the wastegate. This effectively changes the air pressure as seen by the wastegate actuator diaphragm. Solenoids may require small diameter restrictors be installed in the air control lines to limit airflow and even out the on/off nature of their operation.

The wastegate control solenoid can be commanded to run in a variety of frequencies in various gears, engine speeds, or according to various other factors in a deterministic open-loop mode. Or by monitoring manifold pressure in a feedback loop- the engine management system can monitor the efficacy of PWM changes in the boost control solenoid bleed rate at altering boost pressure in the intake manifold, increasing or decreasing the bleed rate to target a particular maximum boost.

The basic algorithm sometimes involves the EMS (engine management system) "learning" how fast the turbocharger can spool and how fast the boost pressure increases. Armed with this knowledge, as long as boost pressure is below a predetermined allowable ceiling, the EMS will open the boost control solenoid to allow the turbocharger to create overboost beyond what the wastegate would normally allow. As overboost reaches the programmable maximum, the EMS begins to decrease the bleed rate through the control solenoid to raise boost pressure as seen at the wastegate actuator diaphragm so the wastegate opens enough to limit boost to the maximum configured level of over-boost.

Stepper motors allow fine control of airflow based on position and speed of the motor, but may have low total airflow capability. Some systems use a solenoid in conjunction with a stepper motor, with the stepper motor allowing fine control and the solenoid coarse control.

Many configurations are possible with 2-, 3-, and 4-port solenoids and stepper motors in series or parallel. Two port solenoid bleed systems with a PID controller tend to be common on factory turbocharged cars.

Advantages

Since less positive pressure can be present at the wastegate actuator as desired boost is approached the wastegate remains closer to a completely closed state. This keeps exhaust gas routed through the turbine and increases energy transferred to the wheels of the turbocharger. Once desired boost is reached, closed loop based systems react by allowing more air pressure to reach the wastegate actuator to stop the further increase in air pressure so desired boost levels are maintained. This reduces turbocharger lag and lowers boost threshold. Boost pressure builds faster when the throttle is depressed quickly and allows boost pressure to build at lower engine RPM than without such a system.

This also allows the use of a much softer spring in the actuator. For instance, a 7 psi (0.48 bar) spring together with a boost controller may still be able to achieve a maximum boost level of well over 15 psi (1.0 bar). The electronic control unit can be programmed to control 7 psi (0.48 bar) psi at half throttle, 12 psi (0.83 bar) at 3/4 throttle, and 15 psi (1.0 bar) at full throttle, or whatever levels the programmer or designer of the control unit intends. This partial throttle control greatly increases driver control over the engine and vehicle.

Limitations and Disadvantages

Even with an electronic controller, actuator springs that are too soft can cause the wastegate to open before desired. Exhaust gas backpressure is still pushing against the wastegate valve itself. This backpressure can overcome the spring pressure without the aid of the actuator at all. Electronic control may still enable control of boost to over double gauge pressure of the spring's rated pressure.

The solenoid and stepper motors also need to be installed in such a way to maximize the advantages of failure modes. For instance, if a solenoid is installed to control boost electronically, it should be installed such that if the solenoid fails in the most common failure mode (probably non-energized position) the boost control falls back to simple wastegate actuator boost levels. It is possible a solenoid or stepper motor could get stuck in a position that lets no boost pressure reach the wastegate, causing boost to quickly rise out of control.

The electronic systems, extra hoses, solenoids and soforth add complexity to the turbocharger system. This runs counter to the "keep it simple" principle as there are more things that can go wrong. It is worth noting that virtually all modern factory turbocharged cars, the same cars with long warranty periods, implement electronic boost control. Manufacturers such as Subaru, Mitsubishi and Saab integrate electronic boost control in all turbo model cars.

Availability and Applications

Electronic boost control systems are available as aftermarket stand-alone systems such as the HKS EVC and VBC, Apex-i AVC-R, or Gizzmo IBC / MS-IBC as a built-in feature of modern factory turbocharged vehicles such as the Subaru Impreza WRX STi and often as built-in features in full aftermarket stand-alone engine management systems such as the Hydra Nemesis, AEM EMS and MegaSquirt.

Dangers in use

Installing a boost controller in a vehicle that is already well tuned, such as a factory turbocharged car, may allow higher boost pressure than tolerable by the engine or turbocharger, reducing life and reliability. Care should be taken to avoid exceeding the limits of any engine system components such as the engine block, fuel injectors, or engine management system. This is as true with boost control as it is with fuel and timing controls, or any number of other engine system modifications.

In particular, users may find the extremely low cost and ease of adding a manual boost controller a particular draw for extra power at low cost compared to more comprehensive modifications. Users should carefully consider how installing any boost controller may affect and interact with existing complex engine management systems. Additional boost levels may not be tolerated by the existing turbocharger, causing faster wear. Fuel injectors or the fuel pump may not be able to deliver additional fuel needed for higher air flow and power of higher boost pressure. Or the engine management system may not be able to properly compensate for fuel or ignition timing, causing knock and/or engine failure.

Past and Future

There are other outdated methods of boost control, such as intake restriction or bleed off. For instance, it is possible to install a large butterfly valve in the intake to restrict airflow as desired boost is approached. It is also possible to actually release large amounts of already compressed air similar to a blowoff valve but on a constant basis to maintain desired boost at the intake manifold. The currently popular exhaust gas bypass via wastegate is quite superior if compared to creating intake restriction or wasting energy by releasing air that has already been compressed. These methods are rarely used in modern system due to the large sacrifices in efficiency, heat, and reliability.

Other methods may come into widespread use in the future, such as variable geometry turbochargers. With a sufficiently large turbine, no wastegate is necessary. Low speed response and faster spool up are then obtained using variable turbine technologies rather than a smaller turbine. These systems may replace or supplement typical wastegates as they develop. Control methods for the variable mechanical controls, such as the principles of closed loop will still apply even if they no longer involve pneumatics.

Cold Air Intake

A cold air intake (CAI) is an aftermarket assembly of parts used to bring relatively cool air into a car's internal-combustion engine.

Most vehicles manufactured from the mid-1970s until the mid-1990s have thermostatic air intake systems that regulate the temperature of the air entering the engine's intake tract, providing warm air when the engine is cold and cold air when the engine is warm to maximize performance, efficiency, and fuel economy. With the advent of advanced emission controls and more advanced fuel injection methods modern vehicles do not have a thermostatic air intake system and the factory installed air intake draws unregulated cold air. Aftermarket cold air intake systems are marketed with claims of increased engine efficiency and performance. The putative principle behind a cold air intake is that cooler air has a higher density, thus containing more oxygen per volume unit than warmer air.

Design Features

Some strategies used in designing aftermarket cold air intakes are:

- Reworking parts of the intake that create turbulence to reduce air resistance.

- Providing a more direct route to the air intake by eliminating muffling devices.

- Shortening the length of the intake.

- Placing the intake duct so as to use the ram-air effect to give positive pressure at speed.

Construction

Intake systems come in many different styles and can be constructed from plastic, metal, rubber (silicone) or composite materials (fiberglass, carbon fiber or Kevlar). The most efficient intake systems utilize an airbox which is sized to complement the engine and will extend the powerband of the engine. The intake snorkel (opening for the intake air to enter the system) must be large enough to ensure sufficient air is available to the engine under all conditions from idle to full throttle.

The most basic cold air intake consists of a long metal or plastic tube leading to a conical air filter. Power may be lost at certain engine speeds and gained at others. Because of the reduced covering, intake noise is usually increased.

Some intakes use heat shields to isolate the air filter from the rest of the engine compartment, providing cooler air from the front or side of the engine bay. This can make a big difference to intake temperatures, especially when the car is moving slowly. Some systems, called "fender mount," move the filter into the fender wall instead. This system draws air up through the fender wall which provides even more isolation and still cooler air.

States adopting strict CARB emission standards California Air Resources Board do not allow the usage of performance air intakes that increase vehicle's pollution. As a result all aftermarket performance air intakes used in such states must be CARB tested and approved, otherwise they are considered to be illegal for road usage and may result in a failed vehicle inspection.

Dual Mass Flywheel

Dual mass flywheel section

A Dual mass flywheel or DMF is a rotating mechanical device that is used to provide continuous energy (rotational energy) in systems where the energy source is not continuous, the same way as a conventional flywheel acts, but damping any violent variation of torque or revolutions that could cause an unwanted vibration. The vibration reduction is achieved by accumulating stored energy in the two flywheel half masses over a period of time but damped by a series of strong springs, doing that at a rate that is compatible with the energy source, and then releasing that energy at a much higher rate over a relatively short time. The compact dual-mass flywheel also includes the whole clutch, (with the pressure plate and the friction disc).

History

Schaeffler torque converter with a pendulum absorber using the same DMF's bent springs.

The Dual Mass Flywheels were developed to address the escalation of torque and power, especially at low revs. The growing concern for the environment and the adoption of more stringent regulations have marked the development of more efficient new engines, lowering the cylinder number to 3 or even 2 cylinders, and allowing the delivery of more torque and power at low revolutions. The counterpart has been an increase in the level of vibration which traditional clutch discs are unable to absorb. This is where the Dual Mass Flywheels play a key role, making these mechanical developments more viable.

The absorption capacity of the vibration depends on the moving parts of the DMF, these parts are subject to wear. Whenever the clutch is replaced, the DMF should be checked for wear. The two key wear characteristics are freeplay and sideplay (rock). These should be measured to determine whether the flywheel is serviceable. The wear limit specifications can be found in vehicle or flywheel manufacturer's published documentation. Other failure modes are severely grooved/damaged clutch mating surface, grease loss, and cracking.

Types

The main type is called a planetary DMF. The planetary gear and the torsional damper are incorporated into the main flywheel. For this purpose, the main flywheel is divided into primary and secondary pinion-connected masses, and between them there are four different types of bent springs:

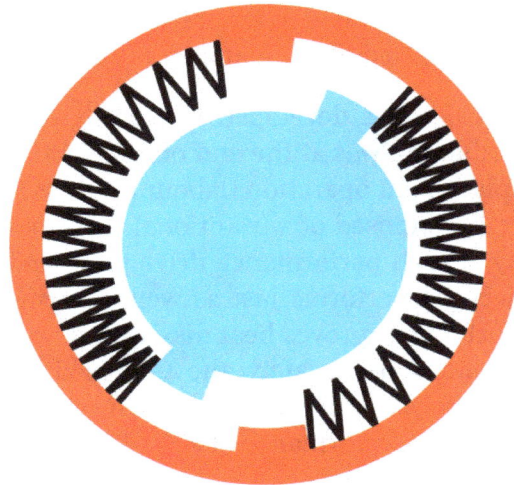

Principle of dual mass flywheel.
Black: absorber springs.
Red: flywheel, crankshaft side
Blue: flywheel, transmission side

Individual Bent Spring

The simplest form of the bent spring is the standard single spring.

One-phase Bent Springs in Parallel

The standard springs are called parallel springs of one phase. These consist of an outer and an inner spring of almost equal lengths and connected in parallel. The individual characteristic curves of the two springs are added to form the characteristic curve of the spring pair.

Two-phase Bent Springs in Parallel

In the case of two-stage spring there are two curved parallel springs, one inside the other, but the internal spring is shorter so that it acts later. The characteristic curve of the outer spring is adapted to increase when the engine is started. The softer outer spring only acts to increase the problematic resonance frequency range. When the torque increases, reaching the maximum value, the internal spring also acts. In this second phase, the inner and outer springs work together. The collaboration of both springs thus ensures good acoustic isolation at all engine speeds.

Three-phase Bent Spring

This curved spring consists of an outer and two inner springs with different elastic characteristics connected in series. This category of bent spring uses the two concepts together: parallel and series connection in order to ensure optimal torsional compensation for each value of torque.

Solutions to Unreliability Issues

Because of expensive unreliability issues with Dual Mass Flywheels there has been a trend to fit solid flywheels from lower power models of the same engine. Some mechanics have even welded the dual mass elements together. These 'fixes' have resulted in instances of gearbox damage and damage to engine crankshafts, as well as refinement and drive-ability issues.

Long established parts maker Valeo has developed a properly engineered 4-piece clutch kit to replace the DMF. The UK introduction was at the end of 2005. It is claimed to provide increased reliability for the expensive replacement operation (labour time cost and the part cost), of the dual mass flywheel. The 4-piece kit is composed of a traditional rigid flywheel and long travel damper consisting of a cover assembly, a high performance drive plate and a release bearing. Rigorous R&D testing showed the 4-piece kit performs just as well in terms of overall clutch operation, transmission protection, comfort and improves heat dissipation and durability. The transmission protection provides full engine and gearbox safety; the 4-piece kit matches the DMF for mechanical vibration reduction and therefore limits premature wear within the gearbox. The long travel damper closely matches the damping performance of a DMF. Fitting times are also comparable.

Choke Valve

A choke valve is a type of valve designed to create a choked flow in a fluid line in an automobile. The viscosity of the fluid passing through the valve is irrelevant in understanding how the mechanism works. The rate of flow is determined only by the ambient pressure on the upstream side of the valve.

In automotive contexts, a choke valve modifies the air pressure in the intake manifold of an internal combustion engine, thereby altering the ratio of fuel and air quantity entering the engine. Choke valves are generally used in naturally aspirated engines with carburetors to supply a richer fuel mixture when starting the engine. Most choke valves in engines are butterfly valves mounted in the manifold above the carburetor jet to produce a higher partial vacuum, which increases the fuel draw.

In heavy industrial or fluid engineering contexts, a choke valve is a particular design of valve that raises and lowers a solid cylinder (called a "plug" or "stem") which is placed around or inside another cylinder that has holes or slots. The design of a choke valve means fluids flowing through the cage are coming from all sides and that the streams of flow (through the holes or slots) collide with each other at the center of the cage cylinder, thereby dissipating the energy of the fluid through "flow impingement". The main advantage of choke valves is that they can be designed to be totally linear in their flow rate.

Automotive

A choke valve is sometimes installed in the carburetor of internal combustion engines. Its purpose is to restrict the flow of air, thereby enriching the fuel-air mixture while starting the engine. Depending on engine design and application, the valve can be activated manually by the operator of the engine (via a lever or pull handle) or automatically by a temperature-sensitive mechanism called an autochoke.

Choke valves are important for naturally-aspirated gasoline engines because small droplets of gasoline do not evaporate well within a cold engine. By restricting the flow of air into the throat of the carburetor, the choke valve reduces the pressure inside the throat, which causes a proportionally-greater amount of fuel to be pushed from the main jet into the combustion chamber during cold-running operation. Once the engine is warm (from combustion), opening the choke valve restores the carburetor to normal operation, supplying fuel and air in the correct stoichiometric ratio for clean, efficient combustion.

Note that the term "choke" is applied to the carburetor's enrichment device even when it works by a totally different method. Commonly, SU carburetors have "chokes" that work by lowering the fuel jet to a narrower part of the needle. Some others work by introducing an additional fuel route to the constant depression chamber.

Chokes were nearly universal in automobiles until fuel injection began to supplant carburetors. Choke valves are still common in other internal-combustion applications, including most small portable engines, motorcycles, small propeller-driven airplanes, riding lawn mowers, and normally-aspirated marine engines.

M21i engine bay original parts

Industrial

Heavy-duty industrial choke valves control the flow to a certain flow coefficient (C_v), which is determined by how far the valve is opened. They are regularly used in the oil industry. For highly-erosive and corrosive purposes, they are often made of tungsten carbide or inconel.

Radiator (Engine Cooling)

A typical engine coolant radiator used in an automobile

Radiators are heat exchangers used for cooling internal combustion engines, mainly in automobiles but also in piston-engined aircraft, railway locomotives, motorcycles, stationary generating plant or any similar use of such an engine.

Internal combustion engines are often cooled by circulating a liquid called engine coolant through the engine block, where it is heated, then through a radiator where it loses heat to the atmosphere, and then returned to the engine. Engine coolant is usually water-based, but may also be oil. It is common to employ a water pump to force the engine coolant to circulate, and also for an axial fan to force air through the radiator.

Automobiles and Motorcycles

Coolant being poured into the radiator of an automobile

In automobiles and motorcycles with a liquid-cooled internal combustion engine, a radiator is connected to channels running through the engine and cylinder head, through which a liquid (coolant) is pumped. This liquid may be water (in climates where water is unlikely to freeze), but is more commonly a mixture of water and antifreeze in proportions appropriate to the climate. Antifreeze itself is usually ethylene glycol or propylene glycol (with a small amount of corrosion inhibitor).

A typical automotive cooling system comprises:

- a series of channels cast into the engine block and cylinder head, surrounding the combustion chambers with circulating liquid to carry away heat;

- a radiator, consisting of many small tubes equipped with a honeycomb of fins to convect heat rapidly, that receives and cools hot liquid from the engine;

- a water pump, usually of the centrifugal type, to circulate the liquid through the system;

- a thermostat to control temperature by varying the amount of liquid going to the radiator;

- a fan to draw fresh air through the radiator.

The radiator transfers the heat from the fluid inside to the air outside, thereby cooling the fluid, which in turn cools the engine. Radiators are also often used to cool automatic transmission fluids, air conditioner refrigerant, intake air, and sometimes to cool motor oil or power steering fluid. Radiators are typically mounted in a position where they receive airflow from the forward movement of the vehicle, such as behind a front grill. Where engines are mid- or rear-mounted, it is common to mount the radiator behind a front grill to achieve sufficient airflow, even though this requires long coolant pipes. Alternatively, the radiator may draw air from the flow over the top of the vehicle or from a side-mounted grill. For long vehicles, such as buses, side airflow is most common for engine and transmission cooling and top airflow most common for air conditioner cooling.

Radiator Construction

Automobile radiators are constructed of a pair of header tanks, linked by a core with many narrow passageways, giving a high surface area relative to volume. This core is usually made of stacked layers of metal sheet, pressed to form channels and soldered or brazed together. For many years radiators were made from brass or copper cores soldered to brass headers. Modern radiators have aluminum cores, and often save money and weight by using plastic headers. This construction is more prone to failure and less easily repaired than traditional materials.

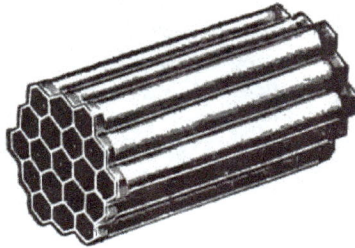

Honeycomb radiator tubes

An earlier construction method was the honeycomb radiator. Round tubes were swaged into hexagons at their ends, then stacked together and soldered. As they only touched at their ends, this formed what became in effect a solid water tank with many air tubes through it.

Some vintage cars use radiator cores made from coiled tube, a less efficient but simpler construction.

Coolant Pump

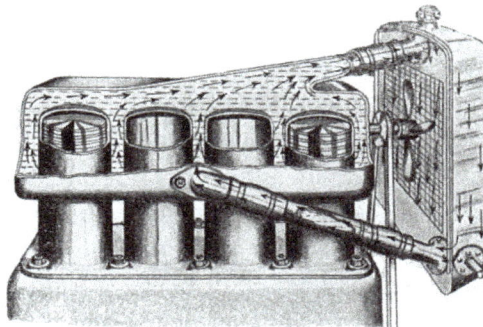

Thermosyphon cooling system of 1937, without circulating pump

Radiators first used downward vertical flow, driven solely by a thermosyphon effect. Coolant is heated in the engine, becomes less dense, and so rises. As the radiator cools the fluid, the coolant becomes denser and falls. This effect is sufficient for low-power stationary engines, but inadequate for all but the earliest automobiles. All automobiles for many years have used centrifugal pumps to circulate the engine coolant because natural circulation has very low flow rates.

Heater

A system of valves or baffles, or both, is usually incorporated to simultaneously operate a small radiator inside the vehicle. This small radiator, and the associated blower fan, is called the heater core, and serves to warm the cabin interior. Like the radiator, the heater core acts by removing heat from the engine. For this reason, automotive technicians often advise operators to turn on the heater and set it to high if the engine is overheating, to assist the main radiator.

Temperature Control

Waterflow Control

Car engine thermostat

The engine temperature on modern cars is primarily controlled by a wax-pellet type of thermostat, a valve which opens once the engine has reached its optimum operating temperature.

When the engine is cold, the thermostat is closed except for a small bypass flow so that the thermostat experiences changes to the coolant temperature as the engine warms up. Engine coolant is directed by the thermostat to the inlet of the circulating pump and is returned directly to the engine, bypassing the radiator. Directing water to circulate only through the engine allows the temperature to reach optimum operating temperature as quickly as possible whilst avoiding localised "hot spots." Once the coolant reaches the thermostat's activation temperature, it opens, allowing water to flow through the radiator to prevent the temperature rising higher.

Once at optimum temperature, the thermostat controls the flow of engine coolant to the radiator so that the engine continues to operate at optimum temperature. Under peak load conditions, such as driving slowly up a steep hill whilst heavily laden on a hot day, the thermostat will be approaching fully open because the engine will be producing near to maximum power while the velocity of air flow across the radiator is low. (The velocity of air flow across the radiator has a major effect on its ability to dissipate heat.) Conversely, when cruising fast downhill on a motorway on a cold night on a light throttle, the thermostat will be nearly closed because the engine is producing little power, and the radiator is able to dissipate much more heat than the engine is producing. Allowing too much flow of coolant to the radiator would result in the engine being over cooled and operating at lower than optimum temperature, resulting in decreased fuel efficiency and increased exhaust emissions. Furthermore, engine durability, reliability, and longevity are sometimes compromised, if any components (such as the crankshaft bearings) are engineered to take thermal expansion into account to fit together with the correct clearances. Another side effect of over-cooling is reduced performance of the cabin heater, though in typical cases it still blows air at a considerably higher temperature than ambient.

The thermostat is therefore constantly moving throughout its range, responding to changes in vehicle operating load, speed and external temperature, to keep the engine at its optimum operating temperature.

On vintage cars you may find a bellows type thermostat, which has a corrugated bellows containing a volatile liquid such as alcohol or acetone. These types of thermostats do not work well at cooling system pressures above about 7 psi. Modern motor vehicles typically run at around 15 psi, which precludes the use of the bellows type thermostat. On direct air-cooled engines this is not a concern for the bellows thermostat that controls a flap valve in the air passages.

Airflow Control

Other factors influence the temperature of the engine, including radiator size and the type of radiator fan. The size of the radiator (and thus its cooling capacity) is chosen such that it can keep the engine at the design temperature under the most extreme conditions a vehicle is likely to encounter (such as climbing a mountain whilst fully loaded on a hot day).

Airflow speed through a radiator is a major influence on the heat it loses. Vehicle speed affects this, in rough proportion to the engine effort, thus giving crude self-regulatory feedback. Where an additional cooling fan is driven by the engine, this also tracks engine speed similarly.

Engine-driven fans are often regulated by a viscous-drive clutch from the drivebelt, which slips and reduces the fan speed at low temperatures. This improves fuel efficiency by not wasting power on driving the fan unnecessarily. On modern vehicles, further regulation of cooling rate is provided by either variable speed or cycling radiator fans. Electric fans are controlled by a thermostatic switch or the engine control unit. Electric fans also have the advantage of giving good airflow and cooling at low engine revs or when stationary, such as in slow-moving traffic.

Before the development of viscous-drive and electric fans, engines were fitted with simple fixed fans that drew air through the radiator at all times. Vehicles whose design required the installation of a large radiator to cope with heavy work at high temperatures, such as commercial vehicles and tractors would often run cool in cold weather under light loads, even with the presence of a thermostat, as the large radiator and fixed fan caused a rapid and significant drop in coolant temperature as soon as the thermostat opened. This problem can be solved by fitting a radiator blind (or radiator shroud) to the radiator that can be adjusted to partially or fully block the airflow through the radiator. At its simplest the blind is a roll of material such as canvas or rubber that is unfurled along the length of the radiator to cover the desired portion. Some older vehicles, like the World War I-era S.E.5 and SPAD S.XIII single-engined fighters, have a series of shutters that can be adjusted from the driver's or pilot's seat to provide a degree of control. Some modern cars have a series of shutters that are automatically opened and closed by the engine control unit to provide a balance of cooling and aerodynamics as needed.

These AEC Regent III RT buses are fitted with radiator blinds, seen here covering the lower half of the radiators.

Coolant Pressure

Because the thermal efficiency of internal combustion engines increases with internal temperature, the coolant is kept at higher-than-atmospheric pressure to increase its boiling point. A calibrated pressure-relief valve is usually incorporated in the radiator's fill cap. This pressure varies between models, but typically ranges from 4 to 30 psi (30 to 200 kPa).

As the coolant expands with increasing temperature, its pressure in the closed system must increase. Ultimately, the pressure relief valve opens, and excess fluid is dumped into an overflow container. Fluid overflow ceases when the thermostat modulates the rate of cooling to keep the temperature of the coolant at optimum. When the engine coolant cools and contracts (as conditions change or when the engine is switched off), the fluid is returned to the radiator through additional valving in the cap.

Engine Coolant

Before World War II, engine coolant was usually plain water. Antifreeze was used solely to control freezing, and this was often only done in cold weather.

Development in high-performance aircraft engines required improved coolants with higher boiling points, leading to the adoption of glycol or water-glycol mixtures. These led to the adoption of glycols for their antifreeze properties.

Since the development of aluminium or mixed-metal engines, corrosion inhibition has become even more important than antifreeze, and in all regions and seasons.

Boiling or Overheating

An overflow tank that runs dry may result in the coolant vaporizing, which can cause localized or general overheating of the engine. Severe damage can result, such as blown headgaskets, cracked cylinder heads or cylinder blocks. Sometimes there will be no warning, because the temperature sensor that provides data for the temperature gauge (either mechanical or electric) is not exposed to the excessively hot coolant, providing a harmfully false reading.

Opening a hot radiator drops the system pressure, which may cause it to boil and eject dangerously hot liquid and steam. Therefore, radiator caps often contain a mechanism that attempts to relieve the internal pressure before the cap can be fully opened.

History

The invention of the automobile water radiator is attributed to Karl Benz. Wilhelm Maybach designed the first honeycomb radiator for the Mercedes 35hp.

Supplementary Radiators

It is sometimes necessary for a car to be equipped with a second, or auxiliary, radiator to increase the cooling capacity, when the size of the original radiator cannot be increased. The second radiator is plumbed in series with the main radiator in the circuit. This was the case when the Audi 100 was first turbocharged creating the 200.

Some engines have an oil cooler, a separate small radiator to cool the engine oil. Cars with an automatic transmission often have extra connections to the radiator, allowing the transmission fluid to transfer its heat to the coolant in the radiator. These may be either oil-air radiators, as for a smaller version of the main radiator. More simply they may be oil-water coolers, where an oil pipe is inserted inside the water radiator. Though the water is hotter than the ambient air, its higher thermal conductivity offers comparable cooling (within limits) from a less complex and thus cheaper and more reliable oil cooler. Less commonly, power steering fluid, brake fluid, and other hydraulic fluids may be cooled by an auxiliary radiator on a vehicle.

Turbo charged or supercharged engines may have an intercooler, which is an air-to-air or air-to-water radiator used to cool the incoming air charge—not to cool the engine.

Aircraft

Aircraft with liquid-cooled piston engines (usually inline engines rather than radial) also require radiators. As airspeed is higher than for cars, these are efficiently cooled in flight, and so do not require large areas or cooling fans. Many high-performance aircraft however suffer extreme overheating problems when idling on the ground - a mere 7 minutes for a Spitfire. This is similar to Formula 1 cars of today, when stopped on the grid with engines running they require ducted air forced into their radiator pods to prevent overheating.

Surface Radiators

Reducing drag is a major goal in aircraft design, including the design of cooling systems. An early technique was to take advantage of an aircraft's abundant airflow to replace the honeycomb core (many surfaces, with a high ratio of surface to volume) by a surface mounted radiator. This uses a single surface blended into the fuselage or wing skin, with the coolant flowing through pipes at the back of this surface. Such designs were seen mostly on World War I aircraft.

As they are so dependent on airspeed, surface radiators are even more prone to overheating when ground-running. Racing aircraft such as the Supermarine S.6B, a racing seaplane with radiators built into the upper surfaces of its floats, have been described as "being flown on the temperature gauge" as the main limit on their performance.

Surface radiators have also been used by a few high-speed racing cars, such as Malcolm Campbell's Blue Bird of 1928.

Pressurized Cooling Systems

Radiator caps for pressurized automotive cooling systems. Of the two valves, one prevents the creation of a vacuum, the other limits the pressure.

It is generally a limitation of most cooling systems that the cooling fluid not be allowed to boil, as the need to handle gas in the flow greatly complicates design. For a water cooled system, this means that the maximum amount of heat transfer is limited by the specific heat capacity of water and the difference in temperature between ambient and 100°C. This provides more effective cooling in the winter, or at higher altitudes where the temperatures are low.

Another effect that is especially important in aircraft cooling is that the specific heat capacity changes with pressure, and this pressure changes more rapidly with altitude than the drop in temperature. Thus, generally, liquid cooling systems lose capacity as the aircraft climbs. This was a major limit on performance during the 1930s when the introduction of turbosuperchargers first allowed convenient travel at altitudes above 15,000 ft, and cooling design became a major area of research.

The most obvious, and common, solution to this problem was to run the entire cooling system under pressure. This maintained the specific heat capacity at a constant value, while the outside air temperature continued to drop. Such systems thus improved cooling capability as they climbed. For most uses, this solved the problem of cooling high-performance piston engines, and almost all liquid-cooled aircraft engines of the World War II period used this solution.

However, pressurized systems were also more complex, and far more susceptible to damage - as the cooling fluid was under pressure, even minor damage in the cooling system like a single rifle-calibre bullet hole, would cause the liquid to rapidly spray out of the hole. Failures of the cooling systems were, by far, the leading cause of engine failures.

Evaporative Cooling

Although it is more difficult to build an aircraft radiator that is able to handle steam, it is by no means impossible. The key requirement is to provide a system that condenses the steam back into liquid before passing it back into the pumps and completing the cooling loop. Such a system can take advantage of the specific heat of vaporization, which in the case of water is five times the specific heat capacity in the liquid form. Additional gains may be had by allowing the steam to become superheated. Such systems, known as evaporative coolers, were the topic of considerable research in the 1930s.

Consider two cooling systems that are otherwise similar, operating at an ambient air temperature of 20°C. An all-liquid design might operate between 30°C and 90°C, offering 60°C of temperature difference to carry away heat. An evaporative cooling system might operate between 80°C and 110°C, which at first glance appears to be much less temperature difference, but this analysis overlooks the enormous amount of heat energy soaked up during the generation of steam, equivalent to 500°C. In effect, the evaporative version is operating between 80°C and 560°C, a 480°C effective temperature difference. Such a system can be effective even with much smaller amounts of water.

The downside to the evaporative cooling system is the area of the condensers required to cool the steam back below the boiling point. As steam is much less dense than water, a correspondingly larger surface area is needed to provide enough airflow to cool the steam back down. The Rolls-Royce Goshawk design of 1933 used conventional radiator-like condensers and this design proved to be a serious problem for drag. In Germany, the Günter brothers developed an alternative design combining evaporative cooling and surface radiators spread all over the aircraft wings, fuselage and even the rudder. Several aircraft were built using their design and set numerous performance records, notably the Heinkel He 119 and Heinkel He 100. However, these systems required numerous pumps to return the liquid from the spread-out radiators and proved to be extremely difficult to keep running properly, and were much more susceptible to battle damage. Efforts to develop this system had generally been abandoned by 1940. The need for evaporative cooling was soon to be negated by the widespread availability of ethylene glycol based coolants, which had a lower specific heat, but a much higher boiling point than water.

Radiator Thrust

An aircraft radiator contained in a duct heats the air passing through, causing the air to expand and gain velocity. This is called the Meredith effect, and high-performance piston aircraft with well-designed low-drag radiators (notably the P-51 Mustang) derive thrust from it. The thrust was significant enough to offset the drag of the duct the radiator was enclosed in and allowed the aircraft to achieve zero cooling drag. At one point, there were even plans to equip the Spitfire with an afterburner, by injecting fuel into the exhaust duct after the radiator and igniting it. Afterburning is achieved by injecting additional fuel into the engine downstream of the main combustion cycle.

Stationary Plant

Engines for stationary plant are normally cooled by radiators in the same way as automobile engines. However, in some cases, evaporative cooling is used via a cooling tower.

References

- Evaluation of Certain Food Additives and Contaminants (Technical Report Series). World Health Organization. p. 105. ISBN 92-4-120909-7.

- Knowling, Michael (November 16, 2007). "Shielded Sucker: A cheap and easy air intake improvement - with no CAI pipe in sight!". AutoSpeed. Retrieved 2013-04-15.

- Edgar, Julian (July 10, 2001). "Siting Cold Air Intakes: Simple testing to find the best place to put the mouth of the cold air intake.". AutoSpeed. Retrieved 2013-04-15.

- A safe and effective propylene glycol based capture liquid for fruit fly traps baited with synthetic lures – page 2|Florida Entomologist. Findarticles.com. Retrieved on 2011-01-01.

- Engine Cooling Testing: Why use a refractometer? Archived July 25, 2011, at the Wayback Machine. posted 2/7/2001 by Michael Reimer

- Allan Browning and David Berry (September / October 2010) "Selecting and maintaining glycol based heat transfer fluids," Facilities Engineering Journal, pages 16-18.

Internal Combustion Engine

Internal combustion engines are heat engines that help in the combustion of a fuel. In these engines, the expansion of the high-temperature and high-pressure gases is produced by combustion; this combustion directly applies force to a particular part of the engine. The section on internal combustion engine offers an insightful focus, keeping in mind the subject matter.

Internal Combustion Engine

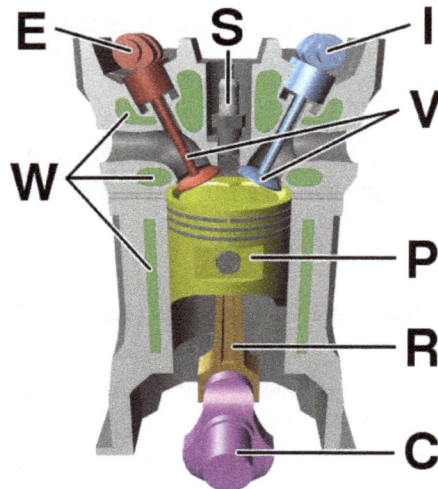

Diagram of a cylinder as found in 4-stroke gasoline engines.:
C – crankshaft.
E – exhaust camshaft.
I – inlet camshaft.
P – piston.
R – connecting rod.
S – spark plug.
V – valves. red: exhaust, blue: intake.
W – cooling water jacket.
gray structure – engine block.

An internal combustion engine (ICE) is a heat engine where the combustion of a fuel occurs with an oxidizer (usually air) in a combustion chamber that is an integral part of the working fluid flow circuit. In an internal combustion engine the expansion of the high-temperature and high-pressure gases produced by combustion apply direct force to some component of the engine. The force is applied typically to pistons, turbine blades, rotor or a nozzle. This force moves the component over a distance, transforming chemical energy into useful mechanical energy.

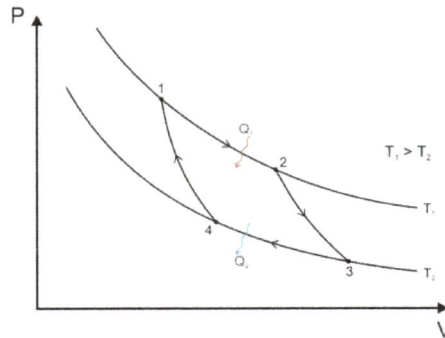

Diagram describing the ideal combustion cycle by Carnot

The first commercially successful internal combustion engine was created by Étienne Lenoir around 1859 and the first modern internal combustion engine was created in 1876 by Nikolaus Otto.

The term internal combustion engine usually refers to an engine in which combustion is intermittent, such as the more familiar four-stroke and two-stroke piston engines, along with variants, such as the six-stroke piston engine and the Wankel rotary engine. A second class of internal combustion engines use continuous combustion: gas turbines, jet engines and most rocket engines, each of which are internal combustion engines on the same principle as previously described. Firearms are also a form of internal combustion engine.

Internal combustion engines are quite different from external combustion engines, such as steam or Stirling engines, in which the energy is delivered to a working fluid not consisting of, mixed with, or contaminated by combustion products. Working fluids can be air, hot water, pressurized water or even liquid sodium, heated in a boiler. ICEs are usually powered by energy-dense fuels such as gasoline or diesel, liquids derived from fossil fuels. While there are many stationary applications, most ICEs are used in mobile applications and are the dominant power supply for vehicles such as cars, aircraft, and boats.

Typically an ICE is fed with fossil fuels like natural gas or petroleum products such as gasoline, diesel fuel or fuel oil. There's a growing usage of renewable fuels like biodiesel for compression ignition engines and bioethanol or methanol for spark ignition engines. Hydrogen is sometimes used, and can be made from either fossil fuels or renewable energy.

History

Various scientists and engineers contributed to the development of internal combustion engines. In 1791, John Barber developed a turbine. In 1794 Thomas Mead patented a gas engine. Also in 1794 Robert Street patented an internal combustion engine, which was also the first to use liquid fuel, and built an engine around that time. In 1798, John Stevens built the first American internal combustion engine. In 1807, Swiss engineer François Isaac de Rivaz built an internal combustion engine ignited by electric spark. In 1823, Samuel Brown patented the first internal combustion engine to be applied industrially.

In 1860, Belgian Jean Joseph Etienne Lenoir produced a gas-fired internal combustion engine. In 1864, Nikolaus Otto patented the first atmospheric gas engine. In 1872, American George Brayton invented the first commercial liquid-fuelled internal combustion engine. In 1876, Nikolaus Otto, working with Gottlieb Daimler and Wilhelm Maybach, patented the compressed charge, four-cycle engine. In 1879, Karl Benz patented a reliable two-stroke gas engine. In 1892, Rudolf Diesel developed the first compressed charge, compression ignition engine. In 1926, Robert Goddard launched the first liquid-fueled rocket. In 1939, the Heinkel He 178 became the world's first jet aircraft.

Etymology

At one time, the word engine (via Old French, from Latin ingenium, "ability") meant any piece of machinery — a sense that persists in expressions such as siege engine. A "motor" (from Latin motor, "mover") is any machine that produces mechanical power. Traditionally, electric motors are not referred to as "Engines"; however, combustion engines are often referred to as "motors." (An electric engine refers to a locomotive operated by electricity.)

In boating an internal combustion engine that is installed in the hull is referred to as an engine, but the engines that sit on the transom are referred to as motors.

Applications

Reciprocating engine as found inside a car

Reciprocating piston engines are by far the most common power source for land and water vehicles, including automobiles, motorcycles, ships and to a lesser extent, locomotives (some are electrical but most use Diesel engines). Rotary engines of the Wankel design are used in some automobiles, aircraft and motorcycles.

Where very high power-to-weight ratios are required, internal combustion engines appear in the form of combustion turbines or Wankel engines. Powered aircraft typically uses an ICE which may be a reciprocating engine. Airplanes can instead use jet engines and helicopters can instead employ turboshafts; both of which are types of turbines. In addition to providing propulsion, airliners may employ a separate ICE as an auxiliary power unit. Wankel engines are fitted to many unmanned aerial vehicles.

Big Diesel generator used for backup power

Combined cycle power plant

ICEs drive some of the large electric generators that power electrical grids. They are found in the form of combustion turbines in combined cycle power plants with a typical electrical output in the range of 100 MW to 1 GW. The high temperature exhaust is used to boil and superheat water to run a steam turbine. Thus, the efficiency is higher because more energy is extracted from the fuel than what could be extracted by the combustion turbine alone. In combined cycle power plants efficiencies in the range of 50% to 60% are typical. In a smaller scale Diesel generators are used for backup power and for providing electrical power to areas not connected to an electric grid.

Small engines (usually 2-stroke gasoline engines) are a common power source for lawnmowers, string trimmers, chain saws, leafblowers, pressure washers, snowmobiles, jet skis, outboard motors, mopeds, and motorcycles.

Classification

There are several possible ways to classify internal combustion engines.

Reciprocating:

    By number of strokes

- Two-stroke engine
- Clerk Cycle 1879
- Day Cycle

- Four-stroke engine (Otto cycle)
- Six-stroke engine

By type of ignition

- Compression-ignition engine
- Spark-ignition engine (commonly found as gasoline engines)

By mechanical/thermodynamical cycle (these 2 cycles do not encompass all reciprocating engines, and are infrequently used):

- Atkinson cycle
- Miller cycle

Rotary:

- Wankel engine

Continuous combustion:

- Gas turbine
- Jet engine
  - Rocket engine
  - Ramjet

The following jet engine types are also gas turbines types:

- Turbojet
- Turbofan
- Turboprop

Reciprocating Engines

Structure

Bare cylinder block of a V8 engine

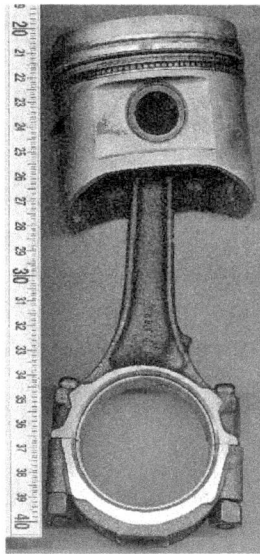

Piston, piston ring, gudgeon pin and connecting rod

The base of a reciprocating internal combustion engine is the engine block, which is typically made of cast iron or aluminium. The engine block contains the cylinders. In engines with more than one cylinder they are usually arranged either in 1 row (straight engine) or 2 rows (boxer engine or V engine); 3 rows are occasionally used (W engine) in contemporary engines, and other engine configurations are possible and have been used. Single cylinder engines are common for motorcycles and in small engines of machinery. Water-cooled engines contain passages in the engine block where cooling fluid circulates (the water jacket). Some small engines are air-cooled, and instead of having a water jacket the cylinder block has fins protruding away from it to cool by directly transferring heat to the air. The cylinder walls are usually finished by honing to obtain a cross hatch, which is better able to retain the oil. A too rough surface would quickly harm the engine by excessive wear on the piston.

The pistons are short cylindrical parts which seal one end of the cylinder from the high pressure of the compressed air and combustion products and slide continuously within it while the engine is in operation. The top wall of the piston is termed its crown and is typically flat or concave. Some two-stroke engines use pistons with a deflector head. Pistons are open at the bottom and hollow except for an integral reinforcement structure (the piston web). When an engine is working the gas pressure in the combustion chamber exerts a force on the piston crown which is transferred through its web to a gudgeon pin. Each piston has rings fitted around its circumference that mostly prevent the gases from leaking into the crankcase or the oil into the combustion chamber. A ventilation system drives the small amount of gas that escape past the pistons during normal operation (the blow-by gases) out of the crankcase so that it does not accumulate contaminating the oil and creating corrosion. In two-stroke gasoline engines the crankcase is part of the air–fuel path and due to the continuous flow of it they do not need a separate crankcase ventilation system.

The cylinder head is attached to the engine block by numerous bolts or studs. It has several functions. The cylinder head seals the cylinders on the side opposite to the pistons; it contains short ducts (the ports) for intake and exhaust and the associated intake valves that open to let the cylinder be filled with fresh air and exhaust valves that open to allow the combustion gases to escape.

However, 2-stroke crankcase scavenged engines connect the gas ports directly to the cylinder wall without poppet valves; the piston controls their opening and occlusion instead. The cylinder head also holds the spark plug in the case of spark ignition engines and the injector for engines that use direct injection. All CI engines use fuel injection, usually direct injection but some engines instead use indirect injection. SI engines can use a carburetor or fuel injection as port injection or direct injection. Most SI engines have a single spark plug per cylinder but some have 2. A head gasket prevents the gas from leaking between the cylinder head and the engine block. The opening and closing of the valves is controlled by one or several camshafts and springs—or in some engines—a desmodromic mechanism that uses no springs. The camshaft may press directly the stem of the valve or may act upon a rocker arm, again, either directly or through a pushrod.

Valve train above a Diesel engine cylinder head. This engine uses rocker arms but no pushrods.

Engine block seen from below. The cylinders, oil spray nozzle and half of the main bearings are clearly visible.

The crankcase is sealed at the bottom with a sump that collects the falling oil during normal operation to be cycled again. The cavity created between the cylinder block and the sump houses a crankshaft that converts the reciprocating motion of the pistons to rotational motion. The crankshaft is held in place relative to the engine block by main bearings, which allow it to rotate. Bulkheads in the crankcase form a half of every main bearing; the other half is a detachable cap. In some cases a single main bearing deck is used rather than several smaller caps. A connecting rod is connected to offset sections of the crankshaft (the crankpins) in one end and to the piston in the other end through the gudgeon pin and thus transfers the force and translates the reciprocating motion of the pistons to the circular motion of the crankshaft. The end of the connecting rod attached to the gudgeon pin is called its small end, and the other end, where it is connected to the crankshaft, the big end. The big end has a detachable half to allow assembly around the crankshaft. It is kept together to the connecting rod by removable bolts.

The cylinder head has an intake manifold and an exhaust manifold attached to the corresponding ports. The intake manifold connects to the air filter directly, or to a carburetor when one is present, which is then connected to the air filter. It distributes the air incoming from these devices to the individual cylinders. The exhaust manifold is the first component in the exhaust system. It collects the exhaust gases from the cylinders and drives it to the following component in the path. The exhaust system of an ICE may also include a catalytic converter and muffler. The final section in the path of the exhaust gases is the tailpipe.

4-Stroke Engines

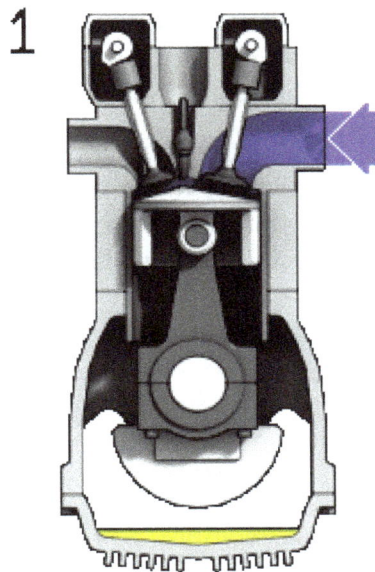

Diagram showing the operation of a 4-stroke SI engine. Labels:
1 - Induction
2 - Compression
3 - Power
4 - Exhaust

The top dead center (TDC) of a piston is the position where it is nearest to the valves; bottom dead center (BDC) is the opposite position where it is furthest from them. A stroke is the movement of a piston from TDC to BDC or vice versa together with the associated process. While an engine is in operation the crankshaft rotates continuously at a nearly constant speed. In a 4-stroke ICE each piston experiences 2 strokes per crankshaft revolution in the following order. Starting the description at TDC, these are:

1. Intake, induction or suction: The intake valves are open as a result of the cam lobe pressing down on the valve stem. The piston moves downward increasing the volume of the combustion chamber and allowing air to enter in the case of a CI engine or an air fuel mix in the case of SI engines that do not use direct injection. The air or air-fuel mixture is called the charge in any case.

2. Compression: In this stroke, both valves are closed and the piston moves upward reducing

the combustion chamber volume which reaches its minimum when the piston is at TDC. The piston performs work on the charge as it is being compressed; as a result its pressure, temperature and density increase; an approximation to this behavior is provided by the ideal gas law. Just before the piston reaches TDC, ignition begins. In the case of a SI engine, the spark plug receives a high voltage pulse that generates the spark which gives it its name and ignites the charge. In the case of a CI engine the fuel injector quickly injects fuel into the combustion chamber as a spray; the fuel ignites due to the high temperature.

3. Power or working stroke: The pressure of the combustion gases pushes the piston downward, generating more work than it required to compress the charge. Complementary to the compression stroke, the combustion gases expand and as a result their temperature, pressure and density decreases. When the piston is near to BDC the exhaust valve opens. The combustion gases expand irreversibly due to the leftover pressure—in excess of back pressure, the gauge pressure on the exhaust port—; this is called the blowdown.

4. Exhaust: The exhaust valve remains open while the piston moves upward expelling the combustion gases. For naturally aspirated engines a small part of the combustion gases may remain in the cylinder during normal operation because the piston does not close the combustion chamber completely; these gases dissolve in the next charge. At the end of this stroke, the exhaust valve closes, the intake valve opens, and the sequence repeats in the next cycle. The intake valve may open before the exhaust valve closes to allow better scavenging.

2-Stroke Engines

The defining characteristic of this kind of engine is that each piston completes a cycle every crankshaft revolution. The 4 processes of intake, compression, power and exhaust take place in only 2 strokes so that it is not possible to dedicate a stroke exclusively for each of them. Starting at TDC the cycle consist of:

1. Power: While the piston is descending the combustion gases perform work on it—as in a 4-stroke engine—. The same thermodynamic considerations about the expansion apply.

2. Scavenging: Around 75° of crankshaft rotation before BDC the exhaust valve or port opens, and blowdown occurs. Shortly thereafter the intake valve or transfer port opens. The incoming charge displaces the remaining combustion gases to the exhaust system and a part of the charge may enter the exhaust system as well. The piston reaches BDC and reverses direction. After the piston has traveled a short distance upwards into the cylinder the exhaust valve or port closes; shortly the intake valve or transfer port closes as well.

3. Compression: With both intake and exhaust closed the piston continues moving upwards compressing the charge and performing a work on it. As in the case of a 4-stroke engine, ignition starts just before the piston reaches TDC and the same consideration on the thermodynamics of the compression on the charge.

While a 4-stroke engine uses the piston as a positive displacement pump to accomplish scavenging taking 2 of the 4 strokes, a 2-stroke engine uses the last part of the power stroke and the first part of the compression stroke for combined intake and exhaust. The work required to displace the charge and exhaust gases comes from either the crankcase or a separate blower. For scavenging, expulsion of burned gas and entry of fresh mix, two main approaches are described: Loop scavenging, and Uniflow scavenging, SAE news published in the 2010s that 'Loop Scavenging' is better under any circumstance than Uniflow Scavenging.

Crankcase Scavenged

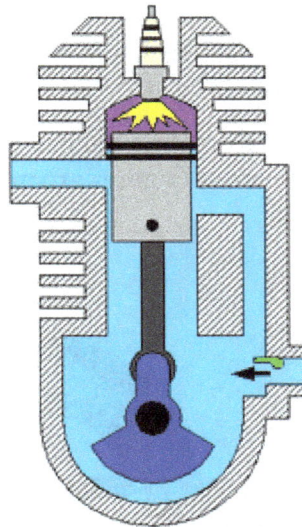

Diagram of a crankcase scavenged 2-stroke engine in operation

Some SI engines are crankcase scavenged and do not use poppet valves. Instead the crankcase and the part of the cylinder below the piston is used as a pump. The intake port is connected to the crankcase through a reed valve or a rotary disk valve driven by the engine. For each cylinder a transfer port connects in one end to the crankcase and in the other end to the cylinder wall. The exhaust port is connected directly to the cylinder wall. The transfer and exhaust port are opened and closed by the piston. The reed valve opens when the crankcase pressure is slightly below intake pressure, to let it be filled with a new charge; this happens when the piston is moving upwards. When the piston is moving downwards the pressure in the crankcase increases and the reed valve closes promptly, then the charge in the crankcase is compressed. When the piston is moving upwards, it uncovers the exhaust port and the transfer port and the higher pressure of the charge in the crankcase makes it enter the cylinder through the transfer port, blowing the exhaust gases. Lubrication is accomplished by adding 2-stroke oil to the fuel in small ratios. Petroil refers to the mix of gasoline with the aforesaid oil. This kind of 2-stroke engines has a lower efficiency than comparable 4-strokes engines and release a more polluting exhaust gases for the following conditions:

- They use a total-loss lubrication system: all the lubricating oil is eventually burned along with the fuel.

- There are conflicting requirements for scavenging: On one side, enough fresh charge needs to be introduced in each cycle to displace almost all the combustion gases but introducing too much of it means that a part of it gets in the exhaust.

- They must use the transfer port(s) as a carefully designed and placed nozzle so that a gas current is created in a way that it sweeps the whole cylinder before reaching the exhaust port so as to expel the combustion gases, but minimize the amount of charge exhausted. 4-stroke engines have the benefit of forcibly expelling almost all of the combustion gases because during exhaust the combustion chamber is reduced to its minimum volume. In crankcase scavenged 2-stroke engines, exhaust and intake are performed mostly simultaneously and with the combustion chamber at its maximum volume.

The main advantage of 2-stroke engines of this type is mechanical simplicity and a higher power-to-weight ratio than their 4-stroke counterparts. Despite having twice as many power strokes per cycle, less than twice the power of a comparable 4-stroke engine is attainable in practice.

In the USA two stroke motorcycle and automobile engines were banned due to the pollution, although many thousands of lawn maintenance engines are in use.

Blower Scavenged

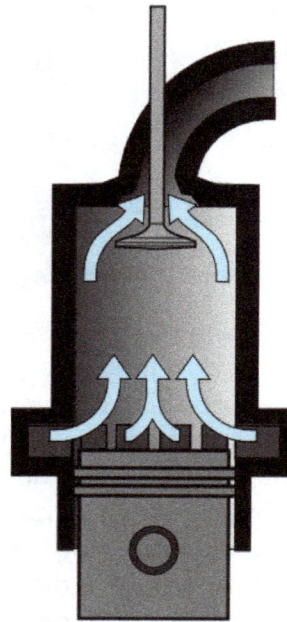

Diagram of uniflow scavenging

Using a separate blower avoids many of the shortcomings of crankcase scavenging, at the expense of increased complexity which means a higher cost and an increase in maintenance requirement. An engine of this type uses ports or valves for intake and valves for exhaust, except opposed piston engines which may also use ports for exhaust. The blower is usually of the Roots-type but other types have been used too. This design is commonplace in CI engines, and has been occasionally used in SI engines.

CI engines that use a blower typically use uniflow scavenging. In this design the cylinder wall contains several intake ports placed uniformly spaced along the circumference just above the position that the piston crown reaches when at BDC. An exhaust valve or several like that of 4-stroke engines is used. The final part of the intake manifold is an air sleeve which feeds the intake ports. The intake ports are placed at an horizontal angle to the cylinder wall (I.e: they are in plane of the piston crown) to give a swirl to the incoming charge to improve combustion. The largest reciprocating IC are low speed CI engines of this type; they are used for marine propulsion or electric power generation and achieve the highest thermal efficiencies among internal combustion engines of any kind. Some Diesel-electric locomotive engines operate on the 2-stroke cycle. The most powerful of them have a brake power of around 4.5 MW or 6,000 HP. The EMD SD90MAC class of locomotives use a 2-stroke engine. The comparable class GE AC6000CW whose prime mover has almost the same brake power uses a 4-stroke engine.

An example of this type of engine is the Wärtsilä-Sulzer RTA96-C turbocharged 2-stroke Diesel, used in large container ships. It is the most efficient and powerful internal combustion engine in the world with a thermal efficiency over 50%. For comparison, the most efficient small four-stroke engines are around 43% thermally-efficient (SAE 900648); size is an advantage for efficiency due to the increase in the ratio of volume to surface area.

Historical Design

Dugald Clerk developed the first two cycle engine in 1879. It used a separate cylinder which functioned as a pump in order to transfer the fuel mixture to the cylinder.

In 1899 John Day simplified Clerk's design into the type of 2 cycle engine that is very widely used today. Day cycle engines are crankcase scavenged and port timed. The crankcase and the part of the cylinder below the exhaust port is used as a pump. The operation of the Day cycle engine begins when the crankshaft is turned so that the piston moves from BDC upward (toward the head) creating a vacuum in the crankcase/cylinder area. The carburetor then feeds the fuel mixture into the crankcase through a reed valve or a rotary disk valve (driven by the engine). There are cast in ducts from the crankcase to the port in the cylinder to provide for intake and another from the exhausst port to the exhaust pipe. The height of the port in relationship to the length of the cylinder is called the "port timing."

On the first upstroke of the engine there would be no fuel inducted into the cylinder as the crankcase was empty. On the downstroke the piston now compresses the fuel mix, which has lubricated the piston in the cylinder and the bearings due to the fuel mix having oil added to it. As the piston moves downward is first uncovers the exhaust, but on the first stroke there is no burnt fuel to exhaust. As the piston moves downward further, it uncovers the intake port which has a duct that runs to the crankcase. Since the fuel mix in the crankcase is under pressure the mix moves through the duct and into the cylinder.

Because there is no obstruction in the cylinder of the fuel to move directly out of the exhaust port prior to the piston rising far enough to close the port, early engines used a high domed piston to slow down the flow of fuel. Later the fuel was "resonated" back into the cylinder using an expansion chamber design. When the piston rose close to TDC a spark ignites the fuel. As the piston is driven downward with power it first uncovers the exhaust port where the burned fuel is expelled under high pressure and then the intake port where the process has been completed and will keep repeating.

Later engines used a type of porting devised by the Deutz company to improve performance. It was called the Schnurle Reverse Flow system. DKW licensed this design for all their motorcycles. Their DKW RT 125 was one of the first motor vehicles to achieve over 100 mpg as a result.

Ignition

Internal combustion engines require ignition of the mixture, either by spark ignition (SI) or compression ignition (CI). Before the invention of reliable electrical methods, hot tube and flame methods were used. Experimental engines with laser ignition have been built.

Spark Ignition Process

Bosch Magneto

Points and Coil Ignition

The spark ignition engine was a refinement of the early engines which used Hot Tube ignition. When Bosch developed the magneto it became the primary system for producing electricity to energize a spark plug. Many small engines still use magneto ignition. Small engines are started by hand cranking using a recoil starter or hand crank . Prior to Charles F. Kettering of Delco's development of the automotive starter all gasoline engined automobiles used a hand crank.

Larger engines typically power their starting motors and Ignition systems using the electrical energy stored in a lead–acid battery. The battery's charged state is maintained by an automotive alternator or (previously) a generator which uses engine power to create electrical energy storage.

The battery supplies electrical power for starting when the engine has a starting motor system, and supplies electrical power when the engine is off. The battery also supplies electrical power during rare run conditions where the alternator cannot maintain more than 13.8 volts (for a common 12V automotive electrical system). As alternator voltage falls below 13.8 volts, the lead-acid storage battery increasingly picks up electrical load. During virtually all running conditions, including normal idle conditions, the alternator supplies primary electrical power.

Some systems disable alternator field (rotor) power during wide open throttle conditions. Disabling the field reduces alternator pulley mechanical loading to nearly zero, maximizing crankshaft power. In this case the battery supplies all primary electrical power.

Gasoline engines take in a mixture of air and gasoline and compress it by the movement of the piston from bottom dead center to top dead center when the fuel is at maximum compression. The reduction in the size of the swept area of the cylinder and taking into account the volume of the combustion chamber is described by a ratio. Early engines had compression ratios of 6 to 1. As compression ratios were increased the efficiency of the engine increased as well.

With early induction and ignition systems the compression ratios had to be kept low. With advances in fuel technology and combustion management high performance engines can run reliably at 12:1 ratio. With low octane fuel a problem would occur as the compression ratio increased as the fuel was igniting due to the rise in temperature that resulted. Charles Kettering developed a lead additive which allowed higher compression ratios.

The fuel mixture is ignited at difference progressions of the piston in the cylinder. At low rpm the spark is timed to occur close to the piston achieving top dead center. In order to produce more power, as rpm rises the spark is advanced sooner during piston movement. The spark occurs while the fuel is still being compressed progressively more as rpm rises.

The necessary high voltage, typically 10,000 volts, is supplied by an induction coil or transformer. The induction coil is a fly-back system, using interruption of electrical primary system current through some type of synchronized interrupter. The interrupter can be either contact points or a power transistor. The problem with this type of ignition is that as RPM increases the available of electrical energy decreases. This is especially as problem since the amount of energy needed to ignite a more dense fuel mixture is higher. The result was often a high rpm misfire.

Capacitor discharge ignition was developed. It produces a rising voltage that is sent to the spark plug. CD system voltages can reach 60,000 volts. CD ignitions use step-up transformers. The step-up transformer uses energy stored in a capacitance to generate electric spark. With either system, a mechanical or electrical control system provides a carefully timed high-voltage to the proper cylinder. This spark, via the spark plug, ignites the air-fuel mixture in the engine's cylinders.

While gasoline internal combustion engines are much easier to start in cold weather than diesel engines, they can still have cold weather starting problems under extreme conditions. For years the solution was to park the car in heated areas. In some parts of the world the oil was actually drained and heated over night and returned to the engine for cold starts. In the early 1950s the gasoline Gasifier unit was developed, where, on cold weather starts, raw gasoline was diverted to the unit where part of the fuel was burned causing the other part to become a hot vapor sent directly to the intake valve manifold. This unit was quite popular until electric engine block heaters became standard on gasoline engines sold in cold climates.

Compression Ignition Process

Diesel, PPC (Partially premixed combustion) and HCCI (Homogeneous charge compression ignition) engines, rely solely on heat and pressure created by the engine in its compression process for ignition. The compression level that occurs is usually twice or more than a gasoline engine. Diesel engines take in air only, and shortly before peak compression, spray a small quantity of diesel fuel into the cylinder via a fuel injector that allows the fuel to instantly ignite. HCCI type engines take in both air and fuel, but continue to rely on an unaided auto-combustion process, due to higher pressures and heat. This is also why diesel and HCCI engines are more susceptible to cold-starting issues, although they run just as well in cold weather once started. Light duty diesel engines with indirect injection in automobiles and light trucks employ glowplugs that pre-heat the combustion chamber just before starting to reduce no-start conditions in cold weather. Most diesels also have a battery and charging system; nevertheless, this system is secondary and is added by manufacturers as a luxury for the ease of starting, turning fuel on and off (which can also be done via a switch or mechanical apparatus), and for running auxiliary electrical components and accessories. Most new engines rely on electrical and electronic engine control units (ECU) that also adjust the combustion process to increase efficiency and reduce emissions.

Lubrication

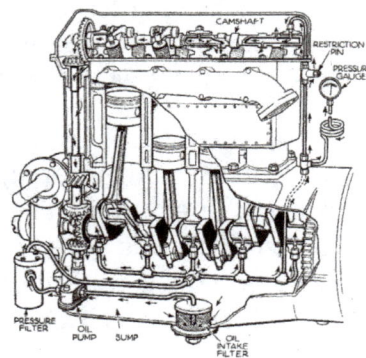

Diagram of an engine using pressurized lubrication

Surfaces in contact and relative motion to other surfaces require lubrication to reduce wear, noise and increase efficiency by reducing the power wasting in overcoming friction, or to make the mechanism work at all. At the very least, an engine requires lubrication in the following parts:

- Between pistons and cylinders
- Small bearings
- Big end bearings
- Main bearings

- Valve gear (The following elements may not be present):

  - Tappets

  - Rocker arms

  - Pushrods

  - Timing chain or gears. Toothed belts do not require lubrication.

In 2-stroke crankcase scavenged engines, the interior of the crankcase, and therefore the crank-shaft, connecting rod and bottom of the pistons are sprayed by the 2-stroke oil in the air-fuel-oil mixture which is then burned along with the fuel. The valve train may be contained in a compart-ment flooded with lubricant so that no oil pump is required.

In a splash lubrication system no oil pump is used. Instead the crankshaft dips into the oil in the sump and due to its high speed, it splashes the crankshaft, connecting rods and bottom of the pistons. The connecting rod big end caps may have an attached scoop to enhance this effect. The valve train may also be sealed in a flooded compartment, or open to the crankshaft in a way that it receives splashed oil and allows it to drain back to the sump. Splash lubrication is common for small 4-stroke engines.

In a forced (also called pressurized) lubrication system, lubrication is accomplished in a closed loop which carries motor oil to the surfaces serviced by the system and then returns the oil to a reservoir. The auxiliary equipment of an engine is typically not serviced by this loop; for instance, an alternator may use ball bearings sealed with its lubricant. The reservoir for the oil is usually the sump, and when this is the case, it is called a wet sump system. When there is a different oil reservoir the crankcase still catches it, but it is continuously drained by a dedicated pump; this is called a dry sump system.

On its bottom, the sump contains an oil intake covered by a mesh filter which is connected to an oil pump then to an oil filter outside the crankcase, from there it is diverted to the crankshaft main bearings and valve train. The crankcase contains at least one oil gallery (a conduit inside a crankcase wall) to which oil is introduced from the oil filter. The main bearings contain a groove through all or half its circumference; the oil enters to these grooves from channels connected to the oil gallery. The crankshaft has drillings which take oil from these grooves and deliver it to the big end bearings. All big end bearings are lubricated this way. A single main bearing may provide oil for 0, 1 or 2 big end bearings. A similar system may be used to lubricate the piston, its gudgeon pin and the small end of its connecting rod; in this system, the connecting rod big end has a groove around the crankshaft and a drilling connected to the groove which distributes oil from there to the bottom of the piston and from then to the cylinder.

Other systems are also used to lubricate the cylinder and piston. The connecting rod may have a nozzle to throw an oil jet to the cylinder and bottom of the piston. That nozzle is in movement rel-ative to the cylinder it lubricates, but always pointed towards it or the corresponding piston.

Typically a forced lubrication systems have a lubricant flow higher than what is required to lubri-cate satisfactorily, in order to assist with cooling. Specifically, the lubricant system helps to move heat from the hot engine parts to the cooling liquid (in water-cooled engines) or fins (in air-cooled engines) which then transfer it to the environment. The lubricant must be designed to be chemically stable and maintain suitable viscosities within the temperature range it encounters in the engine.

Cylinder Configuration

Common cylinder configurations include the straight or inline configuration, the more compact V configuration, and the wider but smoother flat or boxer configuration. Aircraft engines can also adopt a radial configuration, which allows more effective cooling. More unusual configurations such as the H, U, X, and W have also been used.

Multiple cylinder engines have their valve train and crankshaft configured so that pistons are at different parts of their cycle. It is desirable to have the piston's cycles uniformly spaced (this is called even firing) especially in forced induction engines; this reduces torque pulsations and makes inline engines with more than 3 cylinders statically balanced in its primary forces. However, some engine configurations require odd firing to achieve better balance than what is possible with even firing. For instance, a 4-stroke I2 engine has better balance when the angle between the crankpins is 180° because the pistons move in opposite directions and inertial forces partially cancel, but this gives an odd firing pattern where one cylinder fires 180° of crankshaft rotation after the other, then no cylinder fires for 540°. With an even firing pattern the pistons would move in unison and the associated forces would add.

Multiple crankshaft configurations do not necessarily need a cylinder head at all because they can instead have a piston at each end of the cylinder called an opposed piston design. Because fuel inlets and outlets are positioned at opposed ends of the cylinder, one can achieve uniflow scavenging, which, as in the four-stroke engine is efficient over a wide range of engine speeds. Thermal efficiency is improved because of a lack of cylinder heads. This design was used in the Junkers Jumo 205 diesel aircraft engine, using two crankshafts at either end of a single bank of cylinders, and most remarkably in the Napier Deltic diesel engines. These used three crankshafts to serve three banks of double-ended cylinders arranged in an equilateral triangle with the crankshafts at the corners. It was also used in single-bank locomotive engines, and is still used in marine propulsion engines and marine auxiliary generators.

Diesel Cycle

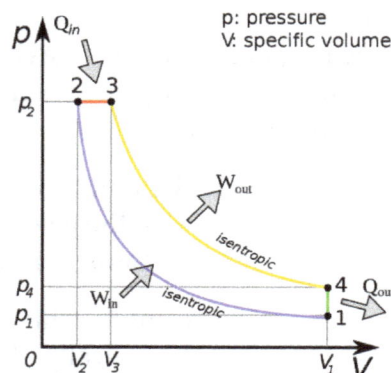

P-v Diagram for the Ideal Diesel cycle. The cycle follows the numbers 1–4 in clockwise direction.

Most truck and automotive diesel engines use a cycle reminiscent of a four-stroke cycle, but with a compression heating ignition system, rather than needing a separate ignition system. This variation is called the diesel cycle. In the diesel cycle, diesel fuel is injected directly into the cylinder so that combustion occurs at constant pressure, as the piston moves.

Otto Cycle

Otto cycle is the typical cycle for most of the cars internal combustion engines, that work using gasoline as a fuel. Otto cycle is exactly the same one that was described for the four-stroke engine. It consists of the same major steps: Intake, compression, ignition, expansion and exhaust.

Five-stroke Engine

In 1879, Nikolaus Otto manufactured and sold a double expansion engine (the double and triple expansion principles had ample usage in steam engines), with two small cylinders at both sides of a low-pressure larger cylinder, where a second expansion of exhaust stroke gas took place; the owner returned it, alleging poor performance. In 1906, the concept was incorporated in a car built by EHV (Eisenhuth Horseless Vehicle Company) CT, USA; and in the 21st century Ilmor designed and successfully tested a 5-stroke double expansion internal combustion engine, with high power output and low SFC (Specific Fuel Consumption).

Six-stroke Engine

The six-stroke engine was invented in 1883. Four kinds of six-stroke use a regular piston in a regular cylinder (Griffin six-stroke, Bajulaz six-stroke, Velozeta six-stroke and Crower six-stroke), firing every three crankshaft revolutions. The systems capture the wasted heat of the four-stroke Otto cycle with an injection of air or water.

The Beare Head and "piston charger" engines operate as opposed-piston engines, two pistons in a single cylinder, firing every two revolutions rather more like a regular four-stroke.

Other Cycles

The very first internal combustion engines did not compress the mixture. The first part of the piston downstroke drew in a fuel-air mixture, then the inlet valve closed and, in the remainder of the downstroke, the fuel-air mixture fired. The exhaust valve opened for the piston upstroke. These attempts at imitating the principle of a steam engine were very inefficient. There are a number of variations of these cycles, most notably the Atkinson and Miller cycles. The diesel cycle is somewhat different.

Split-cycle engines separate the four strokes of intake, compression, combustion and exhaust into two separate but paired cylinders. The first cylinder is used for intake and compression. The compressed air is then transferred through a crossover passage from the compression cylinder into the second cylinder, where combustion and exhaust occur. A split-cycle engine is really an air compressor on one side with a combustion chamber on the other.

Previous split-cycle engines have had two major problems—poor breathing (volumetric efficiency) and low thermal efficiency. However, new designs are being introduced that seek to address these problems.

The Scuderi Engine addresses the breathing problem by reducing the clearance between the piston and the cylinder head through various turbo charging techniques. The Scuderi design requires the use of outwardly opening valves that enable the piston to move very close to the cylinder head without the interference of the valves. Scuderi addresses the low thermal efficiency via firing after top dead centre (ATDC).

Firing ATDC can be accomplished by using high-pressure air in the transfer passage to create sonic flow and high turbulence in the power cylinder.

Combustion Turbines

Jet Engine

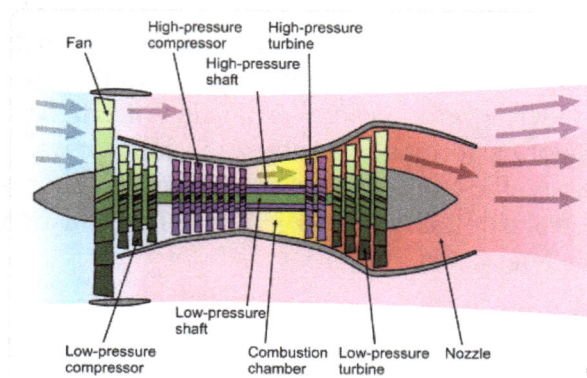

Turbofan Jet Engine

Jet engines use a number of rows of fan blades to compress air which then enters a combustor where it is mixed with fuel (typically JP fuel) and then ignited. The burning of the fuel raises the temperature of the air which is then exhausted out of the engine creating thrust. A modern turbofan engine can operate at as high as 48% efficiency.

There are six sections to a Fan Jet engine:

- Fan
- Compressor
- Combustor
- Turbine
- Mixer
- Nozzle

Gas Turbines

A gas turbine compresses air and uses it to turn a turbine. It is essentially a jet engine which directs its output to a shaft. There are three stages to a turbine: 1) air is drawn through a compressor where the temperature rises due to compression, 2) fuel is added in the combuster, and 3) hot air is exhausted through turbine blades which rotate a shaft connected to the compressor.

Turbine Power Plant

A gas turbine is a rotary machine similar in principle to a steam turbine and it consists of three main components: a compressor, a combustion chamber, and a turbine. The air, after being compressed in the compressor, is heated by burning fuel in it. The heated air and the products of combustion expand in a turbine, producing work output. About ⬚ of the work drives the compressor: the rest (about ⬚) is available as useful work output.

Gas Turbines are among the most efficient internal combustion engines. The General Electric 7HA and 9HA turbine combined cycle electrical plants are rated at over 61% efficiency.

Brayton Cycle

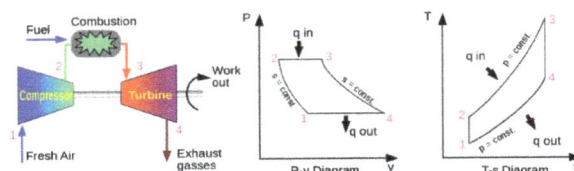

Brayton cycle

A gas turbine is a rotary machine somewhat similar in principle to a steam turbine. It consists of three main components: compressor, combustion chamber, and turbine. The air is compressed by the compressor where a temperature rise occurs. The compressed air is further heated by combustion of injected fuel in the combustion chamber which expands the air. This energy rotates the turbine which powers the compressor via a mechanical coupling. The hot gases are then exhausted to provide thrust.

Gas turbine cycle engines employ a continuous combustion system where compression, combustion, and expansion occur simultaneously at different places in the engine—giving continuous power. Notably, the combustion takes place at constant pressure, rather than with the Otto cycle, constant volume.

Wankel Engines

The Wankel engine (rotary engine) does not have piston strokes. It operates with the same separation of phases as the four-stroke engine with the phases taking place in separate locations in the engine. In thermodynamic terms it follows the Otto engine cycle, so may be thought of as a "four-phase" engine. While it is true that three power strokes typically occur per rotor revolution, due to the 3:1 revolution ratio of the rotor to the eccentric shaft, only one power stroke per shaft revolution actually occurs. The drive (eccentric) shaft rotates once during every power stroke instead of twice (crankshaft), as in the Otto cycle, giving it a greater power-to-weight ratio than piston engines. This type of engine was most notably used in the Mazda RX-8, the earlier RX-7, and other vehicle models. The engine is also use in unmanned aerial vehicles, where the small size and weight and the high power-to-weight ratio are advantages.

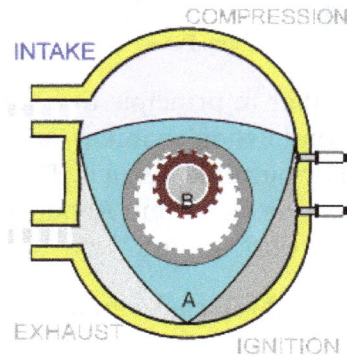

The Wankel rotary cycle. The shaft turns three times for each rotation of the rotor around the lobe and once for each orbital revolution around the eccentric shaft.

Forced Induction

Forced induction is the process of delivering compressed air to the intake of an internal combustion engine. A forced induction engine uses a gas compressor to increase the pressure, temperature and density of the air. An engine without forced induction is considered a naturally aspirated engine.

Forced induction is used in the automotive and aviation industry to increase engine power and efficiency. It particularly helps aviation engines, as they need to operate at high altitude.

Forced induction is achieved by a supercharger, where the compressor is directly powered from the engine shaft or, in the turbocharger, from a turbine powered by the engine exhaust.

Fuels and Oxidizers

All internal combustion engines depend on combustion of a chemical fuel, typically with oxygen from the air (though it is possible to inject nitrous oxide to do more of the same thing and gain a power boost). The combustion process typically results in the production of a great quantity of heat, as well as the production of steam and carbon dioxide and other chemicals at very high temperature; the temperature reached is determined by the chemical make up of the fuel and oxidisers, as well as by the compression and other factors.

Fuels

The most common modern fuels are made up of hydrocarbons and are derived mostly from fossil fuels (petroleum). Fossil fuels include diesel fuel, gasoline and petroleum gas, and the rarer use of propane. Except for the fuel delivery components, most internal combustion engines that are designed for gasoline use can run on natural gas or liquefied petroleum gases without major modifications. Large diesels can run with air mixed with gases and a pilot diesel fuel ignition injection. Liquid and gaseous biofuels, such as ethanol and biodiesel (a form of diesel fuel that is produced from crops that yield triglycerides such as soybean oil), can also be used. Engines with appropriate modifications can also run on hydrogen gas, wood gas, or charcoal gas, as well as from so-called producer gas made from other convenient biomass. Experiments have also been conducted using powdered solid fuels, such as the magnesium injection cycle.

Presently, fuels used include:

- Petroleum:
    - Petroleum spirit (North American term: gasoline, British term: petrol)
    - Petroleum diesel.
    - Autogas (liquified petroleum gas).
    - Compressed natural gas.
    - Jet fuel (aviation fuel)
    - Residual fuel
- Coal:
    - Gasoline can be made from carbon (coal) using the Fischer-Tropsch process
    - Diesel fuel can be made from carbon using the Fischer-Tropsch process
- Biofuels and vegetable oils:
    - Peanut oil and other vegetable oils.
    - Woodgas, from an onboard wood gasifier using solid wood as a fuel
    - Biofuels:
        - Biobutanol (replaces gasoline).
        - Biodiesel (replaces petrodiesel).
        - Dimethyl Ether (replaces petrodiesel).
        - Bioethanol and Biomethanol (wood alcohol) and other biofuels.
        - Biogas
- Hydrogen (mainly spacecraft rocket engines)

Even fluidized metal powders and explosives have seen some use. Engines that use gases for fuel are called gas engines and those that use liquid hydrocarbons are called oil engines; however, gasoline engines are also often colloquially referred to as, "gas engines" ("petrol engines" outside North America).

The main limitations on fuels are that it must be easily transportable through the fuel system to the combustion chamber, and that the fuel releases sufficient energy in the form of heat upon combustion to make practical use of the engine.

Diesel engines are generally heavier, noisier, and more powerful at lower speeds than gasoline engines. They are also more fuel-efficient in most circumstances and are used in heavy road vehicles, some automobiles (increasingly so for their increased fuel efficiency over gasoline engines), ships, railway locomotives, and light aircraft. Gasoline engines are used in most other road vehicles including most cars, motorcycles, and mopeds. Note that in Europe, sophisticated diesel-engined cars have taken over about 45% of the market since the 1990s. There are also engines that run on hydrogen, methanol, ethanol, liquefied petroleum gas (LPG), biodiesel, paraffin and tractor vaporizing oil (TVO).

Hydrogen

Hydrogen could eventually replace conventional fossil fuels in traditional internal combustion engines. Alternatively fuel cell technology may come to deliver its promise and the use of the internal combustion engines could even be phased out.

Although there are multiple ways of producing free hydrogen, those methods require converting combustible molecules into hydrogen or consuming electric energy. Unless that electricity is produced from a renewable source—and is not required for other purposes— hydrogen does not solve any energy crisis. In many situations the disadvantage of hydrogen, relative to carbon fuels, is its storage. Liquid hydrogen has extremely low density (14 times lower than water) and requires extensive insulation—whilst gaseous hydrogen requires heavy tankage. Even when liquefied, hydrogen has a higher specific energy but the volumetric energetic storage is still roughly five times lower than gasoline. However, the energy density of hydrogen is considerably higher than that of electric batteries, making it a serious contender as an energy carrier to replace fossil fuels. The 'Hydrogen on Demand' process creates hydrogen as needed, but has other issues, such as the high price of the sodium borohydride that is the raw material.

Oxidizers

One-cylinder gasoline engine, c. 1910

Since air is plentiful at the surface of the earth, the oxidizer is typically atmospheric oxygen, which has the advantage of not being stored within the vehicle. This increases the power-to-weight and power-to-volume ratios. Other materials are used for special purposes, often to increase power output or to allow operation under water or in space.

- Compressed air has been commonly used in torpedoes.

- Compressed oxygen, as well as some compressed air, was used in the Japanese Type 93 torpedo. Some submarines carry pure oxygen. Rockets very often use liquid oxygen.

- Nitromethane is added to some racing and model fuels to increase power and control combustion.

- Nitrous oxide has been used—with extra gasoline—in tactical aircraft, and in specially equipped cars to allow short bursts of added power from engines that otherwise run on gasoline and air. It is also used in the Burt Rutan rocket spacecraft.

- Hydrogen peroxide power was under development for German World War II submarines. It may have been used in some non-nuclear submarines, and was used on some rocket engines (notably the Black Arrow and the Me-163 rocket plane).

- Other chemicals such as chlorine or fluorine have been used experimentally, but have not been found practical.

Cooling

Cooling is required to remove excessive heat — over heating can cause engine failure, usually from wear(due to heat-induced failure of lubrication), cracking or warping. Two most common forms of engine cooling are air-cooled and water-cooled. Most modern automotive engines are both water and air-cooled, as the water/liquid-coolant is carried to air-cooled fins and/or fans, whereas larger engines may be singularly water-cooled as they are stationary and have a constant supply of water through water-mains or fresh-water, while most power tool engines and other small engines are air-cooled. Some engines (air or water-cooled) also have an oil cooler. In some engines, especially for turbine engine blade cooling and liquid rocket engine cooling, fuel is used as a coolant, as it is simultaneously preheated before injecting it into a combustion chamber.

Starting

Internal Combustion engines must have their cycles started. In reciprocating engines this is accomplished by turning the crankshaft (Wankel Rotor Shaft) which induces the cycles of intake, compression, combustion, and exhaust. The first engines were started with a turn of their flywheels, while the first vehicle (the Daimler Reitwagen) was started with a hand crank. All ICE engined automobiles were started with hand cranks until Charles Kettering developed the electric starter for automobiles.

The most often found methods of starting ICE today is with an electric motor. As diesel engines have become larger another method has come into use as well, that is Air Starters.

Another method of starting is to use compressed air that is pumped into some cylinders of an engine to start it turning.

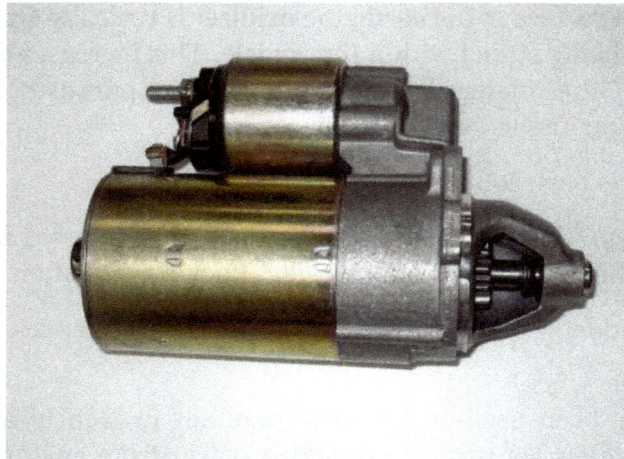

Electric Starter as used in automobiles

With two wheeled vehicles their engines may be started in four ways:

- By pedaling, as on a bicycle

- By pushing the vehicle and then engaging the clutch (Run and Bump Starting)

- By kicking downward on a single pedal, known as Kick Starting

- Electric Starting

There are also starters where a spring is compressed by a crank motion and then used to start an engine. Small engines use a pull rope mechanism called recoil starting as the rope returns to storage after it has been pulled fully out to start the engine.

Turbine engines are frequently started by electric motor, or by air.

Measures of Engine Performance

Engine types vary greatly in a number of different ways:

- energy efficiency

- fuel/propellant consumption (brake specific fuel consumption for shaft engines, thrust specific fuel consumption for jet engines)

- power-to-weight ratio

- thrust to weight ratio

- Torque curves (for shaft engines) thrust lapse (jet engines)

- Compression ratio for piston engines, overall pressure ratio for jet engines and gas turbines

Energy Efficiency

Once ignited and burnt, the combustion products—hot gases—have more available thermal energy than the original compressed fuel-air mixture (which had higher chemical energy). The available energy is manifested as high temperature and pressure that can be translated into work by the engine. In a reciprocating engine, the high-pressure gases inside the cylinders drive the engine's pistons.

Once the available energy has been removed, the remaining hot gases are vented (often by opening a valve or exposing the exhaust outlet) and this allows the piston to return to its previous position (top dead center, or TDC). The piston can then proceed to the next phase of its cycle, which varies between engines. Any heat that is not translated into work is normally considered a waste product and is removed from the engine either by an air or liquid cooling system.

Internal combustion engines are heat engines, and as such their theoretical efficiency can be approximated by idealized thermodynamic cycles. The thermal efficiency of a theoretical cycle cannot exceed that of the Carnot cycle, whose efficiency is determined by the difference between the lower and upper operating temperatures of the engine. The upper operating temperature of an engine is limited by two main factors; the thermal operating limits of the materials, and the auto-ignition resistance of the fuel. All metals and alloys have a thermal operating limit, and there is significant research into ceramic materials that can be made with greater thermal stability and desirable structural properties. Higher thermal stability allows for a greater temperature difference between the lower (ambient) and upper operating temperatures, hence greater thermodynamic efficiency. Also, as the cylinder temperature rises, the engine becomes more prone to auto-ignition. This is caused when the cylinder temperature nears the flash point of the charge. At this point, ignition can spontaneously occur before the spark plug fires, causing excessive cylinder pressures. Auto-ignition can be mitigated by using fuels with high auto-ignition resistance (octane rating), however it still puts an upper bound on the allowable peak cylinder temperature.

The thermodynamic limits assume that the engine is operating under ideal conditions: a frictionless world, ideal gases, perfect insulators, and operation for infinite time. Real world applications introduce complexities that reduce efficiency. For example, a real engine runs best at a specific load, termed its power band. The engine in a car cruising on a highway is usually operating significantly below its ideal load, because it is designed for the higher loads required for rapid acceleration. In addition, factors such as wind resistance reduce overall system efficiency. Engine fuel economy is measured in miles per gallon or in liters per 100 kilometres. The volume of hydrocarbon assumes a standard energy content.

Most iron engines have a thermodynamic limit of 37%. Even when aided with turbochargers and stock efficiency aids, most engines retain an average efficiency of about 18%-20 %. The latest technologies in Formula One engines have seen a boost in thermal efficiency to almost 47%. Rocket engine efficiencies are much better, up to 70%, because they operate at very high temperatures and pressures and can have very high expansion ratios. Electric motors are better still, at around 85 -90 % efficiency or more, but they rely on an external power source (often another heat engine at a power plant subject to similar thermodynamic efficiency limits). However large stationary power plant turbines are typically significantly more efficient and cleaner than small mobile combustion engines in vehicles.

There are many inventions aimed at increasing the efficiency of IC engines. In general, practical engines are always compromised by trade-offs between different properties such as efficiency, weight, power, heat, response, exhaust emissions, or noise. Sometimes economy also plays a role in not only the cost of manufacturing the engine itself, but also manufacturing and distributing the fuel. Increasing the engine's efficiency brings better fuel economy but only if the fuel cost per energy content is the same.

Measures of Fuel Efficiency and Propellant Efficiency

For stationary and shaft engines including propeller engines, fuel consumption is measured by calculating the brake specific fuel consumption, which measures the mass flow rate of fuel consumption divided by the power produced.

For internal combustion engines in the form of jet engines, the power output varies drastically with airspeed and a less variable measure is used: thrust specific fuel consumption (TSFC), which is the mass of propellant needed to generate impulses that is measured in either pound force-hour or the grams of propellant needed to generate an impulse that measures one kilonewton-second.

For rockets, TSFC can be used, but typically other equivalent measures are traditionally used, such as specific impulse and effective exhaust velocity.

Air and Noise Pollution

Air Pollution

Internal combustion engines such as reciprocating internal combustion engines produce air pollution emissions, due to incomplete combustion of carbonaceous fuel. The main derivatives of the process are carbon dioxide CO_2, water and some soot — also called particulate matter (PM). The effects of inhaling particulate matter have been studied in humans and animals and include asthma, lung cancer, cardiovascular issues, and premature death. There are, however, some additional products of the combustion process that include nitrogen oxides and sulfur and some uncombusted hydrocarbons, depending on the operating conditions and the fuel-air ratio.

Not all of the fuel is completely consumed by the combustion process; a small amount of fuel is present after combustion, and some of it reacts to form oxygenates, such as formaldehyde or acetaldehyde, or hydrocarbons not originally present in the input fuel mixture. Incomplete combustion usually results from insufficient oxygen to achieve the perfect stoichiometric ratio. The flame is "quenched" by the relatively cool cylinder walls, leaving behind unreacted fuel that is expelled with the exhaust. When running at lower speeds, quenching is commonly observed in diesel (compression ignition) engines that run on natural gas. Quenching reduces efficiency and increases knocking, sometimes causing the engine to stall. Incomplete combustion also leads to the production of carbon monoxide (CO). Further chemicals released are benzene and 1,3-butadiene that are also hazardous air pollutants.

Increasing the amount of air in the engine reduces emissions of incomplete combustion products, but also promotes reaction between oxygen and nitrogen in the air to produce nitrogen oxides (NOx). NOx is hazardous to both plant and animal health, and leads to the production of ozone (O_3). Ozone is not emitted directly; rather, it is a secondary air pollutant, produced in the atmosphere by the reaction of NOx and volatile organic compounds in the presence of sunlight.

Ground-level ozone is harmful to human health and the environment. Though the same chemical substance, ground-level ozone should not be confused with stratospheric ozone, or the ozone layer, which protects the earth from harmful ultraviolet rays.

Carbon fuels contain sulfur and impurities that eventually produce sulfur monoxides (SO) and sulfur dioxide (SO_2) in the exhaust, which promotes acid rain.

In the United States, nitrogen oxides, PM, carbon monoxide, sulphur dioxide, and ozone, are regulated as criteria air pollutants under the Clean Air Act to levels where human health and welfare are protected. Other pollutants, such as benzene and 1,3-butadiene, are regulated as hazardous air pollutants whose emissions must be lowered as much as possible depending on technological and practical considerations.

NOx, carbon monoxide and other pollutants are frequently controlled via exhaust gas recirculation which returns some of the exhaust back into the engine intake, and catalytic converters, which convert exhaust chemicals to harmless chemicals.

Non-road Engines

The emission standards used by many countries have special requirements for non-road engines which are used by equipment and vehicles that are not operated on the public roadways. The standards are separated from the road vehicles.

Noise Pollution

Significant contributions to noise pollution are made by internal combustion engines. Automobile and truck traffic operating on highways and street systems produce noise, as do aircraft flights due to jet noise, particularly supersonic-capable aircraft. Rocket engines create the most intense noise.

Idling

Internal combustion engines continue to consume fuel and emit pollutants when idling so it is desirable to keep periods of idling to a minimum. Many bus companies now instruct drivers to switch off the engine when the bus is waiting at a terminal.

In England, the Road Traffic Vehicle Emissions Fixed Penalty Regulations 2002 (Statutory Instrument 2002 No. 1808) introduced the concept of a "stationary idling offence". This means that a driver can be ordered "by an authorised person ... upon production of evidence of his authorisation, require him to stop the running of the engine of that vehicle" and a "person who fails to comply ... shall be guilty of an offence and be liable on summary conviction to a fine not exceeding level 3 on the standard scale". Only a few local authorities have implemented the regulations, one of them being Oxford City Council.

Types of Internal Combustion Engine

Two-stroke Engine

Two-stroke engines often have a high power-to-weight ratio, power being available in a narrow range of rotational speeds called the "power band". Compared to four-stroke engines, two-stroke engines have a greatly reduced number of moving parts, and so can be more compact and significantly lighter.

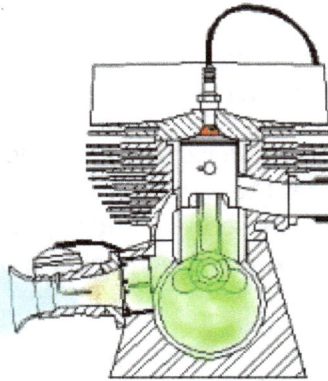

A two-stroke, or two-cycle, engine is a type of internal combustion engine which completes a power cycle with two strokes (up and down movements) of the piston during only one crankshaft revolution. This is in contrast to a "four-stroke engine", which requires four strokes of the piston to complete a power cycle. In a two-stroke engine, the end of the combustion stroke and the beginning of the compression stroke happen simultaneously, with the intake and exhaust (or scavenging) functions occurring at the same time.

History

The first commercial two-stroke engine involving in-cylinder compression is attributed to Scottish engineer Dugald Clerk, who patented his design in 1881. However, unlike most later two-stroke engines, his had a separate charging cylinder. The crankcase-scavenged engine, employing the area below the piston as a charging pump, is generally credited to Englishman Joseph Day. The first truly practical two-stroke engine is attributed to Yorkshireman Alfred Angas Scott, who started producing twin-cylinder water-cooled motorcycles in 1908.

Gasoline (spark ignition) versions are particularly useful in lightweight or portable applications such as chainsaws and motorcycles. However, when weight and size are not an issue, the cycle's potential for high thermodynamic efficiency makes it ideal for diesel compression ignition engines operating in large, weight-insensitive applications, such as marine propulsion, railway locomotives and electricity generation. In a two-stroke engine, the heat transfer from the engine to the cooling system is less than in a four-stroke, which means that two-stroke engines can be more efficient.

Emissions

Crankcase-compression two-stroke engines, such as common small gasoline-powered engines, create more exhaust emissions than four-stroke engines because their two-stroke oil (petroil) lubrication mixture is also burned in the engine, due to the engine's total-loss oiling system.

Applications

Two-stroke petrol engines are preferred when mechanical simplicity, light weight, and high power-to-weight ratio are design priorities. With the traditional lubrication technique of mixing oil into the fuel, they also have the advantage of working in any orientation, as there is no oil reservoir dependent on gravity; this is an essential property for hand-held power tools such as chainsaws.

1966 Saab Sport

A two-stroke minibike

Lateral view of a two-stroke Forty series British Seagull outboard engine, the serial number dates it to 1954/1955

A number of mainstream automobile manufacturers have used two-stroke engines in the past, including the Swedish Saab and German manufacturers DKW, Auto-Union, VEB Sachsenring Automobilwerke Zwickau, and VEB Automobilwerk Eisenach. The Japanese manufacturer Suzuki did the same in the 1970s. Production of two-stroke cars ended in the 1980s in the West, due to increasingly stringent regulation of air pollution. Eastern Bloc countries continued until around 1991, with the Trabant and Wartburg in East Germany. Two-stroke engines are still found in a variety of small propulsion applications, such as outboard motors, high-performance, small-capacity motorcycles, mopeds, and dirt bikes, underbones, scooters, tuk-tuks, snowmobiles, karts, ultra-light airplanes, and model airplanes and other model vehicles. They are also common in power tools used outdoors, such as lawnmowers, chainsaws, and weed-wackers.

With direct fuel injection and a sump-based lubrication system, a two-stroke engine produces air pollution no worse than a four-stroke, and it can achieve higher thermodynamic efficiency. Therefore, the cycle has historically also been used in large diesel engines, most notably large industrial and marine engines, as well as some trucks and heavy machinery. There are several experimental designs intended for automobile use: for instance, Lotus of Norfolk, UK, has a prototype direct-injection two-stroke engine intended for alcohol fuels called the Omnivore which it is demonstrating in a version of the Exige.

Different Two-stroke Design Types

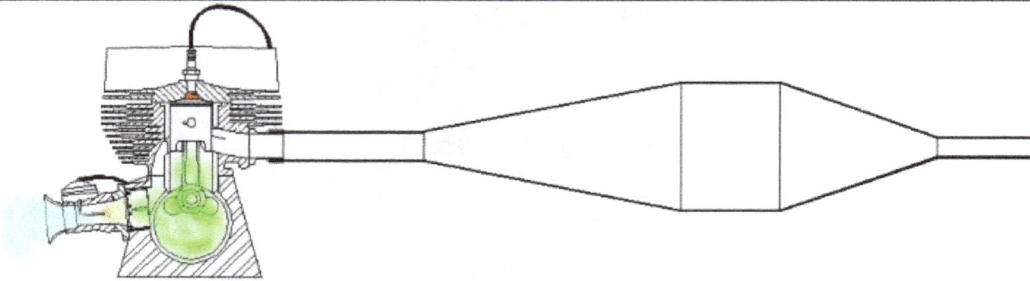

A two-stroke engine, in this case with an expansion chamber illustrates the effect of a reflected pressure wave on the fuel charge. This is important for maximum charge pressure (volumetric efficiency) and fuel economy. It is used on most high-performance engine designs.

Although the principles remain the same, the mechanical details of various two-stroke engines differ depending on the type. The design types vary according to the method of introducing the charge to the cylinder, the method of scavenging the cylinder (exchanging burnt exhaust for fresh mixture) and the method of exhausting the cylinder.

Piston-controlled Inlet Port

Piston port is the simplest of the designs and the most common in small two-stroke engines. All functions are controlled solely by the piston covering and uncovering the ports as it moves up and down in the cylinder. In the 1970s, Yamaha worked out some basic principles for this system. They found that, in general, widening an exhaust port increases the power by the same amount as raising the port, but the power band does not narrow as it does when the port is raised. However, there is a mechanical limit to the width of a single exhaust port, at about 62% of the bore diameter for reasonable ring life. Beyond this, the rings will bulge into the exhaust port and wear quickly. A maximum is 70% of bore width is possible in racing engines, where rings are changed every few races. Intake duration is between 120 and 160 degrees. Transfer port time is set at a minimum of 26 degrees. The strong low pressure pulse of a racing two-stroke expansion chamber can drop the pressure to -7 PSI when the piston is at bottom dead center, and the transfer ports nearly wide open. One of the reasons for high fuel consumption in 2-strokes is that some of the incoming pressurized fuel/air mixture is forced across the top of the piston, where it has a cooling action, and straight out the exhaust pipe. An expansion chamber with a strong reverse pulse will stop this out-going flow. A fundamental difference from typical four-stroke engines is that the two-stroke's crankcase is sealed and forms part of the induction process in gasoline and hot bulb engines. Diesel two-strokes often add a Roots blower or piston pump for scavenging.

Reed Inlet Valve

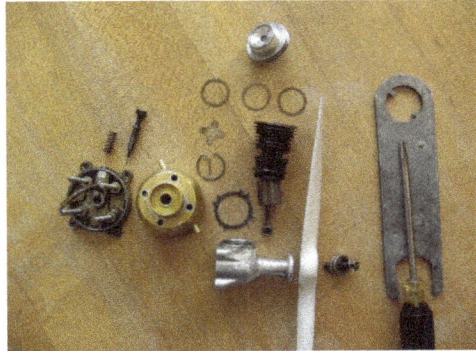

A Cox Babe Bee 0.049 cubic inch (0.8 cubic cm) reed valve engine, disassembled, uses glow plug ignition. The mass is 64 grams.

The reed valve is a simple but highly effective form of check valve commonly fitted in the intake tract of the piston-controlled port. They allow asymmetric intake of the fuel charge, improving power and economy, while widening the power band. They are widely used in motorcycle, ATV and marine outboard engines.

Rotary Inlet Valve

The intake pathway is opened and closed by a rotating member. A familiar type sometimes seen on small motorcycles is a slotted disk attached to the crankshaft which covers and uncovers an opening in the end of the crankcase, allowing charge to enter during one portion of the cycle (aka Disc Valve).

Another form of rotary inlet valve used on two-stroke engines employs two cylindrical members with suitable cutouts arranged to rotate one within the other - the inlet pipe having passage to the crankcase only when the two cutouts coincide. The crankshaft itself may form one of the members, as in most glow plug model engines. In another embodiment, the crank disc is arranged to be a close-clearance fit in the crankcase, and is provided with a cutout which lines up with an inlet passage in the crankcase wall at the appropriate time, as in the Vespa motor scooter.

The advantage of a rotary valve is it enables the two-stroke engine's intake timing to be asymmetrical, which is not possible with piston port type engines. The piston port type engine's intake timing opens and closes before and after top dead center at the same crank angle, making it symmetrical, whereas the rotary valve allows the opening to begin earlier and close earlier.

Rotary valve engines can be tailored to deliver power over a wider speed range or higher power over a narrower speed range than either piston port or reed valve engine. Where a portion of the rotary valve is a portion of the crankcase itself, it is particularly important that no wear is allowed to take place.

Cross-flow-scavenged

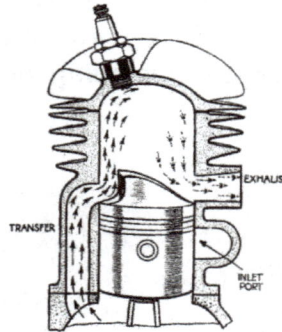

Deflector piston with cross-flow scavenging

In a cross-flow engine, the transfer and exhaust ports are on opposite sides of the cylinder, and a deflector on the top of the piston directs the fresh intake charge into the upper part of the cylinder, pushing the residual exhaust gas down the other side of the deflector and out the exhaust port. The deflector increases the piston's weight and exposed surface area, affecting piston cooling and also making it difficult to achieve an efficient combustion chamber shape. This design has been superseded since the 1960s by the loop scavenging method (below), especially for motorbikes, although for smaller or slower engines, such as lawn mowers, the cross-flow-scavenged design can be an acceptable approach.

Loop-scavenged

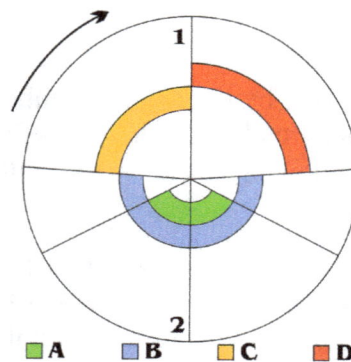

The two-stroke cycle

    Top dead center (TDC)

    Bottom dead center (BDC)

A: Intake/scavenging

B: Exhaust

C: Compression

D: Expansion (power)

This method of scavenging uses carefully shaped and positioned transfer ports to direct the flow of fresh mixture toward the combustion chamber as it enters the cylinder. The fuel/air mixture strikes the cylinder head, then follows the curvature of the combustion chamber, and then is deflected downward.

This not only prevents the fuel/air mixture from traveling directly out the exhaust port, but also creates a swirling turbulence which improves combustion efficiency, power and economy. Usually, a piston deflector is not required, so this approach has a distinct advantage over the cross-flow scheme (above).

Often referred to as "Schnuerle" (or "Schnürle") loop scavenging after the German inventor of an early form in the mid-1920s, it became widely adopted in that country during the 1930s and spread further afield after World War II.

Loop scavenging is the most common type of fuel/air mixture transfer used on modern two-stroke engines. Suzuki was one of the first manufacturers outside of Europe to adopt loop-scavenged two-stroke engines. This operational feature was used in conjunction with the expansion chamber exhaust developed by German motorcycle manufacturer, MZ and Walter Kaaden.

Loop scavenging, disc valves and expansion chambers worked in a highly coordinated way to significantly increase the power output of two-stroke engines, particularly from the Japanese manufacturers Suzuki, Yamaha and Kawasaki. Suzuki and Yamaha enjoyed success in grand Prix motorcycle racing in the 1960s due in no small way to the increased power afforded by loop scavenging.

An additional benefit of loop scavenging was the piston could be made nearly flat or slightly dome shaped, which allowed the piston to be appreciably lighter and stronger, and consequently to tolerate higher engine speeds. The "flat top" piston also has better thermal properties and is less prone to uneven heating, expansion, piston seizures, dimensional changes and compression losses.

SAAB built 750 and 850 cc 3-cylinder engines based on a DKW design that proved reasonably successful employing loop charging. The original SAAB 92 had a two-cylinder engine of comparatively low efficiency. At cruising speed, reflected wave exhaust port blocking occurred at too low a frequency. Using the asymmetric three-port exhaust manifold employed in the identical DKW engine improved fuel economy.

The 750 cc standard engine produced 36 to 42 hp, depending on the model year. The Monte Carlo Rally variant, 750 cc (with a filled crankshaft for higher base compression), generated 65 hp. An 850 cc version was available in the 1966 SAAB Sport (a standard trim model in comparison to the deluxe trim of the Monte Carlo). Base compression comprises a portion of the overall compression ratio of a two-stroke engine. Work published at SAE in 2012 points that loop scavenging is under every circumstance more efficient than cross-flow scavenging.

Uniflow-scavenged

In a uniflow engine, the mixture, or "charge air" in the case of a diesel, enters at one end of the cylinder controlled by the piston and the exhaust exits at the other end controlled by an exhaust valve or piston. The scavenging gas-flow is therefore in one direction only, hence the name uniflow. The valved arrangement is common in on-road, off-road and stationary two-stroke engines (Detroit

Diesel), certain small marine two-stroke engines (Gray Marine), certain railroad two-stroke diesel locomotives (Electro-Motive Diesel) and large marine two-stroke main propulsion engines (Wärtsilä). Ported types are represented by the opposed piston design in which there are two pistons in each cylinder, working in opposite directions such as the Junkers Jumo 205 and Napier Deltic. The once-popular split-single design falls into this class, being effectively a folded uniflow. With advanced angle exhaust timing, uniflow engines can be supercharged with a crankshaft-driven (piston or Roots) blower.

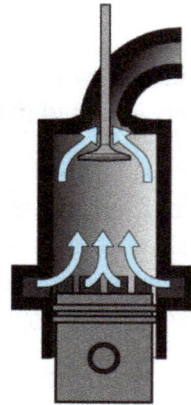

Uniflow scavenging

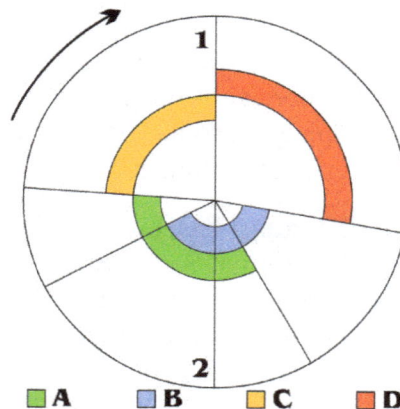

The uniflow two-stroke cycle

  Top dead center (TDC)

  Bottom dead center (BDC)

  A: Intake (effective scavenging, 135°–225°; necessarily symmetric about BDC; Diesel injection is usually initiated at 4° before TDC)

  B: Exhaust

  C: Compression

  D: Expansion (power)

Stepped Piston Engine

The piston of this engine is "top-hat" shaped; the upper section forms the regular cylinder, and the lower section performs a scavenging function. The units run in pairs, with the lower half of one piston charging an adjacent combustion chamber.

This system is still partially dependent on total loss lubrication (for the upper part of the piston), the other parts being sump lubricated with cleanliness and reliability benefits. The piston weight is only about 20% heavier than a loop-scavenged piston because skirt thicknesses can be less. Bernard Hooper Engineering Ltd. (BHE) is one of the more recent engine developers using this approach.

Power Valve Systems

Many modern two-stroke engines employ a power valve system. The valves are normally in or around the exhaust ports. They work in one of two ways: either they alter the exhaust port by closing off the top part of the port, which alters port timing, such as Ski-doo R.A.V.E, Yamaha YPVS, Honda RC-Valve, Kawasaki K.I.P.S., Cagiva C.T.S. or Suzuki AETC systems, or by altering the volume of the exhaust, which changes the resonant frequency of the expansion chamber, such as the Suzuki SAEC and Honda V-TACS system. The result is an engine with better low-speed power without sacrificing high-speed power. However, as power valves are in the hot gas flow they need regular maintenance to perform well.

Direct Injection

Direct injection has considerable advantages in two-stroke engines, eliminating some of the waste and pollution caused by carbureted two-strokes where a proportion of the fuel/air mixture entering the cylinder goes directly out, unburned, through the exhaust port. Two systems are in use, low-pressure air-assisted injection, and high pressure injection.

Since the fuel does not pass through the crankcase, a separate source of lubrication is needed.

Two-stroke Diesel Engine

Brons two-stroke V8 Diesel engine driving a N.V. Heemaf generator.

Diesel engines rely solely on the heat of compression for ignition. In the case of Schnuerle ported and loop-scavenged engines, intake and exhaust happens via piston-controlled ports. A uniflow diesel engine takes in air via scavenge ports, and exhaust gases exit through an overhead poppet valve. Two-stroke diesels are all scavenged by forced induction. Some designs use a mechanically driven Roots blower, whilst marine diesel engines normally use exhaust-driven turbochargers, with electrically driven auxiliary blowers for low-speed operation when exhaust turbochargers are unable to deliver enough air.

Marine two-stroke diesel engines directly coupled to the propeller are able to start and run in either direction as required. The fuel injection and valve timing is mechanically readjusted by using a different set of cams on the camshaft. Thus, the engine can be run in reverse to move the vessel backwards.

Lubrication

Most small petrol two-stroke engines cannot be lubricated by oil contained in their crankcase and sump, since the crankcase is already being used to pump fuel-air mixture into the cylinder. Traditionally, the moving parts (both rotating crankshaft and sliding piston) were lubricated by a premixed fuel-oil mixture (at a ratio between 16:1 and 100:1). As late as the 1970s, petrol stations would often have a separate pump to deliver such a premix fuel to motorcycles. Even then, in many cases, the rider would carry a bottle of their own two-stroke oil.

Two-stroke oils which became available worldwide in the 1970s are specifically designed to mix with petrol and be burnt in the combustion chamber without leaving undue unburnt oil or ash. This led to a marked reduction in spark plug fouling, which had previously been a factor in two-stroke engines.

More recent two-stroke engines might pump lubrication from a separate tank of two-stroke oil. The supply of this oil is controlled by the throttle position and engine speed. Examples are found in Yamaha's PW80 (Pee-wee), a small, 80cc two-stroke dirt bike designed for young children, and many two-stroke snowmobiles. The technology is referred to as auto-lube. This is still a total-loss system with the oil being burnt the same as in the pre-mix system; however, given that the oil is not properly mixed with the fuel when burned in the combustion chamber, it translates into a slightly more efficient lubrication. This lubrication method also pays dividends in terms of user friendliness by eliminating the user's need to mix the gasoline at every refill, makes the motor much less susceptible to atmospheric conditions (Ambient temperature, elevation) and ensures proper engine lubrication, with less oil at light loads (such as idle) and more oil at high loads (such as full throttle). Some companies, such as Bombardier, had some oil pump designs have no oil injected at idle to reduce smoke levels, as the loading on the engine parts was light enough to not require additional lubrication beyond the low levels that the fuel provides. Ultimately oil injection is still the same as premixed gasoline in that the oil is burnt in the combustion chamber (albeit not as completely as pre-mix) and the gas is still mixed with the oil, although not as thoroughly as in pre-mix. In addition, this method requires extra mechanical parts to pump the oil from the separate tank to the carburetor or throttle body. In applications where performance, simplicity and/or dry weight are significant considerations, the pre-mix lubrication method is almost always used. For example, a two-stroke engine in a motocross bike pays major consideration to performance, simplicity and weight. Chainsaws and brush cutters must be as light as possible to reduce user fatigue and hazard, especially when used in a professional work environment.

All two-stroke engines running on a petrol/oil mix will suffer oil starvation if forced to rotate at speed with the throttle closed, e.g. motorcycles descending long hills and perhaps when decelerating gradually from high speed by changing down through the gears. Two-stroke cars (such as those that were popular in Eastern Europe in the mid-20th century) were in particular danger and were usually fitted with freewheel mechanisms in the powertrain, allowing the engine to idle when the throttle was closed, requiring the use of the brakes in all slowing situations.

Large two-stroke engines, including diesels, normally use a sump lubrication system similar to four-stroke engines. The cylinder must still be pressurized, but this is not done from the crankcase, but by an ancillary Roots-type blower or a specialized turbocharger (usually a turbo-compressor system) which has a "locked" compressor for starting (and during which it is powered by the engine's crankshaft), but which is "unlocked" for running (and during which it is powered by the engine's exhaust gases flowing through the turbine).

Two-stroke Reversibility

For the purpose of this discussion, it is convenient to think in motorcycle terms, where the exhaust pipe faces into the cooling air stream, and the crankshaft commonly spins in the same axis and direction as do the wheels i.e. "forward". Some of the considerations discussed here apply to four-stroke engines (which cannot reverse their direction of rotation without considerable modification), almost all of which spin forward, too.

Regular gasoline two-stroke engines will run backwards for short periods and under light load with little problem, and this has been used to provide a reversing facility in microcars, such as the Messerschmitt KR200, that lacked reverse gearing. Where the vehicle has electric starting, the motor will be turned off and restarted backwards by turning the key in the opposite direction. Two-stroke golf carts have used a similar kind of system. Traditional flywheel magnetos (using contact-breaker points, but no external coil) worked equally well in reverse because the cam controlling the points is symmetrical, breaking contact before top dead center (TDC) equally well whether running forwards or backwards. Reed-valve engines will run backwards just as well as piston-controlled porting, though rotary valve engines have asymmetrical inlet timing and will not run very well.

There are serious disadvantages to running many engines backwards under load for any length of time, and some of these reasons are general, applying equally to both two-stroke and four-stroke engines. This disadvantage is accepted in most cases where cost, weight and size are major considerations. The problem comes about because in "forwards" running the major thrust face of the piston is on the back face of the cylinder which, in a two-stroke particularly, is the coolest and best-lubricated part. The forward face of the piston in a trunk engine is less well-suited to be the major thrust face since it covers and uncovers the exhaust port in the cylinder, the hottest part of the engine, where piston lubrication is at its most marginal. The front face of the piston is also more vulnerable since the exhaust port, the largest in the engine, is in the front wall of the cylinder. Piston skirts and rings risk being extruded into this port, so it is always better to have them pressing hardest on the opposite wall (where there are only the transfer ports in a crossflow engine) and there is good support. In some engines, the small end is offset to reduce thrust in the intended rotational direction and the forward face of the piston has been made thinner and lighter to compensate - but when running backwards, this weaker forward face suffers increased mechanical stress it was not designed to resist. This can be avoided by the use of crossheads and also using thrust bearings to isolate the engine from end loads.

Large two-stroke ship diesels are sometimes made to be reversible. Like four-stroke ship engines (some of which are also reversible) they use mechanically operated valves, so require additional camshaft mechanisms. These engine use crossheads to eliminate sidethrust on the piston and isolate the under-piston space from the crankcase.

On top of other considerations, the oil-pump of a modern two-stroke may not work in reverse, in which case the engine will suffer oil starvation within a short time. Running a motorcycle engine backwards is relatively easy to initiate, and in rare cases, can be triggered by a back-fire. It is not advisable.

Model airplane engines with reed-valves can be mounted in either tractor or pusher configuration without needing to change the propeller. These motors are compression ignition, so there are no ignition timing issues and little difference between running forward and running backward.

Power-to-weight Ratio

Power-to-weight ratio (or specific power or power-to-mass ratio) is a calculation commonly applied to engines and mobile power sources to enable the comparison of one unit or design to another. Power-to-weight ratio is a measurement of actual performance of any engine or power source. It is also used as a measurement of performance of a vehicle as a whole, with the engine's power output being divided by the weight (or mass) of the vehicle, to give a metric that is independent of the vehicle's size. Power-to-weight is often quoted by manufacturers at the peak value, but the actual value may vary in use and variations will affect performance.

The inverse of power-to-weight, weight-to-power ratio (power loading) is a calculation commonly applied to aircraft, cars, and vehicles in general, to enable the comparison of one vehicle's performance to another. Power-to-weight ratio is equal to thrust per unit mass multiplied by the velocity of any vehicle.

Power-to-weight (Specific Power)

The power-to-weight ratio (Specific Power) formula for an engine (power plant) is the power generated by the engine divided by the mass. ("Weight" in this context is a colloquial term for "mass", what an engineer means by the "power to weight ratio" of an electric motor is not infinite in a zero gravity environment.)

A typical turbocharged V8 diesel engine might have an engine power of 250 kW (340 hp) and a mass of 380 kg (840 lb), giving it a power-to-weight ratio of 0.65 kW/kg (0.40 hp/lb).

Examples of high power-to-weight ratios can often be found in turbines. This is because of their ability to operate at very high speeds. For example, the Space Shuttle's main engines used turbopumps (machines consisting of a pump driven by a turbine engine) to feed the propellants (liquid oxygen and liquid hydrogen) into the engine's combustion chamber. The original liquid hydrogen turbopump is similar in size to an automobile engine (weighing approximately 352 kilograms (775 lb)) and produces 72,000 hp (53.6 MW) for a power-to-weight ratio of 153 kW/kg (93 hp/lb).

Engine Power

The actual useful power of any traction engine can be calculated using a dynamometer to measure torque and rotational speed, with peak power sustained when the transmission and/or operator keeps the product of torque and rotational speed maximised. For jet engines there is often a cruise speed and power can be usefully calculated there, for rockets there is typically no cruise speed, so it is less meaningful.

Peak power of a traction engine occurs at a rotational speed higher than the speed when torque is maximised and at or below the maximum rated rotational speed - Max RPM. A rapidly falling torque curve would correspond with sharp torque and power curve peaks around their maxima at similar rotational speed, for example a small, lightweight engine with a large turbocharger. A slowly falling or near flat torque curve would correspond with a slowly rising power curve up to a maximum at a rotational speed close to Max RPM, for example a large, heavy multi-cylinder engine suitable for cargo/hauling. A falling torque curve could correspond with a near flat power curve across rotational speeds for smooth handling at different vehicle speeds.

Examples

Heat Engines and Heat Pumps

Thermal energy is made up from molecular kinetic energy and latent phase energy. Heat engines are able to convert thermal energy in the form of a temperature gradient between a hot source and a cold sink into other desirable mechanical work. Heat pumps take mechanical work to regener-ate thermal energy in a temperature gradient. Care should be made when interpreting propulsive power, especially for jet engines and rockets, deliverable from heat engines to a vehicle.

Heat Engine/Heat pump type	Peak Power Output		Power-to-weight ratio		Example Use
	SI	English	SI	English	
Wärtsilä RTA96-C 14-cylinder two-stroke Turbo Diesel engine	80,080 kW	108,920 hp	0.03 kW/kg	0.02 hp/lb	Emma Mærsk container ship
Suzuki 538 cc V2 4-stroke gas (petrol) outboard Otto engine	19 kW	25 hp	0.27 kW/kg	0.16 hp/lb	Runabout boats
DOE/NASA/0032-28 Mod 2 502 cc gas (petrol) Stirling engine	62.3 kW	83.5 hp	0.30 kW/kg	0.18 hp/lb	Chevrolet Celebrity[·] 1985
GM 6.6 L Duramax LMM (LYE option) V8 Turbo Diesel engine	246 kW	330 hp	0.65 kW/kg	0.40 hp/lb	Chevrolet Kodiak[·], GMC Topkick[·]
Junkers Jumo 205A opposed-piston two-stroke Diesel engine	647 kW	867 hp	1.1 kW/kg	0.66 hp/lb	Ju 86C-1 airliner, B&V Ha 139 floatplane
GE LM2500+ marine turboshaft Brayton gas turbine	30,200 kW	40,500 hp	1.31 kW/kg	0.80 hp/lb	GTS Millennium cruiseship, QM2 ocean liner
Mazda 13B-MSP Renesis 1.3 L Wankel engine	184 kW	247 hp	1.5 kW/kg	0.92 hp/lb	Mazda RX-8[·]

Engine	SI	English	kW/kg	hp/lb	Example Use
PW R-4360 71.5 L 28-cylinder supercharged Radial engine	3,210 kW	4,300 hp	1.83 kW/kg	1.11 hp/lb	B-50 Superfortress, Convair B-36
					C-97 Stratofreighter, C-119 Flying Boxcar
					Hughes H-4 Hercules "Spruce Goose"
Wright R-3350 54.57 L 18-c s/c Turbo-compound Radial engine	2,535 kW	3,400 hp	2.09 kW/kg	1.27 hp/lb	B-29 Superfortress, Douglas DC-7
					C-97 S/f prototype, Kaiser-Frazer C-119F
O.S. Engines 49-PI Type II 4.97 cc UAV Wankel engine	0.934 kW	1.252 hp	2.8 kW/kg	1.7 hp/lb	Model aircraft, Radio-controlled aircraft
JetCat SPT10-RX-H UAV turboshaft	9 kW	12 hp	3.67 kW/kg	2.24 hp/lb	Model aircraft, Radio-controlled aircraft
GE LM6000 marine turboshaft Brayton gas turbine	44,700 kW	59,900 hp	5.67 kW/kg	3.38 hp/lb	Peaking power plant
GE CF6-80C2 Brayton high-bypass turbofan jet engine					Boeing 747[.], 767, Airbus A300
BMW V10 3L P84/5 2005 gas (petrol) Otto engine	690 kW	925 hp	7.5 kW/kg	4.6 hp/lb	Williams FW27 car[.], Formula One auto racing
BMW i4 1.490L M12 engine 1987 gas (petrol) Otto engine	1030 kW	1,400 hp	8.25 kW/kg	5.07 hp/lb	Arrows A10 car[.], Formula One auto racing
GE90-115B Brayton turbofan jet engine	83,164 kW	111,526 hp	10.0 kW/kg	6.10 hp/lb	Boeing 777
PWR RS-24 (SSME) Block II H_2 Brayton turbopump	63,384 kW	85,000 hp	138 kW/kg	84 hp/lb	Space Shuttle (STS-110 and later) [.]
PWR RS-24 (SSME) Block I H_2 Brayton turbopump	53,690 kW	72,000 hp	153 kW/kg	93 hp/lb	Space Shuttle

Electric Motors/Electromotive Generators

An electric motor uses electrical energy to provide mechanical work, usually through the interaction of a magnetic field and current-carrying conductors. By the interaction of mechanical work on an electrical conductor in a magnetic field, electrical energy can be generated.

Electric motor type	Weight		Peak Power Output		Power-to-weight ratio		Example Use
	SI	English	SI	English	kW/kg	hp/lb	
Panasonic MSMA202S1G AC servo motor	6.5 kg	14 lb	2 kW	2.7 hp	0.31 kW/kg	0.19 hp/lb	Conveyor belts, Robotics
Toshiba 660 MVA water cooled 23kV AC turbo generator	1,342 t	2,959,000 lb	660 MW	890,000 hp	0.49 kW/kg	0.30 hp/lb	Bayswater, Eraring Coal Power stations

	SI	English	SI	English	SI	English	
Canopy Tech. Cypress 32 MW 15 kV AC PM generator	33,557 kg	73,981 lb	32 MW	43,000 hp	0.95 kW/kg	0.58 hp/lb	Electric Power stations
Toyota Brushless AC Nd Fe B PM motor	36.3 kg	80 lb	50 kW	67 hp	1.37 kW/kg	0.84 hp/lb	Toyota Prius[*] 2004
Himax HC6332-250 Brushless DC motor	0.45 kg	0.99 lb	1.7 kW	2.3 hp	3.78 kW/kg	2.30 hp/lb	Radio controlled cars
Hi-Pa Drive HPD40 Brushless DC wheel hub motor	25 kg	55 lb	120 kW	160 hp	4.8 kW/kg	2.92 hp/lb	Mini QED HEV, Ford F150 HEV
ElectriFly GPMG4805 Brushless DC	1.48 kg	3.3 lb	8.4 kW	11.3 hp	5.68 kW/kg	3.45 hp/lb	Radio-controlled aircraft
YASA-400 Brushless AC	24 kg	53 lb	165 kW	221 hp	6.875 kW/kg	4.18 hp/lb	Electric Vehicle, Drive eO
ElectriFly GPMG5220 Brushless DC	0.133 kg	0.29 lb	1.035 kW	1.388 hp	7.78 kW/kg	4.73 hp/lb	Radio-controlled aircraft
Remy HVH250-090-POC3 Brushless DC	33.5 kg	74 lb	297 kW	398 hp	8.87 kW/kg	5.39 hp/lb	Electric Vehicle
EMRAX268 Brushless AC	19.9 kg	44 lb	200 kW	270 hp	10.05 kW/kg	6.12 hp/lb	Battery Electric Air Plane

Fluid Engines and Fluid Pumps

Fluids (liquid and gas) can be used to transmit and/or store energy using pressure and other fluid properties. Hydraulic (liquid) and pneumatic (gas) engines convert fluid pressure into other desirable mechanical or electrical work. Fluid pumps convert mechanical or electrical work into movement or pressure changes of a fluid, or storage in a pressure vessel.

Fluid Powerplant type	Dry Weight		Peak Power Output		Power-to-weight ratio	
	SI	English	SI	English	SI	English
PlatypusPower Q2/200 hydroelectric turbine	43 kg	95 lb	2 kW	2.7 hp	0.047 kW/kg	0.029 hp/lb
PlatypusPower PP20/200 hydroelectric turbine	330 kg	728 lb	20 kW	27 hp	0.060 kW/kg	0.037 hp/lb
Atlas Copco LZL 35 pneumatic motor	20 kg	44.1 lb	6.5 kW	8.7 hp	0.33 kW/kg	0.20 hp/lb
Atlas Copco LZB 14 pneumatic motor	0.30 kg	0.66 lb	0.16 kW	0.22 hp	0.53 kW/kg	0.33 hp/lb
Bosch 0 607 954 307 pneumatic motor	0.32 kg	0.71 lb	0.1 kW	0.13 hp	0.31 kW/kg	0.19 hp/lb
Atlas Copco LZB 46 pneumatic motor	1.2 kg	2.65 lb	0.84 kW	1.13 hp	0.7 kW/kg	0.43 hp/lb
Bosch 0 607 957 307 pneumatic motor	1.7 kg	3.7 lb	0.74 kW	0.99 hp	0.44 kW/kg	0.26 hp/lb

SAI GM7 radial piston hydraulic motor	300 kg	661 lb	250 kW	335 hp	0.83 kW/kg	0.50 hp/lb
SAI GM3 radial piston hydraulic motor	15 kg	33 lb	15 kW	20 hp	1 kW/kg	0.61 hp/lb
Denison GOLD CUP P14 axial piston hydraulic motor	110 kg	250 lb	384 kW	509 hp	3.5 kW/kg	2.0 hp/lb
Denison TB vane pump	7 kg	15 lb	40.2 kW	53.9 hp	5.7 kW/kg	3.6 hp/lb

Thermoelectric Generators and Electrothermal Actuators

A variety of effects can be harnessed to produce thermoelectricity, thermionic emission, pyroelectricity and piezoelectricity. Electrical resistance and ferromagnetism of materials can be harnessed to generate thermoacoustic energy from an electric current.

Thermoelectric Powerplant type	Dry Weight		Peak Power Output		Power-to-weight ratio		Example Use
Teledyne ^{238}Pu GPHS-RTG 1980	56 kg	123 lb	285 W	0.39 hp	5.09 W/kg	0.003 hp/lb	Galileo probe, New Horizons probe
Boeing ^{238}Pu MMRTG MSL	44.1 kg	97.2 lb	123 W	0.16 hp	2.79 W/kg	0.002 hp/lb	Mars Science Laboratory
HZ-20 thermoelectric module	0.115 kg	0.254 lb	19 W	0.025 hp	165 W/kg	0.098 hp/lb	Hi-Z Technology Inc.

Electrochemical (Galvanic) and Electrostatic Cell Systems

(Closed Cell) Batteries

All electrochemical cell batteries deliver a changing voltage as their chemistry changes from "charged" to "discharged". A nominal output voltage and a cutoff voltage are typically specified for a battery by its manufacturer. The output voltage falls to the cutoff voltage when the battery becomes "discharged". The nominal output voltage is always less than the open-circuit voltage produced when the battery is "charged". The temperature of a battery can affect the power it can deliver, where lower temperatures reduce power. Total energy delivered from a single charge cycle is affected by both the battery temperature and the power it delivers. If the temperature lowers or the power demand increases, the total energy delivered at the point of "discharge" is also reduced.

Battery discharge profiles are often described in terms of a factor of battery capacity. For example, a battery with a nominal capacity quoted in ampere-hours (Ah) at a C/10 rated discharge current (derived in amperes) may safely provide a higher discharge current - and therefore higher power-to-weight ratio - but only with a lower energy capacity. Power-to-weight ratio for batteries is therefore less meaningful without reference to corresponding energy-to-weight ratio and cell temperature. This relationship is known as Peukert's law.

Battery type	Volts	Temp.	Energy-to-weight ratio	Power-to-weight ratio
Energizer 675 Mercury Free Zinc-air battery	1.4V	21 °C	1,645 kJ/kg to 0.9 V	1.65 W/kg 2.24 mA
GE Durathon™ NaMx A2 UPS Molten salt battery	54.2V	- 40 - 65 °C	342 kJ/kg to 37.8 V	15.8 W/kg C/6 (76 A)

Panasonic R03 AAA Zinc–carbon battery	1.5 V	20±2 °C	47 kJ/kg 20 mA to 0.9 V	3.3 W/kg 20 mA
			88 kJ/kg 150 mA to 0.9 V	24 W/kg 150 mA
Eagle-Picher SAR-10081 60Ah 22-cell Nickel–hydrogen battery	27.7 V	10 °C	192 kJ/kg C/2 to 22 V	23 W/kg C/2
			165 kJ/kg C/1 to 22 V	46 W/kg C/1
ClaytonPower 400Ah Lithium-ion battery	12V		617 kJ/kg	85.7 W/kg C/1 (175 A)
Energizer 522 Prismatic Zn–MnO$_2$ Alkaline battery	9 V	21 °C	444 kJ/kg 25 mA to 4.8 V	4.9 W/kg 25 mA
			340 kJ/kg 100 mA to 4.8 V	19.7 W/kg 100 mA
			221 kJ/kg 500 mA to 4.8 V	99 W/kg 500 mA
Panasonic HHR900D 9.25Ah Nickel–metal hydride battery	1.2 V	20 °C	209.65 kJ/kg to 0.7 V	11.7 W/kg C/5
				58.2 W/kg C/1
				116 W/kg 2C
URI 1418Ah replaceable anode Aluminium–air battery model	244.8 V	60 °C	4680 kJ/kg	130.3 W/kg (142 A)
LG Chemical/CPI E2 6Ah LiMn$_2$O$_4$ Lithium-ion polymer battery	3.8 V	25 °C	530.1 kJ/kg C/2 to 3.0 V	71.25 W/kg
			513 kJ/kg 1C to 3.0 V	142.5 W/kg
Saft 45E Fe Super-Phosphate Lithium iron phosphate battery	3.3 V	25 °C	581 kJ/kg C to 2.5 V	161 W/kg
			560 kJ/kg 1.14 C to 2.0 V	183 W/kg
			0.73 kJ/kg 2.27 C to 1.5 V	367 W/kg
Energizer CH35 C 1.8Ah Nickel–cadmium battery	1.2 V	21 °C	152 kJ/kg C/10 to 1 V	4 W/kg C/10
			147.1 kJ/kg 5C to 1 V	200 W/kg 5 C
Firefly Energy Oasis FF12D1-G31 6-cell 105Ah VRLA battery	12 V	25 °C	142 kJ/kg C/10 to 7.2 V	4 W/kg C/10
		-1 8 °C	7 kJ/kg CCA to 7.2V	234 W/kg CCA (625A)
		0 °C	9 kJ/kg CA to 7.2 V	300 W/kg CA (800 A)
Panasonic CGA103450A 1.95Ah LiCoO$_2$ Lithium-ion battery	3.7 V	20 °C	666 kJ/kg C/5.3 to 2.75 V	35 W/kg C/5.3
		0 °C	633 kJ/kg C/1 to 2.75 V	176 W/kg C/1
		20 °C	655 kJ/kg C/1 to 2.75 V	182 W/kg C/1
		20 °C	641 kJ/kg 2C to 2.75 V	356 W/kg 2C

Electric Fuel Battery Corp. UUV 120Ah Zinc–air fuel cell			630 kJ/kg	500 W/kg C/1
Sion Power 2.5Ah Li–S Lithium-ion battery	2.15 V	25 °C	1260 kJ/kg	70 W/kg C/5
			1209 kJ/kg	672 W/kg 2C
Stanford Prussian Blue durable Potassium-ion battery	1.35 V	room	54 kJ/kg	13.8 W/kg C/1
			50 kJ/kg	138 W/kg 10C
			39 kJ/kg	693 W/kg 50C
Maxell / Yuasa / AIST Nickel–metal hydride lab prototype		45 °C		980 W/kg
Toshiba SCiB cell 4.2Ah Li_2TiO_3 Lithium-ion battery	2.4 V	25 °C	242 kJ/kg	67.2 W/kg C/1
			218 kJ/kg	4000 W/kg 12C
Ionix Power Systems $LiMn_2O_4$ Lithium-ion battery lab model		lab	270 kJ/kg	1700 W/kg
		lab	29 kJ/kg	4900 W/kg
A123 Systems 26650 Cell 2.3Ah $LiFePO_4$ Lithium ion battery	3.3 V	-20 °C	347 kJ/kg C/1 to 2V	108 W/kg C/1
		0 °C	371 kJ/kg C/1 to 2 V	108 W/kg C/1
		25 °C	390 kJ/kg C/1 to 2 V	108 W/kg C/1
		25 °C	390 kJ/kg 27C to 2 V	3300 W/kg 27C
		25 °C	57 kJ/kg 32C to 2 V	5657 W/kg 32C
Saft VL 6Ah Lithium-ion battery	3.65 V	-20 °C	154 kJ/kg 30C to 2.5 V	41.4 W/kg 30C (180 A)
			182 kJ/kg 1C to 2.5 V	67.4 W/kg 1C
		25 °C	232 kJ/kg 1C to 2.5 V	64.4 W/kg 1C
			233 kJ/kg 58.3C to 2.5 V	3757 W/kg 58.3C (350A)
			34 kJ/kg 267C to 2.5 V	17176 W/kg 267C (1.6kA)
			4.29 kJ/kg 333C to 2.5 V	21370 W/kg 333C (2kA)

Electrostatic, Electrolytic and Electrochemical Capacitors

Capacitors store electric charge onto two electrodes separated by an electric field semi-insulating (dielectric) medium. Electrostatic capacitors feature planar electrodes onto which electric charge accumulates. Electrolytic capacitors use a liquid electrolyte as one of the electrodes and the electric double layer effect upon the surface of the dielectric-electrolyte boundary to increase the amount of charge stored per unit volume. Electric double-layer capacitors extend both electrodes with a nanopourous material such as activated carbon to significantly increase the surface area upon which electric charge can accumulate, reducing the dielectric medium to nanopores and a very thin high permittivity separator.

While capacitors tend not to be as temperature sensitive as batteries, they are significantly capacity constrained and without the strength of chemical bonds suffer from self-discharge. Power-to-weight ratio of capacitors is usually higher than batteries because charge transport units within the cell are smaller (electrons rather than ions), however energy-to-weight ratio is conversely usually lower.

Capacitor type	Capacity	Volts	Temp.	Energy-to-weight ratio	Power-to-weight ratio
ACT Premlis Lithium ion capacitor	2000 F	4.0 V	25 °C	54 kJ/kg to 2.0 V	44.4 W/kg @ 5 A
				31 kJ/kg to 2.0 V	850 W/kg @ 10 A
Nesccap Electric double-layer capacitor	5000 F	2.7 V	25 °C	19.58 kJ/kg to 1.35 V	5.44 W/kg C/1 (1.875 A)
				5.2 kJ/kg to 1.35 V	5,200 W/kg @ 2,547A
EEStor EESU barium titanate supercapacitor	30.693 F	3500 V	85 °C	1471.98 kJ/kg	80.35 W/kg C/5
				1471.98 kJ/kg	8,035 W/kg 20 C
General Atomics 3330CMX2205 High Voltage Capacitor	20.5 mF	3300 V	? °C	2.3 kJ/kg	6.8 MW/kg @ 100 kA

Fuel Cell Stacks and Flow Cell Batteries

Fuel cells and flow cells, although perhaps using similar chemistry to batteries, have the distinction of not containing the energy storage medium or fuel. With a continuous flow of fuel and oxidant, available fuel cells and flow cells continue to convert the energy storage medium into electric energy and waste products. Fuel cells distinctly contain a fixed electrolyte whereas flow cells also require a continuous flow of electrolyte. Flow cells typically have the fuel dissolved in the electrolyte.

Fuel cell type	Dry weight	Power-to-weight ratio	Example Use
Redflow Power+BOS ZB600 10kWh ZBB	900 kg	5.6 W/kg (9.3 W/kg peak)	Rural Grid support
Ceramic Fuel Cells BlueGen MG 2.0 CHP SOFC	200 kg	10 W/kg	
		15 W/kg CHP	
MTU Friedrichshafen 240 kW MCFC HotModule 2006	20,000 kg	12 W/kg	
Smart Fuel Cell Jenny 600S 25W DMFC	1.7 kg	14.7 W/kg	Portable military electronics
UTC Power PureCell 400 kW PAFC	27,216 kg	14.7 W/kg	
GEFC 50V50A-VRB Vanadium redox battery	80 kg	31.3 W/kg (125 W/kg peak)	
Ballard Power Systems Xcellsis HY-205 205 kW PEMFC	2,170 kg	94.5 W/kg	Mercedes-Benz Citaro O530BZ[·]
UTC Power/NASA 12 kW AFC	122 kg	98 W/kg	Space Shuttle orbiter[·]
Ballard Power Systems FCgen-1030 1.2 kW CHP PEMFC	12 kg	100 W/kg	Residential cogeneration
Ballard Power Systems FCvelocity-HD6 150 kW PEMFC	400 kg	375 W/kg	Bus and heavy duty
NASA Glenn Research Center 50 W SOFC	0.071 kg	700 W/kg	
Honda 2003 43 kW FC Stack PEMFC[·]	43 kg	1000 W/kg	Honda FCX Clarity[·]

| Lynntech, Inc. PEMFC lab prototype | 0.347 kg | 1,500 W/kg | |

Photovoltaics

Photovoltaic Panel type	Power-to-weight ratio
Thyssen Solartec 128W Nanocrystalline Si Triplejunction PV module	6 W/kg
Suntech/UNSW HiPerforma PLUTO220-Udm 220W Ga-F22 Polycrystalline Si PV module	13.1 W/kg STP
	9.64 W/kg nominal
Global Solar PN16015A 62W CIGS polycrystalline thin film PV module	40 W/kg
Able (AEC) PUMA 6 kW GaInP2/GaAs/Ge-on-Ge Triplejunction PV array	65 W/kg
Current spacecraft grade	~77 W/kg
ITO/InP on Kapton foil	2000 W/kg

Vehicles

Power-to-weight ratios for vehicles are usually calculated using curb weight (for cars) or wet weight (for motorcycles), that is, excluding weight of the driver and any cargo. This could be slightly misleading, especially with regard to motorcycles, where the driver might weigh 1/3 to 1/2 as much as the vehicle itself. In the sport of competitive cycling athlete's performance is increasingly being expressed in VAMs and thus as a power-to-weight ratio in W/kg. This can be measured through the use of a bicycle power-meter or calculated from measuring incline of a road climb and the rider's time to ascend it.

Utility and Practical Vehicles

Most vehicles are designed to meet passenger comfort and cargo carrying requirements. Different designs trade off power-to-weight ratio to increase comfort, cargo space, fuel economy, emissions control, energy security and endurance. Reduced drag and lower rolling resistance in a vehicle design can facilitate increased cargo space without increase in the (zero cargo) power-to-weight ratio. This increases the role flexibility of the vehicle. Energy security considerations can trade off power (typically decreased) and weight (typically increased), and therefore power-to-weight ratio, for fuel flexibility or drive-train hybridisation. Some utility and practical vehicle variants such as hot hatches and sports-utility vehicles reconfigure power (typically increased) and weight to provide the perception of sports car like performance or for other psychological benefit. Rail locomotives require high mass to maintain adhesive traction on the rails, therefore improving the power-to-weight ratio by reducing mass is not necessarily beneficial. However choice of rail locomotive traction system (i.e. AC VFD over DC) can support improved power-to-weight ratio by reducing mass for the same adhesion.

Notable Low Ratio

Vehicle	Power	Vehicle Weight	Power to Weight ratio
Benz Patent Motorwagen 954 cc 1886	560 W / 0.75 bhp	265 kg / 584 lb	2.1 W/kg / 779 lb/hp

Stephenson's Rocket 0-2-2 steam locomotive with tender 1829	15 kW / 20 bhp	4,320 kg / 9524 lb	3.5 W/kg / 476 lb/hp
CBQ Zephyr streamliner diesel locomotive with railcars 1934	492 kW / 660 bhp	94 t / 208,000 lb	5.21 W/kg / 315 lb/hp
Alberto Contador's Verbier climb 2009 Tour de France on Specialized bike	420 W / 0.56 bhp	62 kg / 137 lb	6.7 W/kg / 245 lb/hp
Force Motors Minidor Diesel 499 cc auto rickshaw	6.6 kW / 8.8 bhp	700 kg / 1543 lb	9 W/kg / 175 lb/hp
PRR Q2 4-4-6-4 steam locomotive with tender 1944	5,956 kW / 7,987 bhp	475.9 t / 1,049,100 lb	12.5 W/kg / 131 lb/hp
Mercedes-Benz Citaro O530BZ H_2 fuel cell bus 2002	205 kW / 275 bhp	14,500 kg / 32,000 lb	14.1 W/kg / 116 lb/hp
TGV BR Class 373 high-speed Eurostar Trainset 1993	12,240 kW / 16,414 bhp	816 t / 1,798,972 lb	15 W/kg / 110 lb/hp
General Dynamics M1 Abrams Main battle tank 1980	1,119 kW / 1500 bhp	55.7 t / 122,800 lb	20.1 W/kg / 81.9 lb/hp
BR Class 43 high-speed diesel electric locomotive 1975	1,678 kW / 2,250 bhp	70.25 t / 154,875 lb	23.9 W/kg / 69 lb/hp
GE AC6000CW diesel electric locomotive 1996	4,660 kW / 6,250 bhp	192 t / 423,000 lb	24.3 W/kg / 68 lb/hp
BR Class 55 Napier Deltic diesel electric locomotive 1961	2,460 kW / 3,300 bhp	101 t / 222,667 lb	24.4 W/kg / 68 lb/hp
International CXT 2004	164 kW / 220 bhp	6,577 kg / 14500 lb	25 W/kg / 66 lb/hp
Ford Model T 2.9 L flex-fuel 1908	15 kW / 20 bhp	540 kg / 1,200 lb	28 W/kg / 60 lb/hp
TH!NK City 2008	30 kW / 40 bhp	1038 kg / 2,288 lb	28.9 W/kg / 56.9 lb/hp
Messerschmitt KR200 Kabinenroller 191 cc 1955	6 kW / 8.2 bhp	230 kg / 506 lb	30 W/kg / 50 lb/hp
Wright Flyer 1903	9 kW / 12 bhp	274 kg / 605 lb	33 W/kg / 50 lb/hp
Tata Nano 624 cc 2008	26 kW / 35 bhp	635 kg / 1,400 lb	41.0 W/kg / 40 lb/hp
Bombardier JetTrain high-speed gas turbine-electric locomotive 2000	3,750 kW / 5,029 bhp	90,750 kg / 200,000 lb	41.2 W/kg / 39.8 lb/hp
Suzuki MightyBoy 543 cc 1988	23 kW / 31 bhp	550 kg / 1,213 lb	42 W/kg / 39 lb/hp
Mitsubishi i MiEV 2009	47 kW / 63 bhp	1,080 kg / 2,381 lb	43.5 W/kg / 37.8 lb/hp
Holden FJ 2,160 cc 1953	44.7 kW / 60 bhp	1,021 kg / 2,250 lb	43.8 W/kg / 37.5 lb/hp
Chevrolet Kodiak/GMC Topkick LYE 6.6 L 2005	246 kW / 330 bhp	5126 kg / 11,300 lb	48 W/kg / 34.2 lb/hp
DOE/NASA/0032-28 Chevrolet Celebrity 502 cc ASE Mod II 1985	62.3 kW / 83.5 bhp	1,297 kg / 2,860 lb	48.0 W/kg / 34.3 lb/hp
Suzuki Alto 796 cc 2000	35 kW / 46 bhp	720 kg / 1,587 lb	49 W/kg / 35 lb/hp

| Land Rover Defender 2.4 L 1990 | 90 kW / 121 bhp | 1,837 kg / 4,050 lb | 49 W/kg / 33 lb/hp |

Common power

Vehicle	Power	Vehicle Weight	Power to Weight ratio
Toyota Prius 1.8 L 2010 (petrol only)	73 kW / 98 bhp	1,380 kg / 3,042 lb	53 W/kg / 31 lb/hp
Bajaj Platina Naked 100 cc 2006	6 kW / 8 bhp	113 kg / 249 lb	53 W/kg / 31 lb/hp
Subaru R2 type S 2003	47 kW / 63 bhp	830 kg / 1,830 lb	57 W/kg / 29 lb/hp
Ford Fiesta ECOnetic 1.6 L TDCi 5dr 2009	66 kW / 89 bhp	1,155 kg / 2,546 lb	57 W/kg / 29 lb/hp
Volvo C30 1.6D DRIVe S/S 3dr Hatch 2010	80 kW / 108 bhp	1,347 kg / 2,970 lb	59.4 W/kg / 27.5 lb/hp
Ford Focus ECOnetic 1.6 L TDCi 5dr Hatch 2009	81 kW / 108 bhp	1,357 kg / 2,992 lb	59.7 W/kg / 27 lb/hp
Ford Focus 1.8 L Zetec S TDCi 5dr Hatch 2009	84 kW / 113 bhp	1,370 kg / 3,020 lb	61 W/kg / 27 lb/hp
Honda FCX Clarity 4 kg Hydrogen 2008	100 kW / 134 bhp	1,600 kg / 3,528 lb	63 W/kg / 26 lb/hp
Hummer H1 6.6 L V8 2006	224 kW / 300 bhp	3,559 kg / 7,847 lb	63 W/kg / 26 lb/hp
Audi A2 1.4 L TDI 90 type S 2003	66 kW / 89 bhp	1,030 kg / 2,270 lb	64 W/kg / 25 lb/hp
Opel/Vauxhall/Holden/Chevrolet Astra 1.7 L CTDi 125 2010	92 kW / 123 bhp	1,393 kg / 3,071 lb	66 W/kg / 24.9 lb/hp
Mini (new) Cooper 1.6D 2007	81 kW / 108 bhp	1,185 kg / 2,612 lb	68 W/kg / 24 lb/hp
Toyota Prius 1.8 L 2010 (electric boost)	100 kW / 134 bhp	1,380 kg / 3,042 lb	72 W/kg / 23 lb/hp
Ford Focus 2.0 L Zetec S TDCi 5dr Hatch 2009	100 kW / 134 bhp	1,370 kg / 3,020 lb	73 W/kg / 23 lb/hp
General Motors EV1 electric car Gen II 1998	102.2 kW / 137 bhp	1,400 kg / 3,086 lb	73 W/kg / 23 lb/hp
Toyota Venza I4 2.7 L FWD 2009	136 kW / 182 bhp	1,706 kg / 3,760 lb	80 W/kg / 20.7 lb/hp
Ford Focus 2.0 L Zetec S 5dr Hatch 2009	107 kW / 143 bhp	1,327 kg / 2,926 lb	81 W/kg / 20 lb/hp
Fiat Grande Punto 1.6 L Multijet 120 2005	88 kW / 118 bhp	1,075 kg / 2,370 lb	82 W/kg / 20 lb/hp
Mini (classic) 1275GT 1969	57 kW / 76 bhp	686 kg / 1,512 lb	83 W/kg / 20 lb/hp
Opel/Vauxhall/Holden/Chevrolet Astra 2.0 L CTDi 160 2010	118 kW / 158 bhp	1,393 kg / 3,071 lb	85 W/kg / 19.4 lb/hp
Ford Focus 2.0 auto 2007	104.4 kW / 140 bhp	1,198 kg / 2,641 lb	87.1 W/kg / 19 lb/hp
Subaru Legacy/Liberty 2.0R 2005	121 kW / 162 bhp	1,370 kg / 3,020 lb	88 W/kg / 19 lb/hp

Subaru Outback 2.5i 2008	130.5 kW / 175 bhp	1,430 kg / 3,153 lb	91 W/kg / 18 lb/hp
Smart Fortwo 1.0 L Brabus 2009	72 kW / 97 bhp	780 kg / 1,720 lb	92 W/kg / 18 lb/hp
Toyota Venza V6 3.5 L AWD 2009	200 kW / 268 bhp	1,835 kg / 4,045 lb	109 W/kg / 15 lb/hp
Toyota Venza I4 2.7 L FWD 2009 with Lotus mass reduction	136 kW / 182 bhp	1,210 kg / 2,667 lb	112.2 W/kg / 14.7 lb/hp
Toyota Hilux V6 DOHC 4 L 4×2 Single Cab Pickup ute 2009	175 kW / 235 bhp	1,555 kg / 3,428 lb	112.5 W/kg / 14.6 lb/hp
Toyota Venza V6 3.5 L FWD 2009	200 kW / 268 bhp	1,755 kg / 3,870 lb	114 W/kg / 14.4 lb/hp

Performance Luxury, Roadsters and Mild Sports

Increased engine performance is a consideration, but also other features associated with luxury vehicles. Longitudinal engines are common. Bodies vary from hot hatches, sedans (saloons), coupés, convertibles and roadsters. Mid-range dual-sport and cruiser motorcycles tend to have similar power-to-weight ratios.

Vehicle	Power	Vehicle Weight	Power to Weight ratio
Honda Accord sedan V6 2011	202 kW / 271 bhp	1630 kg / 3593 lb	124 W/kg / 13.26 lb/hp
Mini (new) Cooper 1.6T S JCW 2008	155 kW / 208 bhp	1205 kg / 2657 lb	129 W/kg / 13 lb/hp
Mazda RX-8 1.3 L Wankel 2003	173 kW / 232 bhp	1309 kg / 2888 lb	132 W/kg / 12 lb/hp
Holden Statesman/Caprice / Buick Park Avenue / Daewoo Veritas 6 L V8 2007	270 kW / 362 bhp	1891 kg / 4170 lb	143 W/kg / 12 lb/hp
Kawasaki KLR650 Gasoline DualSport 650 cc	26 kW / 35 bhp	182 kg / 401 lb	143 W/kg / 11 lb/hp
NATO HTC M1030M1 Diesel/Jet fuel DualSport 670 cc	26 kW / 35 bhp	182 kg / 401 lb	143 W/kg / 11 lb/hp
Harley-Davidson FLSTF Softail Fat Boy Cruiser 1,584 cc 2009	47 kW / 63 bhp	324 kg / 714 lb	145 W/kg / 11.3 lb/hp
BMW 7 Series 760Li 6 L V12 2006	327 kW / 439 bhp	2250 kg / 4960 lb	145 W/kg / 11 lb/hp
Subaru Impreza WRX STi 2.0 L 2008	227 kW / 304 bhp	1530 kg / 3373 lb	148 W/kg / 11 lb/hp
Honda S2000 roadster 1999	183.88 kW / 240 bhp	1250 kg / 2723 lb	150 W/kg / 11 lb/hp
GMH HSV Clubsport / GMV VXR8 / GMC CSV CR8 / Pontiac G8 6 L V8 2006	317 kW / 425 bhp	1831 kg / 4037 lb	173 W/kg / 9.5 lb/hp
Tesla Roadster 2011	215 kW / 288 bhp	1235 kg / 2723 lb	174 W/kg / 9.5 lb/hp

Sports Vehicles and Aircraft

Power-to-weight ratio is an important vehicle characteristic that affects the acceleration and handling - and therefore the driving enjoyment - of any sports vehicle. Aircraft also depend on high power-to-weight ratio to achieve sufficient lift.

Vehicle	Power	Vehicle Weight	Power to Weight ratio
Lotus Elise SC 2008	163 kW / 218 bhp	910 kg / 2006 lb	179 W/kg / 9.20 lb/hp
Ferrari Testarossa 1984	291 kW / 390 bhp	1506 kg / 3320 lb	193 W/kg / 8.51 lb/hp
Citroën DS3 WRC rally car 2011	235 kW / 315 bhp	1200 kg / 2,645.5 lb	196 W/kg / 8.40 lb/hp
Artega GT	220 kW / 300 bhp	1100 kg / 2425 lb	200 W/kg / 8.08 lb/hp
Lotus Exige GT3 2006	202.1 kW / 271 bhp	980 kg / 2160 lb	206 W/kg / 7.97 lb/hp
Chevrolet Corvette C6 2008	321 kW / 430 bhp	1441 kg / 3177 lb	223 W/kg / 7.39 lb/hp
Nissan GT-R R35 3.6L Turbo V6	406 kW / 545 bhp	1779 kg / 3922 lb	228 W/kg / 7.20 lb/hp
Dodge Charger SRT Hellcat 6.2L Hemi V8	527 kW / 707 bhp	2075 kg / 4575 lb	254 W/kg / 6.47 lb/hp
Chevrolet Corvette C6 Z06	376 kW / 505 bhp	1421 kg / 3133 lb	265 W/kg / 6.2 lb/hp
Porsche 911 GT2 2007	390 kW / 523 bhp	1440 kg / 3200 lb	271 W/kg / 6.1 lb/hp
Lamborghini Murciélago LP 670-4 SV 2009	493 kW / 661 bhp	1550 kg / 3417 lb	318 W/kg / 5.17 lb/hp
Mercedes-Benz C-Coupé DTM touring car 2012	343 kW / 460 bhp	1110 kg / 2,447 lb	309 W/kg / 5.32 lb/hp
Sector111 Drakan Spyder	321 kW / 430 bhp	907 kg / 2000 lb	354 W/kg / 4.65 lb/hp
McLaren F1 GT 1997	467.6 kW / 627 bhp	1220 kg / 2690 lb	403 W/kg / 4.3 lb/hp
BAC Mono 2011	213 kW / 285 bhp	540 kg / 1190 lb	394 W/kg / 4.18 lb/hp
Porsche 918 Spyder	661 kW / 887 bhp	1656 kg / 3650 lb	399 W/kg / 4.16 lb/hp
Lancia Delta S4 group B 1985	350 kW / 480 bhp	890 kg / 1,962 lb	393 W/kg / 4.08 lb/hp
Ariel Atom 3S 2014	272 kW / 365 bhp	639 kg / 1400 lb	426 W/kg / 3.84 lb/hp
Bombardier Dash 8 Q400 turboprop airliner	7,562 kW / 10,142 bhp	17,185 kg / 37,888 lb	440 W/kg / 3.7 lb/hp
Ferrari LaFerrari	708 kW / 950 bhp	1585 kg / 3495 lb	447 W/kg / 3.68 lb/hp

McLaren P1 2013	673 kW / 903 bhp	1490 kg / 3280 lb	452 W/kg / 3.63 lb/hp
Supermarine Spitfire Fighter aircraft 1936	1,096 kW / 1,470 bhp	2,309 kg / 5,090 lb	475 W/kg / 3.46 lb/hp
Messerschmitt Bf 109 Fighter aircraft 1935	1,085 kW / 1,455 bhp	2,247 kg / 4,954 lb	483 W/kg / 3.40 lb/hp
Thunderbolt Land speed record car	3504 kW / 4700 bhp	7 t / 15432 lb	500 W/kg / 3.28 lb/hp
Ferrari FXX 2005	597 kW / 801 bhp	1155 kg / 2546 lb	517 W/kg / 3.18 lb/hp
Polaris Industries Assault Snowmobile 2009	115 kW / 154 bhp	221 kg / 487 lb	523 W/kg / 3.16 lb/hp
Audi R10 TDI Le Mans Prototype 2006	485 kW / 650 bhp	925 kg / 2,039 lb	524 W/kg / 3.13 lb/hp
Ultima GTR 720 2006	536.9 kW / 720 bhp	920 kg / 2183 lb	583 W/kg / 3.03 lb/hp
Honda CBR1000RR 2009	133 kW / 178 bhp	199 kg / 439 lb	668 W/kg / 2.46 lb/hp
Ariel Atom 500 V8 2011	372 kW / 500 bhp	550 kg / 1212 lb	676.3 W/kg / 2.47 lb/hp
BMW S1000RR 2009	144 kW / 193 bhp	207.7 kg / 458 lb	693.3 W/kg / 2.37 lb/hp
Peugeot 208 T16 Pikes Peak 2013	652 kW / 875 bhp	875 kg / 1930 lb	745 W/kg / 2.21 lb/hp
Koenigsegg One:1 2015	1000 kW / 1341 bhp	1310 kg / 2888 lb	763 W/kg / 2.15 lb/hp
Nissan R90C Group C 1990	746 kW / 1000 bhp	900 kg / 1984 lb	829 W/kg / 1.98 lb/hp
Ducati 1199 Panigale R (WSB) 2012	151 kW / 202 bhp	165 kg / 364 lb	915 W/kg / 1.80 lb/hp
KillaCycle Drag racing electric motorcycle	260 kW / 350 bhp	281 kg / 619 lb	925 W/kg / 1.77 lb/hp
MTT Turbine Superbike 2008	213.3 kW / 286 bhp	227 kg / 500 lb	940 W/kg / 1.75 lb/hp
Vyrus 987 C3 4V V supercharged motorcycle 2010	157.3 kW / 211 bhp	158 kg / 348.3 lb	996 W/kg / 1.65 lb/hp
Kawasaki H2R Motorcycle 2015	223 kW / 300 bhp	216 kg / 476 lb	1032 W/kg / 1.43 lb/hp
BMW Williams FW27 Formula One 2005	690 kW / 925 bhp	600 kg / 1323 lb	1150 W/kg / 1.58 lb/hp
Honda RC211V MotoGP 2004-6	176.73 kW / 237 bhp	148 kg / 326 lb	1194 W/kg / 1.37 lb/hp
Boeing 747-300[dead link] at Mach 0.84 cruise, 35,000 ft altitude	245 MW / 328,656 bhp	178.1 t / 392,800 lb	1376 W/kg / 1.20 lb/hp
John Force Racing Funny Car NHRA Drag Racing 2008	5,963.60 kW / 8,000 bhp	1043 kg / 2,300 lb	5717 W/kg / 0.30 lb/hp

Human

Power to weight ratio is important in cycling, since it determines acceleration and the speed during hill climbs. Since a cyclist's power to weight output decreases with fatigue, it is normally discussed with relation to the length of time that he or she maintains that power. A professional cyclist can produce over 20 W/kg as a 5-second maximum.

Wankel Engine

A cut-away of a Wankel engine shown at the Deutsches Museum in Munich, Germany

The Mazda RX-8, a sports car powered by a Wankel engine

Norton Classic air-cooled twin-rotor motorcycle

The Wankel engine is a type of internal combustion engine using an eccentric rotary design to convert pressure into rotating motion. In contrast to the more common reciprocating piston designs, the Wankel engine delivers advantages of simplicity, smoothness, compactness, high revolutions per minute, and a high power-to-weight ratio. The engine is commonly referred to as a rotary engine, although this name applies also to other completely different designs. All parts rotate moving in one direction, as opposed to the common reciprocating piston engine which has pistons violently changing direction. The four-stroke cycle occurs in a moving combustion chamber between the inside of an oval-like epitrochoid-shaped housing, and a rotor that is similar in shape to a Reuleaux triangle with sides that are somewhat flatter.

Concept and Design

The design was conceived by German engineer Felix Wankel. Wankel received his first patent for the engine in 1929, began development in the early 1950s at NSU, and completed a working prototype in 1957. NSU subsequently licensed the design to companies around the world, who have continually added improvements. The engines produced are of spark ignition, with compression ignition engines only in research projects.

The Wankel engine has the advantages of compact design and low weight over the most commonly used internal combustion engine employing reciprocating pistons. These advantages have given rotary engine applications in a variety of vehicles and devices, including: automobiles, motorcycles, racing cars, aircraft, go-karts, jet skis, snowmobiles, chain saws, and auxiliary power units. The point of power to weight has been reached of under one pound weight per horsepower output.

History

Early Developments

The first DKM Wankel engine designed by Felix Wankel, the DKM 54 (Drehkolbenmotor), at the Deutsches Museum in Bonn, Germany: the rotor and its housing spin.

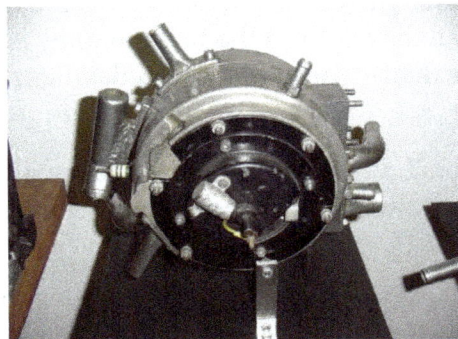

The first KKM Wankel Engine designed by Hanns Dieter Paschke, the NSU KKM 57P (Kreiskolbenmotor), at Autovision und Forum, Germany: the rotor housing is static.

In 1951, NSU Motorenwerke AG in Germany began development of the engine with two models being developed. The first, the DKM motor, was developed by the engineer Felix Wankel. The second, the KKM motor, was developed by Hanns Dieter Paschke, which was adopted forming the modern Wankel engine. The Wankel engine design used today was not designed by Felix Wankel. Titling the engine "the Paschke engine" could be considered to be more apt.

The basis of the DKM type of motor is that both the rotor and the housing spin around on separate axes. The DKM motor reached higher revolutions per minute and was more naturally balanced. However, the engine needed to be stripped to change the spark plugs and contained more parts. The KKM engine is simpler, having a fixed housing.

The first working prototype, DKM 54, produced 21 horsepower and ran on February 1, 1957, at the NSU research and development department Versuchsabteilung TX. The KKM 57 (the Wankel rotary engine, Kreiskolbenmotor) was constructed by NSU engineer Hanns Dieter Paschke in 1957 without the knowledge of Felix Wankel, who later remarked "you have turned my race horse into a plow mare".

Licenses Issued

In 1960, NSU, the firm that employed the two inventors, and the US firm Curtiss-Wright, signed a joint agreement. NSU were to concentrate on low- and medium-powered Wankel engine development and Curtiss-Wright developing high-powered engines, including aircraft engines of which Curtiss-Wright had decades of experience designing and producing. Curtiss-Wright recruited Max Bentele to head their design team.

Many manufacturers signed license agreements for development, attracted by the smoothness, quiet running, and reliability emanating from the uncomplicated design. Amongst them were Alfa Romeo, American Motors, Citroen, Ford, General Motors, Mazda, Mercedes-Benz, Nissan, Porsche, Rolls-Royce, Suzuki, and Toyota. In the United States in 1959, under license from NSU, Curtiss-Wright pioneered improvements in the basic engine design. In Britain, in the 1960s, Rolls Royce's Motor Car Division pioneered a two-stage diesel version of the Wankel engine.

Citroën did much research, producing the M35, GS Birotor and RE-2 (fr) Helicopter using engines produced by Comotor, a joint venture of Citroën and NSU. General Motors seemed to have concluded the Wankel engine was slightly more expensive to build than an equivalent reciprocating engine. General Motors claimed to have solved the fuel economy issue, but failed in obtaining in a concomitant way to acceptable exhaust emissions. Mercedes-Benz fitted a Wankel engine in their C111 concept car.

Deere & Company designed a version that was capable of using a variety of fuels. The design was proposed as the power source for United States Marine Corps combat vehicles and other equipment in the late 1980s.

In 1961, the Soviet research organization of NATI, NAMI, and VNIImotoprom commenced development creating experimental engines with different technologies. Soviet automobile manufacturer AvtoVAZ also experimented in Wankel engine design without a license, introducing a limited number of engines in some cars.

Despite much research and development throughout the world, only Mazda has produced Wankel engines in large quantities.

Developments for Motorcycles

In Britain, Norton Motorcycles developed a Wankel rotary engine for motorcycles, based on the Sachs air-cooled rotor Wankel that powered the DKW/Hercules W-2000 motorcycle, this two-rotor engine was included in their Commander and F1. Norton improved on the Sachs's air cooling, introducing a plenum chamber. Suzuki also made a production motorcycle powered by a Wankel engine, the RE-5, using ferroTiC alloy apex seals and an NSU rotor in a successful attempt to prolong the engine's life.

Developments for Cars

Mazda and NSU signed a study contract to develop the Wankel engine in 1961 and competed to bring the first Wankel-powered automobile to market. Although Mazda produced an experimental Wankel that year, NSU was first with a Wankel automobile for sale, the sporty NSU Spider in 1964; Mazda countered with a display of two- and four-rotor Wankel engines at that year's Tokyo Motor Show. In 1967, NSU began production of a Wankel-engined luxury car, the Ro 80. However, NSU had not produced reliable apex seals on the rotor, unlike Mazda and Curtiss-Wright. NSU had problems with apex seals' wear, poor shaft lubrication, and poor fuel economy, leading to frequent engine failures, not solved until 1972, which led to large warranty costs curtailing further NSU Wankel engine development. This premature release of the new Wankel engine gave a poor reputation for all makes and even when these issues were solved in the last engines produced by NSU in the second half of the '70s, sales did not recover. Audi, after the takeover of NSU, built in 1979 a new KKM 871 engine with side intake ports and 750 cc per chamber, 170 HP at 6,500 rpm, and 220 Nm at 3,500 rpm. The engine was installed in an Audi 100 hull they named "Audi 200", but the engine was not mass-produced.

Mazda's first Wankel engine, at the Mazda Museum in Hiroshima, Japan

Mazda, however, claimed to have solved the apex seal problem, and operated test engines at high speed for 300 hours without failure. After years of development, Mazda's first Wankel engine car was the 1967 Cosmo 110S. The company followed with a number of Wankel ("rotary" in the company's terminology) vehicles, including a bus and a pickup truck. Customers often cited the cars' smoothness of operation. However, Mazda chose a method to comply with hydrocarbon emission standards that, while less expensive to produce, increased fuel consumption. Unfortunately for Mazda, this was introduced immediately prior to a sharp rise in fuel prices. Curtiss-Wright produced the RC2-60 engine which was comparable to a V8 engine in performance and fuel consumption. Unlike NSU, by 1966 Curtiss-Wright had solved the rotor sealing issue with seals lasting 100,000 miles.

Mazda later abandoned the Wankel in most of their automotive designs, continuing to use the engine in their sports car range only, producing the RX-7 until August 2002. The company normally used two-rotor designs. A more advanced twin-turbo three-rotor engine was fitted in the 1991 Eunos Cosmo sports car. In 2003, Mazda introduced the Renesis engine fitted in the RX-8. The Renesis engine relocated the ports for exhaust from the periphery of the rotary housing to the sides, allowing for larger overall ports, better airflow, and further power gains. Some early Wankel engines had also side exhaust ports, the concept being abandoned because of carbon buildup in ports and the sides of the rotor. The Renesis engine solved the problem by using a keystone scraper side seal, and approached the thermal distortion difficulties by adding some parts made of ceramics. The Renesis is capable of 238 hp (177 kW) with improved fuel economy, reliability, and lower emissions than previous Mazda rotary engines, all from a nominal 1.3 L displacement. However, this was not enough to meet more stringent emissions standards. Mazda ended production of their Wankel engine in 2012 after the engine failed to meet the improved Euro 5 emission standards, leaving no automotive company selling a Wankel-powered vehicle. The company is continuing development of the next generation of Wankel engines, the SkyActiv-R with a new rear wheel drive sports car model announced in October 2015 although with no launch date given. Mazda states that the SkyActiv-R solves the three key issues with previous rotary engines: fuel economy, emissions and reliability. Mazda announced the introduction of the series-hybrid Mazda2 EV car using a Wankel engine as a range extender, however no date of introduction has been announced.

1972 GM Rotary engine cutaway shows twin-rotors

American Motors (AMC) was so convinced "... that the rotary engine will play an important role as a powerplant for cars and trucks of the future....", that the chairman, Roy D. Chapin Jr., of the smallest U.S. automaker signed an agreement in February 1973, after a year's negotiations, to build Wankels for both passenger cars and Jeeps, as well as the right to sell any rotary engines it produced to other companies. American Motors' president, William Luneburg, did not expect dramatic development through to 1980, however Gerald C. Meyers, AMC's vice president of the engineering product group, suggested that AMC should buy the engines from Curtiss-Wright before developing its own Wankel engines and predicted a total transition to rotary power by 1984. Plans called for the engine to be used in the AMC Pacer, but development was pushed back. American Motors designed the unique Pacer around the engine. By 1974, AMC had decided to purchase the General Motors Wankel instead of building an engine in-house. Both General Motors and AMC confirmed the relationship would benefit in marketing the new engine, with AMC claiming that the General Motors' Wankel achieved good fuel economy. However, General Motors' engines had not reached production when the Pacer was launched onto the market. The 1973 oil crisis played a part in frustrating the uptake of the Wankel engine. Rising fuel prices and talk about proposed US emission standards legislation also added to the concerns.

By 1974, General Motors R&D had not succeeded in producing a Wankel engine meeting both the emission requirements and good fuel economy, leading the company to consider cancelling the project. As General Motors managers were cancelling the Wankel project, the R&D team released only partly the results of their most recent research, which claimed to have solved the fuel economy problem, and building reliable engines with a lifespan above 530,000 miles. These findings were not taken into account when the cancellation order was issued. The cancellation of General Motors' Wankel project required AMC to reconfigure the Pacer to house its venerable AMC straight-6 engine driving the rear-wheels.

In 1974, the Soviets created a special engine design bureau, which, in 1978, designed an engine designated as "VAZ-311". In 1980, the company commenced delivery of the VAZ-411 twin-rotor Wankel engine in VAZ-2106s and Lada cars, with about 200 manufactured. Most of the production went to the security services. The next models were the VAZ-4132 and VAZ-415. Aviadvigatel, the Soviet aircraft engine design bureau, is known to have produced Wankel engines with electronic injection for aircraft and helicopters, though little specific information has surfaced.

Ford conducted research in Wankel engines, resulting in patents granted: GB 1460229, 1974, method for fabricating housings; US 3833321 1974, side plates coating; US 3890069, 1975, housing coating; CA 1030743, 1978: Housings alignment; CA 1045553, 1979, Reed-Valve assembly. Mr. Henry Ford II 1972 statement regarding the production of a Ford Wankel engine was: 'The Rotary probably won't replace the piston in my lifetime". (Harris Edward Dark, 'the Wankel rotary engine, introduction and guide'. Indiana University Press 1974, pag 80, ISBN 0-253-19021-5).

Design

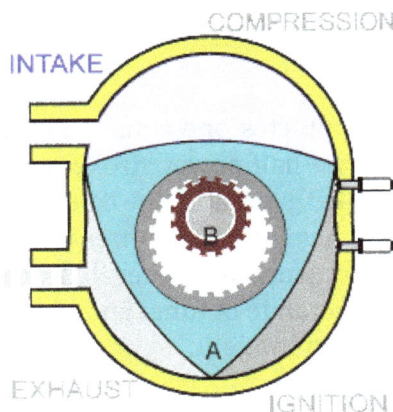

The Wankel KKM motorcycle: The "A" marks one of the three apices of the rotor. The "B" marks the eccentric shaft and the white portion is the lobe of the eccentric shaft. The shaft turns 3 times for each rotation of the rotor around the lobe and once for each orbital revolution around the eccentric shaft.

In the Wankel engine, the four strokes of an Otto cycle piston engine occur in the space between a three-sided symmetric rotor and the inside of a housing. In each rotor of the Wankel engine, the oval-like epitrochoid-shaped housing surrounds a rotor which is triangular with bow-shaped flanks (often confused with a Reuleaux triangle, a three-pointed curve of constant width, but with the bulge in the middle of each side a bit more flattened). The theoretical shape of the rotor between the fixed corners is the result of a minimization of the volume of the geometric combustion chamber and a maximization of the compression ratio, respectively. The symmetric curve connecting two arbitrary apexes of the rotor is maximized in the direction of the inner housing shape with the constraint that it not touch the housing at any angle of rotation (an arc is not a solution of this optimization problem).

The central drive shaft, called the "eccentric shaft" or "E-shaft", passes through the center of the rotor and is supported by fixed bearings. The rotors ride on eccentrics (analogous to crankpins) integral to the eccentric shaft (analogous to a crankshaft). The rotors both rotate around the eccentrics and make orbital revolutions around the eccentric shaft. Seals at the corners of the rotor seal against the periphery of the housing, dividing it into three moving combustion chambers. The rotation of each rotor on its own axis is caused and controlled by a pair of synchronizing gears A fixed gear mounted on one side of the rotor housing engages a ring gear attached to the rotor and ensures the rotor moves exactly 1/3 turn for each turn of the eccentric shaft. The power output of the engine is not transmitted through the synchronizing gears. The force of gas pressure on the rotor (to a first approximation) goes directly to the center of the eccentric, part of the output shaft...

The easiest way to visualize the action of the engine in the animation at left is to look not at the rotor itself, but the cavity created between it and the housing. The Wankel engine is actually a variable-volume progressing-cavity system. Thus, there are three cavities per housing, all repeating the same cycle. Points A and B on the rotor and E-shaft turn at different speeds—point B circles three times as often as point A does, so that one full orbit of the rotor equates to three turns of the E-shaft.

As the rotor rotates orbitally revolving, each side of the rotor is brought closer to and then away from the wall of the housing, compressing and expanding the combustion chamber like the strokes of a piston in a reciprocating piston engine. The power vector of the combustion stage goes through the center of the offset lobe.

While a four-stroke piston engine completes one combustion stroke per cylinder for every two rotations of the crankshaft (that is, one-half power stroke per crankshaft rotation per cylinder), each combustion chamber in the Wankel generates one combustion stroke per driveshaft rotation, i.e. one power stroke per rotor orbital revolution and three power strokes per rotor rotation. Thus, the power output of a Wankel engine is generally higher than that of a four-stroke piston engine of similar engine displacement in a similar state of tune; and higher than that of a four-stroke piston engine of similar physical dimensions and weight.

Wankel engines generally can sustain much higher engine revolutions than reciprocating engines of similar power output. This is due to the smoothness inherent in circular motion, and the absence of highly stressed parts such as crankshafts, camshafts or connecting rods. Eccentric shafts do not have the stress related contours of crankshafts. The maximum revolutions of a rotary engine is limited by tooth load on the synchronizing gears. Hardened steel gears are used for extended operation above 7000 or 8000 rpm. Mazda Wankel engines in auto racing are operated above 10,000 rpm. In aircraft they are used conservatively, up to 6500 or 7500 rpm. However, as gas pressure participates in seal efficiency, racing a Wankel engine at high rpm under no load conditions can destroy the engine.

National agencies that tax automobiles according to displacement and regulatory bodies in automobile racing variously consider the Wankel engine to be equivalent to a four-stroke piston engine of 1.5 to 2 times the displacement. Some racing series have banned the Wankel altogether.

Engineering

Apex seals, left NSU Ro 80 Serie and Research and right Mazda 12A and 13B

Left Mazda old L10A camber axial cooling, middle Audi NSU EA871 axial water cooling only hot bow, right Diamond Engines Wankel radial cooling only in the hot bow

Felix Wankel managed to overcome most of the problems that made previous rotary engines fail by developing a configuration with vane seals that had a tip radius equal to the amount of "oversize" of the rotor housing form, as compared to the theoretical epitrochoid, to minimize radial apex seal motion plus introducing a cylindrical gas-loaded apex pin which abutted all sealing elements to seal around the three planes at each rotor apex.

In the early days, special, dedicated production machines had to be built for different housing dimensional arrangements. However, patented design such as U.S. Patent 3,824,746, G. J. Watt, 1974, for a "Wankel Engine Cylinder Generating Machine", U.S. Patent 3,916,738, "Apparatus for machining and/or treatment of trochoidal surfaces" and U.S. Patent 3,964,367, "Device for machining trochoidal inner walls", and others, solved the issue.

Rotary engines have a problem not found in reciprocating four-stroke engines in that the block housing has intake, compression, combustion, and exhaust occurring at fixed locations around the housing. In contrast, reciprocating engines perform these four strokes in one chamber, so that extremes of "freezing" intake and "flaming" exhaust are averaged and shielded by a boundary layer from overheating working parts. The use of Heat Pipes in an Air Cooled Wankel was proposed by the University of Florida to overcome this uneven heating of the block housing. Pre-heating of certain housing sections with exhaust gas improved performance and fuel economy, also reducing wear and emissions.

The boundary layer shields and the oil film act as thermal insulation, leading to a low temperature of the lubricating film (maximum ~200 °C or 392 °F on a water-cooled Wankel engine. This gives a more constant surface temperature. The temperature around the spark plug is about the same as the temperature in the combustion chamber of a reciprocating engine. With circumferential or axial flow cooling, the temperature difference remains tolerable.

During research in the 1950s and 1960s problems arose. For a while, engineers were faced with what they called "chatter marks" and "devil's scratch" in the inner epitrochoid surface. They discovered that the origin was in the apex seals reaching a resonating vibration, and solved the problem by reducing the thickness and weight of apex seals. Scratches disappeared after the introduction of more compatible materials for seals and housing coatings. Another early problem of the build-up of cracks in the stator surface near the plug hole was eliminated by installing the spark plugs in a separate metal insert/ copper sleeve in the housing instead of plug being screwed directly into the block housing. Toyota found that substituting a glow-plug for the leading site spark plug improved low rpm, part load, specific fuel consumption by 7% and also emissions and idle. A later alternative solution to spark plug boss cooling was provided with a variable coolant velocity scheme for water-cooled rotaries, which has had widespread use, being patented by Curtiss-Wright, with the last-listed for better air-cooled engine spark plug boss cooling. These approaches did not require a high conductivity copper insert, but did not preclude its use. Ford tested a rotary engine with the plugs placed in the side plates, instead of the usual placement in the housing working surface (CA 1036073, 1978).

Four-stroke reciprocating engines are less suitable for use with hydrogen fuel. The hydrogen can misfire on hot parts like the exhaust valve and spark plugs. Another problem concerns the hydrogenate attack on the lubricating film in reciprocating engines. In a Wankel engine, this problem is circumvented by using a ceramic apex seal against a ceramic surface; there is no oil film to suffer hydrogenate attack. The piston shell must be lubricated and cooled with oil. This substantially increases the lubricating oil consumption in a four-stroke hydrogen engine.

Increasing the displacement and power of a rotary engine by adding more rotors to a basic design is simple, but a limit may exist in the number of rotors, as power output is channeled through the last rotor shaft, with all the stresses of the whole engine present at this point. For engines with more than two rotors, the approach of coupling two bi-rotor sets by a serrate coupling between the two rotor sets has been tested successfully.

Research in the United Kingdom under the SPARCS (Self-Pressurising-Air Rotor Cooling System) project, found that idle stability and economy was obtained by supplying an ignitable mix to only one rotor in a multi rotor engine in a forced-air cooled rotor, similar to the Norton air cooled designs.

The Wankel engine's drawbacks of inadequate lubrication and cooling in ambient temperatures, short engine lifespan, high emissions and low fuel efficiencies were addressed by Norton rotary engine specialist David Garside, developing three patented systems in 2016.

- SPARCS.

- Compact-SPARCS.

- CREEV (Compound Rotary Engine for Electric Vehicles)

SPARCS and Compact-SPARCS provides superior heat rejection and efficient thermal balancing to optimise lubrication. This results in reduced engine wear prolonging engine life. As described in Unmanned Systems Technology Magazine "SPARCS uses a sealed rotor cooling circuit consisting of a circulating centrifugal fan and a heat exchanger to reject the heat. This is self-pressurised by capturing the blow-by past the rotor side gas seals from the working chambers." CREEV is a 'exhaust reactor' that consumes unburnt exhaust products delivering lower emissions and improved fuel efficiency. All three patents are currently licensed to UK based engineers, AIE (UK) Ltd.

Materials

Unlike a piston engine, where the cylinder is heated by the combustion process and then cooled by the incoming charge, Wankel rotor housings are constantly heated on one side and cooled on the other, leading to high local temperatures and unequal thermal expansion. While this places high demands on the materials used, the simplicity of the Wankel makes it easier to use alternative materials, such as exotic alloys and ceramics. With water cooling in a radial or axial flow direction, with the hot water from the hot bow heating the cold bow, the thermal expansion remains tolerable; top engine temperature has been reduced to 129 °C (264 °F), with a maximum temperature difference between engine parts of 18 °C (64 °F) by the use of heat pipes around the housing and in side plates as a cooling means.

Among the alloys cited for Wankel housing use are A-132, Inconel 625, and 356 treated to T6 hardness. Several materials have been used for plating the housing working surface, Nikasil being one. Citroen, Mercedes-Benz, Ford, A P Grazen and others applied for patents in this field. For the apex seals, the choice of materials has evolved along with the experience gained, from carbon alloys, to steel, ferrotic, and other materials. The combination between housing plating and apex and side seals materials was determined experimentally, to obtain the best duration of both seals and housing cover. For the shaft, steel alloys with little deformation on load are preferred, the use of Maraging steel has been proposed for this.

Lead is a solid lubricant with leaded gasoline linked to a reduced wear of seals and housings. Leaded gasoline was the predominant type available in the first years of the Wankel engine's development. The first engines had the oil supply calculated with consideration of gasoline's lubricating qualities. Leaded gasoline was phased out, with Wankel engines needing an increased mix of oil in the gasoline to provide lubrication to critical engine parts. Experienced users advise, even in engines with electronic fuel injection, adding at least 1% of oil directly to gasoline as a safety measure in case the pump supplying oil to combustion chamber related parts fails or sucks in air. A SAE paper by David Garside extensively describes Norton's choices of materials and cooling fins.

Several approaches involving solid lubricants were tested, and even the addition of MoS2, one cc per liter of fuel is advised (LiquiMoly). Many engineers agree that the addition of oil to gasoline as in old two-stroke engines is a safer approach for engine reliability than an oil pump injecting into the intake system or directly to the parts requiring lubrication. A combined oil-in-fuel plus oil metering pump is always possible.

Sealing

Early engine designs had a high incidence of sealing loss, both between the rotor and the housing and also between the various pieces making up the housing. Also, in earlier model Wankel engines, carbon particles could become trapped between the seal and the casing, jamming the engine and requiring a partial rebuild. It was common for very early Mazda engines to require rebuilding after 50,000 miles (80,000 km). Further sealing problems arose from the uneven thermal distribution within the housings causing distortion and loss of sealing and compression. This thermal distortion also caused uneven wear between the apex seal and the rotor housing, evident on higher mileage engines. The problem was exacerbated when the engine was stressed before reaching operating temperature. However, Mazda rotary engines solved these initial problems. Current engines have nearly 100 seal-related parts.

The problem of clearance for hot rotor apexes passing between the axially closer side housings in the cooler intake lobe areas was dealt with by using an axial rotor pilot radially inboard of the oils seals, plus improved inertia oil cooling of the rotor interior (C-W US 3261542, C. Jones, 5/8/63, US 3176915, M. Bentele, C. Jones. A.H. Raye. 7/2/62), and slightly "crowned" apex seals (different height in the center and in the extremes of seal).

Modern Wankel engines have fully sealed mainshaft cases. Many engines do not require oil changes, as the oil is not contaminated by the combustion process.

Fuel Economy and Emissions

The shape of the Wankel combustion chamber is more resistant to preignition operating on lower-octane rating gasoline than a comparable piston engine. The combustion chamber shape may also lead to relatively incomplete combustion of the air-fuel charge. This would result in a larger amount of unburned hydrocarbons released into the exhaust. The exhaust is, however, relatively low in NOx emissions, as combustion temperatures are lower than in other engines, and also because of some inherent exhaust gas recirculation (EGR) in early engines. Sir Harry Ricardo showed in the 1920s that for every 1% increase in the proportion of exhaust gas in the admission mix, there is a 7 °C reduction in flame temperature. This allowed Mazda to meet the United States Clean Air Act of 1970 in 1973, with a simple and inexpensive "thermal reactor", which is an enlarged chamber in the exhaust manifold. By decreasing the air-fuel ratio, unburned hydrocarbons (HC) in the exhaust would support combustion in the thermal reactor. Piston-engine cars required expensive catalytic converters to deal with both unburned hydrocarbons and NOx emissions. This inexpensive solution increased fuel consumption, which was already a weak point for the Wankel engine, at the same time that the oil crisis of 1973 raised the price of gasoline. Toyota discovered that injection of air into the exhaust port zone improved fuel economy and emissions. The best results were obtained with holes in the side plates, doing it in the exhaust duct had no noticeable influence. The use of a three-stage catalysts, with air supplied in the middle, as for 2-Stroke piston engines, also proved good.

Mazda improved the fuel efficiency of the thermal reactor system by 40% by the time of introduction of the RX-7 in 1978. However, Mazda eventually shifted to the catalytic converter system. According to the Curtiss-Wright research, the factor that controls the amount of unburned HC in the exhaust is the rotor surface temperature, with higher temperatures producing less HC. Curtiss-Wright showed also that the rotor can be widened, keeping the rest of engine's architecture unchanged, thus reducing friction losses and increasing displacement and power output. The limiting factor for this widening being mechanical considerations, especially shaft deflection at high rotative speeds. Quenching is the dominant source of HC at high speeds, and leakage at low speeds.

Automobile Wankel rotary engines are capable of high-speed operation. However, it was shown that an early opening of the intake port, longer intake ducts, and a greater rotor eccentricity can increase torque at low rpm. The shape and positioning of the recess in the rotor, which forms most of the combustion chamber, influences emissions and fuel economy. The results in terms of fuel economy and exhaust emissions varies depending on the shape of the combustion recess which is determined by the placement of spark plugs per chamber of an individual engine.

Mazda's RX-8 car with the Renesis engine, fuel economy met California State requirements, including California's low emissions vehicle (LEV) standards. This was achieved by a number of innovations. The exhaust ports, which in earlier Mazda rotaries were located in the rotor housings, were moved to the sides of the combustion chamber. This solved the problem of the earlier ash buildup in the engine, and thermal distortion problems of side intake and exhaust ports. A scraper seal was added in the rotor sides, and by use of some ceramic-made parts in the engine. This approach allowed Mazda to eliminate overlap between intake and exhaust port openings, while simultaneously increasing the exhaust port area. The side port trapped the unburned fuel in the chamber, decreased the oil consumption, and improved the combustion stability in the low-speed and light load range. The HC emissions from the side exhaust port Wankel engine are 35–50% less than those from the peripheral exhaust port Wankel engine, because of near zero intake and exhaust port opening overlap. Peripheral ported rotary engines have a better mean effective pressure, especially at high rpm and with a rectangular shaped intake port. However, the RX-8 was not improved to meet EuroV emission regulations and was discontinued in 2012.

Mazda is still continuing development of the next generation of Wankel engines. The company is researching engine laser ignition, eliminating conventional spark plugs, direct fuel injection and sparkless HCCI ignition. These lead to greater rotor eccentricity, equating to a longer stroke in a reciprocating engine, for improved elasticity and low revolutions per minute torque. Research by T Kohno proved that installing a glow-plug in the combustion chamber improved 7% part load and low revolutions per minute fuel economy. These innovations promise to improve fuel consumption and emissions. To improve fuel efficiency further, Mazda is looking at using the Wankel as a range extender in series-hybrid cars announcing a prototype, the Mazda2 EV, for press evaluation in November 2013. This configuration improves fuel efficiency and emissions. As a further advantage, running a Wankel engine at a constant speed gives greater engine life. Keeping to a near constant, or narrow band, of revolutions eliminates, or vastly reduces, many of the disadvantages of the Wankel engine.

In 2015 a new system to reduce emissions and increase fuel efficiency with Wankel Engines was developed by UK based engineers AIE (UK) Ltd following a licensing agreement to utilise patents from Norton rotary engine creator, David Garside. The CREEV system (Compound Rotary Engine for Electric Vehicles) uses a secondary rotor to extract energy from the exhaust, consuming unburnt exhaust products while expansion occurs in the secondary rotor stage, thus reducing overall emissions and fuel costs by recouping exhaust energy that would otherwise be lost. By expanding the exhaust gas to near atmospheric pressure, Garside also ensured the engine exhaust would remain cooler and quieter. AIE (UK) Ltd are now utilising this patent to develop hybrid power units for automobiles and unmanned aerial vehicles.

Laser Ignition

As the rotor's apex seals pass over the spark plug hole, a small amount of compressed charge can be lost from the charge chamber to the exhaust chamber, entailing fuel in the exhaust, reducing efficiency, and giving high emissions. The spark plug needs to be located outside the combustion chamber to enable the apex of the rotor to sweep past. These points may be overcome by using laser ignition, eliminating traditional spark plugs, which may give a narrow slit in the motor housing the rotor apex seals can fully cover with no loss of compression from adjacent chambers. This concept had a precedent in the Glow Plug used by Toyota (SAE paper 790435), and the SAE paper 930680, by D Hixon et al, on 'Catalytic Glow Plugs in the JDTI Stratified Charge Rotary Engine'. The laser plug can fire its spark through the narrow slit. Laser spark plugs can fire deep into the combustion chamber using multiple sparks. Direct fuel injection of which the Wankel engine is suited, combined with laser ignition in single or multiple laser plugs, has shown to enhance the motor even further reducing the disadvantages.

Homogeneous Charge Compression Ignition (HCCI)

Homogeneous charge compression ignition (HCCI) is where the fuel/air intake is a pre-mixed lean air-fuel mixture then compressed to the point of auto-ignition. Electronic spark ignition is eliminated. Gasoline engines combine homogeneous charge (HC) with spark ignition (SI), abbreviated as HCSI. Diesel engines combine stratified charge (SC) with compression ignition (CI), abbreviated as SCCI. HCCI engines achieve gasoline engine-like emissions with compression ignition engine-like efficiency. HCCI engines achieve low levels of nitrogen oxide emissions (NO x) without a catalytic converter. However, unburned hydrocarbon and carbon monoxide emissions still require treatment to reach automotive emission regulations.

Mazda have undertaken research on HCCI ignition for its SkyActiv-R rotary engine project using research from its SkyActiv Generation 2 program. A constraint of rotary engines is the need to locate the spark plug outside the combustion chamber to enable the rotor to sweep past. Mazda confirmed this has been solved in the SkyActiv-R project. Rotaries generally have high compression ratios leaning the design to ease of homogeneous charge compression ignition (HCCI) adoption.

Compression Ignition Rotary

Rolls Royce R6 two stage rotary compression ignition engine

There has been research into compression ignition engines and the burning of diesel heavy fuel in rotaries using spark ignition. The basic design parameters of Wankel preclude obtaining a Compression Ratio higher than 15:1 or 17:1 in a practical engine, but attempts are continuously made to produce a compression ignition Wankel. The Rolls-Royce and Yanmar compression ignition approach was a two stage unit, one rotor acting as compressor, combustion taking place in the other. Conversion of an standard 294 cc per chamber Spark Ignition unit into heavy fuel was described in SAE paper 930682, by L Louthan. SAE paper 930683, by D Eiermann, resulted in the Wankel SuperTec line of compression ignition rotary engines.

Compression ignition engine research is being undertaken by Pratt & Whitney Rocketdyne who were commissioned by DARPA to develop a compression ignition Wankel engine for use in a prototype VTOL flying car called the "Transformer". The engine, based on an earlier unmanned aerial vehicle Wankel diesel concept called "Endurocore". plans to utilize Wankel rotors of varying sizes on a shared eccentric shaft to increase efficiency. The engine is claimed to be a 'full-compression, full-expansion, compression ignition-cycle engine'. An October 28, 2010 patent from Pratt & Whitney Rocketdyne, describes a Wankel engine superficially similar to Rolls-Royce's earlier prototype that required an external air compressor to achieve high enough compression for compression ignition -cycle combustion. The design differs from Rolls-Royce's compression ignition rotary mainly by proposing an injector both in the exhaust passage between the combustor rotor and expansion rotor stages, and an injector in the expansion rotor's expansion chamber, for 'afterburning'.

The British company Rotron, who specialise in unmanned aerial vehicle (UAV) applications of Wankel engines have designed and built a unit to operate on heavy fuel for NATO purposes. The engines uses spark ignition. The prime innovation is flame propagation, ensuring the flame burns smoothly across the whole combustion chamber. The fuel is pre-heated to 98 degrees Celsius before injection into the combustion chamber. Four spark plugs are utilised aligned in two pairs. Two spark plugs ignite the fuel charge at the front of the rotor as it moves into the combustion section of the housing. As the rotor moves the fuel charge, the second two fire a fraction of second behind the first pair of plugs igniting near the rear of the rotor at the back of the fuel charge. The drive shaft is water cooled which has a cooling effect on the internals of the rotor. Cooling water also flows around the external of the engine through a gap in the housing, hence cooling the engine from outside and inside eliminating hot spots.

Advantages

NSU Wankel Spider, the first line of cars sold with a rotor Wankel engine

Mazda Cosmo, the first series two rotor Wankel engine sports car

Prime advantages of the Wankel engine are:

- A far higher power to weight ratio than a piston engine (it is approximately one third of the weight of a piston engine of equivalent power output)

- It is approximately one third of the size of a piston engine of equivalent power output

- No reciprocating parts

- Able to reach higher revolutions per minute than a piston engine

- Operates with almost no vibration

- Not prone to engine-knock

- Cheaper to mass-produce as the engine contains fewer parts

- Superior breathing, filling the combustion charge in 270 degrees of mainshaft rotation rather than 180 degrees in a piston engine

- Supplies torques for about two thirds of the combustion cycle rather than one quarter for a piston engine

- Wider speed range gives greater adaptability

- It can use fuels of wider octane ratings

- Does not suffer from "scale effect" to limit its size

- On some Wankel engines the sump oil remains uncontaminated by the combustion process requiring no oil changes. The oil in the mainshaft is totally sealed from the combustion process. The oil for Apex seals and crankcase lubrication is separate. In piston engines the crankcase oil is contaminated by combustion blow-by through the piston rings.

Wankel engines are considerably lighter and simpler, containing far fewer moving parts than piston engines of equivalent power output. Valves or complex valve trains are eliminated by using simple ports cut into the walls of the rotor housing. Since the rotor rides directly on a large bearing on the output shaft, there are no connecting rods and no crankshaft. The elimination of reciprocating mass and the elimination of the most highly stressed and failure prone parts of piston engines gives the Wankel engine high reliability, a smoother flow of power, and a high power-to-weight ratio.

The surface-to-volume-ratio in the moving combustion chamber is so complex that a direct comparison cannot be made between a reciprocating piston engine and a Wankel engine. The flow velocity and the heat losses behave quite differently. Surface temperatures behave absolutely differently; the film of oil in the Wankel engine acts as insulation. Engines with a higher compression ratio have a worse surface-to-volume ratio. The surface-to-volume ratio of a reciprocating piston diesel engine is much poorer than a reciprocating piston gasoline engine, however diesel engines have a higher efficiency factor. Hence, comparing power outputs is a realistic metric. A reciprocating piston engine with equal power to a Wankel will be approximately twice the displacement. When comparing the power-to-weight ratio, physical size or physical weight to a similar power output piston engine, the Wankel is superior.

A four-stroke cylinder produces a power stroke only every other rotation of the crankshaft, with three strokes being pumping losses. This doubles the real surface-to-volume ratio for the four-stroke reciprocating piston engine and the displacement increased. The Wankel, therefore, has higher volumetric efficiency and lower pumping losses through the absence of choking valves. Because of the quasi-overlap of the power strokes that cause the smoothness of the engine and the avoidance of the four-stroke cycle as in a reciprocating engine, the Wankel engine is very quick to react to power increase changes giving a quick delivery of power when the demand arises, especially at higher rpm's. This difference is more pronounced when compared to four-cylinder reciprocating engines and less pronounced when compared to higher cylinder counts.

In addition to the removal of internal reciprocating stresses by virtue of the complete removal of reciprocating internal parts typically found in a piston engine, the Wankel engine is constructed with an iron rotor within a housing made of aluminium, which has a greater coefficient of thermal expansion. This ensures that even a severely overheated Wankel engine cannot seize, as would be likely to occur in an overheated piston engine. This is a substantial safety benefit of use in aircraft. In addition, the absence of valves and valve trains again increases safety. GM tested an Iron Rotor and Iron Housing in their prototype Wankel engines, that worked at higher temperatures with lower specific fuel consumption.

A further advantage of the Wankel engine for use in aircraft is that a Wankel engine generally has a smaller frontal area than a piston engine of equivalent power, allowing a more aerodynamic nose to be designed around the engine. A cascading advantage is the smaller size and less weight of the Wankel engine also allows for savings in airframe construction costs, compared to piston engines of comparable power output.

Wankel engines that operate within their original design parameters are almost immune to catastrophic failure. A Wankel engine that loses compression, cooling or oil pressure will lose a large amount of power and fail over a short period of time. It will, however, usually continue to produce some power during that time, allowing for a safer landing when used in aircraft. Piston engines under the same circumstances are prone to seizing or breaking parts that almost certainly results in catastrophic failure of the engine and instant, total loss of power. For this reason, Wankel engines are very well suited to snowmobiles, which often take users into remote places where a failure could result in frostbite or death, and aircraft, where abrupt failure is likely to lead to a crash or forced landing in a remote place.

From the combustion chamber shape and features, the fuel ON requirements of Wankel engines are lower than in reciprocating piston engines. The maximum road octane number requirements were 82 for a peripheral intake port wankel engine, and less than 70 for a side inlet port engine. From the point of view of oil refiners this may be an industrial advantage in fuel production costs.

Due to a 50% longer stroke duration than a reciprocating four-cycle engine, there is more time to complete the combustion. This leads to greater suitability for direct fuel injection and stratified charge operation. A Wankel rotary engine has stronger flows of air-fuel mixture and a longer operating cycle than a reciprocating engine, realizing concomitantly thorough mixing of hydrogen and air. The result is a homogeneous mixture, and no hot spots in the engine, which is crucial for hydrogen combustion.

Disadvantages

Many of the disadvantages are in ongoing research with some advances greatly reducing negative aspects of the engine. However, the current disadvantages of the Wankel engine in production are the following:

- Rotor sealing. This is still a minor problem as the engine housing has vastly different temperatures in each separate chamber section. The different expansion coefficients of the materials gives a far from perfect sealing. Additionally, both sides of the seals are being exposed to fuel, and the design does not allow for a dedicated lubrication system, as in two-stroke engines. In comparison, a piston engine has all functions of a cycle in the same chamber giving a more stable temperature for piston rings to act against; additionally, only one side of the piston in a (four-stroke) piston engine is being exposed to fuel, allowing for oil to lubricate the cylinders from the other side. To overcome the differences in temperatures between different regions of housing and side and intermediary plates, and the associated thermal dilatation inequities, the use of a heat pipe, transporting heat from the hot to the cold parts of engine, has been shown to reduce, in a small displacement, charge cooled rotor, air-cooled housing wankel engine, the maximal engine temperature from 231 °C to 129 °C, and the maximum difference from a hotter to a colder region of engine, from 159 °C to 18 °C.

- Apex seal lifting. Centrifugal force pushes the apex seal onto the housing surface forming a firm seal. Gaps can develop between the apex seal and troichoid housing in light-load operation when imbalances in centrifugal force and gas pressure occur. In low engine-rpm ranges, or under low-load conditions, gas pressure in the combustion chamber can cause the seal to lift off the surface, resulting in combustion gas leaking into the next chamber. Mazda has identified this problem and has developed a solution. By changing the shape of the troichoid housing, the seals remain flush to the housing. This points to using the engine at sustained higher revolutions eliminating apex seal lift off, in applications such as an electricity generator. In vehicles this leads to series-hybrid applications of the engine.

- Slow combustion. The combustion is slow as the combustion chamber is long, thin, and moving. Flame travel is done almost exclusively in the direction of rotor movement, adding to the Quenching that is the main source of unburned HC at high rpm. The trailing side of the combustion chamber naturally produces a "squeeze stream" that prevents the flame from reaching the chamber trailing edge. Fuel injection in which fuel is injected towards the leading edge of the combustion chamber can minimize the amount of unburnt fuel in the exhaust. Kawasaki proposed a triangular tail extension of the plug hole, pointing to the combustion chamber trailing side to solve this.

- Bad fuel economy. This is due to seals leakages, and the "difficult" shape of the combustion chamber, with poor combustion behavior and mean effective pressure at part load, low rpm. Meeting the emissions regulations requirements sometimes mandates a fuel-air ratio that is not the best for fuel economy. Acceleration and deceleration as in direct drive average driving conditions also affect fuel economy. Running the engine at a constant speed and load eliminates excess fuel consumption.

- Poor emissions. As unburnt fuel is in the exhaust stream, emissions requirements are difficult to meet. This problem may be overcome by implementing direct fuel injection into the combustion chamber. The Freedom Motors Rotapower Wankel engine, which is not yet in production, met the ultra low California emissions standards. The Mazda Renesis engine, with both intake and exhaust side ports, suppressed the loss of unburned mix to exhaust formerly induced by port overlap.

Although in two dimensions the seal system of a Wankel looks to be even simpler than that of a corresponding multi-cylinder piston engine, in three dimensions the opposite is true. As well as the rotor apex seals evident in the conceptual diagram, the rotor must also seal against the chamber ends.

Piston rings are not perfect seals: each has a gap to allow for expansion. The sealing at the Wankel apexes is less critical, as leakage is between adjacent chambers on adjacent strokes of the cycle, rather than to the crankcase. Although sealing has improved over the years, the less than effective sealing of the Wankel, which is mostly due to lack of lubrication, is still a factor reducing its efficiency. Comparison tests have shown that the Mazda rotary powered RX-8 sports car may use more fuel than a heavier vehicle powered by larger displacement V-8 engines for similar performance results.

The fuel-air mixture cannot be pre-stored as there are consecutive intake cycles. The Wankel engine has a 50% longer stroke duration than a piston engine. The four Otto cycles last 1080° for a Wankel engine (three revolutions of the output shaft) versus 720° for a four-stroke reciprocating piston engine, but the four strokes are still the same proportion of the total.

There are various methods of calculating the engine displacement of a Wankel. The Japanese regulations for calculating displacements for engine ratings use the volume displacement of one rotor face only, and the auto industry commonly accepts this method as the standard for calculating the displacement of a rotary. When compared by specific output, however, the convention results in large imbalances in favor of the Wankel motor, an early approach was rating displacement of each rotor as two times the chamber.

Wankel rotary engine and piston engine displacement and corresponding power output can more accurately be compared by displacement per revolution of the eccentric shaft. A calculation of this form dictates that a two-rotor Wankel displacing 654 cc per face will have a displacement of 1.3 liters per every rotation of the eccentric shaft (only two total faces, one face per rotor going through a full power stroke) and 2.6 liters after two revolutions (four total faces, two faces per rotor going through a full power stroke). The results are directly comparable to a 2.6-liter piston engine with an even number of cylinders in a conventional firing order, which will likewise displace 1.3 liters through its power stroke after one revolution of the crankshaft, and 2.6 liters through its power strokes after two revolutions of the crankshaft. A Wankel rotary engine is still a four-stroke engine and pumping losses from non-power strokes still apply, but the absence of throttling valves and a 50% longer stroke duration result in a significantly lower pumping loss compared to a four-stroke reciprocating piston engine. Measuring a Wankel rotary engine in this way more accurately explains its specific output, as the volume of its air fuel mixture put through a complete power stroke per revolution is directly responsible for torque and thus power produced.

The trailing side of the rotary engine's combustion chamber develops a squeeze stream which pushes back the flamefront. With the conventional one or two-spark-plug system and homogenous mixture, this squeeze stream prevents the flame from propagating to the combustion chamber's trailing side in the mid and high engine speed ranges, Mazda engineers described the full process. Kawasaki addressed this problem in their US patent US 3848574, and Toyota obtained a 7% economy improvement by placing a glow-plug in the leading site and using Reed-Valves in intake ducts. This poor combustion in the trailing side of chamber is one of the reasons why there is more carbon monoxide and unburnt hydrocarbons in a Wankel's exhaust stream. A side-port exhaust, as is used in the Mazda Renesis, avoids one of the causes of this because the unburned mixture cannot escape. The Mazda 26B avoided this issue through a three spark-plug ignition system. (At the 24 Hours of Le Mans endurance race in 1991 the 26B had significantly lower fuel consumption than the competing reciprocating piston engines. All competitors had the same amount of fuel available due to the Le Mans limited fuel quantity rule.)

A peripheral intake port gives the highest mean effective pressure, however, side intake porting produces a more steady idle, as it helps to prevent blow-back of burned gases into the intake ducts which cause "misfirings": alternating cycles where the mixture ignites and fails to ignite; peripheral porting (PP) gives the best mean effective pressure throughout the rpm range, but PP was linked also to worse idle stability and part-load performance. Early work from Toyota led to the addition of a fresh air supply to the exhaust port and proved also that a Reed-valve in the intake port or ducts improved the low rpm and partial load performance of wankel engines, by preventing blow-back of exhaust gas into the intake port and ducts, and reducing the misfiring-inducing high EGR, at the cost of a small loss of power at top rpm; this is according to David W. Garside, the developer of the Norton rotary engine, who proposed that an earlier opening of the intake port before top dead center (TDC) and longer intake ducts improved low rpm torque and elasticity of wankel engines, also described in Kenichi Yamamoto's books. Elasticity is also improved with a greater rotor eccentricity, analogous to a longer stroke in a reciprocating engine. Wankel engines operate better with a low pressure exhaust system, higher exhaust back pressure reducing mean effective pressure, more severely in peripheral intake port engines. The Mazda RX-8 Renesis engine improved performance by doubling the exhaust port area respect to earlier designs, and there is specific work about the effect of intake and exhaust piping configuration on wankel engines performance.

All Mazda-made Wankel rotaries, including the Renesis found in the RX-8, burn a small quantity of oil by design, metered into the combustion chamber to preserve the apex seals. Owners must periodically add small amounts of oil, thereby increasing running costs. Some sources (rotaryeng.

net) claim that better results come with the use of an oil-in-fuel mixture rather than an oil metering pump. Liquid cooled engines require a mineral multigrade oil for cold starts, and wankel engines need a warm-up time before full load operation as reciprocating engines do. All engines exhibit oil loss, however the rotary engine is engineered with a sealed motor, unlike a piston engine that has a film of oil that splashes on the walls of the cylinder to lubricate them, hence an oil "control" ring. No-oil-loss engines have been developed, eliminating much of the oil lubrication problems.

Applications

Automobile Racing

Mazda 787B

In the racing world, Mazda has had substantial success with two-rotor, three-rotor, and four-rotor cars. Private racers have also had considerable success with stock and modified Mazda Wankel-engine cars.

The Sigma MC74 powered by a Mazda 12A engine was the first engine and only team from outside Western Europe or the United States to finish the entire 24 hours of the 24 Hours of Le Mans race, in 1974. Mazda is the only team from outside Western Europe or the United States to have won Le Mans outright and the only non-piston engine ever to win Le Mans, which the company accomplished in 1991 with their four-rotor 787B (2,622 cc or 160 cu in—actual displacement, rated by FIA formula at 4,708 cc or 287 cu in).

Formula Mazda Racing features open-wheel race cars with Mazda Wankel engines, adaptable to both oval tracks and road courses, on several levels of competition. Since 1991, the professionally organized Star Mazda Series has been the most popular format for sponsors, spectators, and upward bound drivers. The engines are all built by one engine builder, certified to produce the prescribed power, and sealed to discourage tampering. They are in a relatively mild state of racing tune, so that they are extremely reliable and can go years between motor rebuilds.

The Malibu Grand Prix chain, similar in concept to commercial recreational kart racing tracks, operates several venues in the United States where a customer can purchase several laps around a track in a vehicle very similar to open wheel racing vehicles, but powered by a small Curtiss-Wright rotary engine.

In engines having more than two rotors, or two rotor race engines intended for high-rpm use, a multi-piece eccentric shaft may be used, allowing additional bearings between rotors. While this approach does increase the complexity of the eccentric shaft design, it has been used successfully in the Mazda's production three-rotor 20B-REW engine, as well as many low volume production race engines. The C-111-2 4 Rotor Mercedes-Benz eccentric shaft for the KE Serie 70, Typ DB M950 KE409 is made in one piece. Mercedes-Benz used split bearings.

Motorcycle Engines

Norton Interpol2 prototype

The small size and attractive power to weight ratio of the Wankel engine appealed to motorcycle manufacturers. The first Wankel-engined motorcycle was the 1960 'IFA/MZ KKM 175W' built by German motorcycle manufacturer MZ licensed from NSU.

In 1972, Yamaha introduced the RZ201 at the Tokyo Motor Show, a prototype with a Wankel engine, weighing 220 kg and producing 60 hp from a twin-rotor 660 cc engine (US patent N3964448). In 1972 Kawasaki presented its two-rotor Kawasaki X99 rotary engine prototype (US patents N 3848574 &3991722). Both Yamaha and Kawasaki claimed to have solved the problems in early Wankels, namely: poor fuel economy, high exhaust emissions, and poor engine longevity, but neither prototype reached production.

In 1974 Hercules produced W-2000 Wankel motorcycles, but low production numbers meant the project was unprofitable, and production ceased in 1977.

From 1975 to 1976, Suzuki produced its RE5 single-rotor Wankel motorcycle. It was a complex design, with both liquid cooling and oil cooling, and multiple lubrication and carburetor systems. It worked well and was smooth, but being rather heavy and having a modest 62 bhp power output, it did not sell well.

Dutch motorcycle importer and manufacturer Van Veen produced small quantities of their dual-rotor Wankel-engined OCR-1000 between 1978 and 1980, using surplus Comotor engines.

In the early 1980s, using earlier work at BSA, Norton produced the air-cooled twin-rotor Classic, followed by the liquid-cooled Commander and the Interpol2 (a police version). Subsequent Norton Wankel bikes included the Norton F1, F1 Sports, RC588, Norton RCW588, and NRS588. Norton has proposed a new 588 cc twin-rotor model called the "NRV588" and a 700 cc version called the "NRV700". A former mechanic at Norton, Brian Crighton, started developing his own rotary engined motorcycles line named "Roton", whose products won several local Australian races.

Despite successes in racing, no motorcycles powered by Wankel engines have been produced for sale to the general public for road use since 1992.

The two different design approaches, taken by Suzuki and BSA may usefully be compared. Even before Suzuki produced the RE5, in Birmingham BSA's research engineer David Garside, was developing a twin-rotor Wankel motorcycle. BSA's collapse put a halt on development, but Garside's machine eventually reached production as the Norton Classic.

Wankel engines run very hot on the ignition and exhaust side of the engine's trochoid chamber, whereas the intake and compression parts are cooler. Suzuki opted for a complicated oil-cooling and water cooling system, with Garside reasoning that provided the power did not exceed 80 bhp, air-cooling would suffice. Garside cooled the interior of the rotors with filtered ram-air. This very hot air was cooled in a plenum contained within the semi-monocoque frame and afterwards, once mixed with fuel, fed into the engine. This air was quite oily after running through the interior of the rotors, and thus was used to lubricate the rotor tips. The exhaust pipes become very hot, with Suzuki opting for a finned exhaust manifold, twin-skinned exhausted pipes with cooling grilles, heatproof pipe wrappings and silencers with heat shields. Garside simply tucked the pipes out of harm's way under the engine, where heat would dissipate in the breeze of the vehicle's forward motion. Suzuki opted for complicated multi-stage carburation, whilst Garside choose simple carburetors. Suzuki had three lube systems, whilst Garside had a single total-loss oil injection system which was fed to both the main bearings and the intake manifolds. Suzuki chose a single rotor that was fairly smooth, however with rough patches at 4,000 rpm; Garside opted for a turbine-smooth twin-rotor motor. Suzuki mounted the massive rotor high in the frame; Garside put his rotors as low as possible to lower the center of gravity of the motorcycle.

Although it was said to handle well, The result was that the Suzuki was heavy, overcomplicated, expensive to manufacture, and (at 62 bhp) a little short on power. Garside's design was simpler, smoother, lighter and (at 80 bhp) significantly more powerful.

Aircraft Engines

Diamond DA20 with Diamond Engines Wankel

Sikorsky Cypher Unmanned aerial vehicle (UAV) powered with a UEL AR801 Wankel engine

ARV Super2 with the British MidWest AE110 twin-rotor Wankel engine

In principle, a Wankel engine should be ideal for light aircraft, as it is light, compact, almost vibrationless and has a high power-to-weight ratio. Further aviation benefits of a Wankel engine include:

1. Rotors cannot seize, since rotor casings expand more than rotors;

2. A Wankel engine is less prone to the serious condition known as "engine-knock", which can destroy the plane's engines in mid-flight.

3. A Wankel is not susceptible to "shock-cooling" during descent;

4. A Wankel does not require an enriched mixture for cooling at high power;

5. Having no reciprocating parts, there is less vulnerability to damage when the engines revolves higher than the designed maximum running operation. The limit to the revolutions is the strength of the main bearings.

Unlike the case with some cars and motorcycles, a Wankel aero-engine will be sufficiently warm before full power is asked of it because of the time taken for pre-flight checks. A Wankel aero-engine spends most of its operational time at high power outputs, with little idling. This makes ideal the use of peripheral ports. An advantage is that modular engines with more than two rotors are feasible. If icing of any intake tracts is an issue, there is plenty of waste engine heat available to prevent icing.

The first Wankel rotary-engine aircraft was the experimental Lockheed Q-Star civilian version of the United States Army's reconnaissance QT-2, basically a powered Schweizer sailplane, in the late 1960s. The plane was powered by a 185 hp (138 kW) Curtiss-Wright RC2-60 Wankel rotary engine; the same engine model was also flown in a Cessna Cardinal and other airplanes and a helicopter. In Germany in the mid-1970s, a pusher ducted fan airplane powered by a modified NSU multi-rotor Wankel engine was developed in both civilian and military versions, Fanliner and Fantrainer.

In roughly the same timeframe as the first experiments with full-scale aircraft powered with Wankel engines, model aircraft-sized versions were pioneered by a combine of the well-known Japanese O.S. Engines firm and the then-extant German Graupner aeromodeling products firm, under license from NSU/Auto-Union. By 1968 the first prototype air-cooled, single-rotor glow plug-ignition, methanol fueled 4.9 cm^3 displacement OS/Graupner model Wankel engine was running, and was produced in at least two differing versions from 1970 to the present day, solely by the O.S. firm since Graupner's demise in 2012.

Aircraft Wankels have been taken up with the advantages over other engines being exploited. Wankels are increasingly being found in roles where the compact size, high power-to-weight ratio and quiet operation is important, notably in drones and unmanned aerial vehicles. Many companies and hobbyists adapt Mazda rotary engines (taken from automobiles) to aircraft use; others, including Wankel GmbH itself, manufacture Wankel rotary engines dedicated for this purpose. One such use are the "Rotapower" engines in the Moller Skycar M400. Another example of purpose built aircraft rotaries are Austro Engine's 55 hp (40.4 kW) AE50R (certified) and 75 hp (55 kW) AE75R (under development) both appr. 2 hp/kg.

Wankel engines are also becoming increasingly popular in homebuilt experimental aircraft, such as the ARV Super2 which can be re-engined with the British MidWest AE series aero-engine. Most are Mazda 12A and 13B automobile engines, converted to aviation use. This is a very cost-effective alternative to certified aircraft engines, providing engines ranging from 100 to 300 horsepower (220 kW) at a fraction of the cost of traditional engines. These conversions first took place in the early 1970s. With a number of these engines mounted on aircraft, as of 10 December 2006 the National Transportation Safety Board has only seven reports of incidents involving aircraft with Mazda engines, and none of these were a failure due to design or manufacturing flaws.

Peter Garrison, contributing editor for Flying magazine, has said that "In my opinion, however, the most promising engine for aviation use is the Mazda rotary." Mazdas have indeed worked well when converted for use in homebuilt aircraft. However, the real challenge in aviation is producing FAA-certified alternatives to the standard reciprocating engines that power most small general aviation aircraft. Mistral Engines, based in Switzerland, developed purpose-built rotaries for factory and retrofit installations on certified production aircraft. The G-190 and G-230-TS rotary engines were already flying in the experimental market, and Mistral Engines hoped for FAA and JAA certification by 2011. As of June 2010, G-300 rotary engine development ceased, with the company citing a need for cash flow to complete development.

Mistral claims to have overcome the challenges of fuel consumption inherent in the rotary, at least to the extent that the engines are demonstrating specific fuel consumption within a few points of reciprocating engines of similar displacement. While fuel burn is still marginally higher than traditional engines, it is outweighed by other beneficial factors.

At the price of increased complication for a high pressure diesel type injection system, fuel consumption in the same range as small pre-chamber automotive and industrial diesels has been demonstrated with Curtiss-Wright's stratified charge multi-fuel engines, while preserving the aforementioned Wankel rotary advantages Unlike a piston and overhead valve engine, there are no valves which can float at higher rpm causing loss of performance. The Wankel is a more effective design at high revolutions with no reciprocating parts, far fewer moving parts and no cylinder head.

The French company Citroën had developed Wankel powered RE-2 (fr) helicopter in the 1970s.

Since Wankel engines operate at a relatively high rotational speed, at 6'000 rpm of output shaft, the Rotor makes only 2'000 turns, with relatively low torque, propeller driven aircraft must use a propeller speed reduction unit to maintain propellers within the designed speed range. Experimental aircraft with Wankel engines use propeller speed reduction units, for instance the MidWest twin-rotor engine has a 2.95:1 reduction gearbox. The rotational shaft speed of a Wankel engine is high compared to reciprocating piston designs. Only the eccentric shaft spins fast, while the rotors turn at exactly one-third of the shaft speed. If the shaft is spinning at 7,500 rpm, the rotors are turning at a much slower 2,500 rpm.

Pratt & Whitney Rocketdyne have been commissioned by DARPA to develop a diesel Wankel engine for use in a prototype VTOL flying car called the "Transformer". The engine, based on an earlier unmanned aerial vehicle Wankel diesel concept called "Endurocore".

The sailplane manufacturer Schleicher uses Wankel engines in their self-launching models ASK-21 Mi, ASH-26E, ASH-25 M/Mi, ASH-30 Mi, ASH-31 Mi, ASW-22 BLE, and ASG-32 Mi.

In 2013 e-Go aeroplanes, based in Cambridge, United Kingdom, announced that their new single-seater canard aircraft, the winner of a design competition to meet the new UK single-seat deregulated category, will be powered by a Wankel engine from Rotron Power, a specialist manufacturer of advanced rotary engines for unmanned aeronautical vehicle (UAV) applications. The first sale was 2016. The aircraft is expected to deliver 100 kts cruise speed from a 30 hp Wankel engine, with a fuel economy of 75 mpg using standard motor gasoline (MOGAS), developing 22 kW (30 hp).

The DA36 E-Star, an aircraft designed by Siemens, Diamond Aircraft and EADS, employs a series hybrid powertrain with the propeller being turned by a Siemens 70 kW (94 hp) electric motor. The aim is to reduce fuel consumption and emissions by up to 25 percent. An onboard 40 hp (30 kW) Austro Engines Wankel rotary engine and generator provides the electricity. A propeller speed reduction unit is eliminated. The electric motor uses electricity stored in batteries, with the generator engine off, to take off and climb reducing sound emissions. The series-hybrid powertrain using the Wankel engine reduces the weight of the plane by 100 kilograms from its predecessor. The DA36 E-Star first flew in June 2013, making this the first ever flight of a series-hybrid powertrain. Diamond Aircraft state that the technology using Wankel engines is scalable to a 100-seat aircraft.

Range Extender

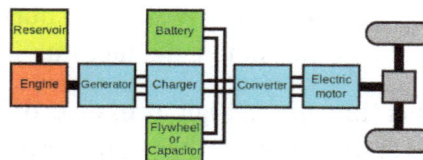

Structure of a series-hybrid vehicle. The grey square represents a differential gear. An alternative arrangement (not shown) is to have electric motors at two or four wheels.

Mazda2 EV prototype

Due to the compact size and the high power to weight ratio of a Wankel engine, a number have been proposed for electric vehicles as range extenders to supplement when electric battery levels are low, with a number of concept cars incorporating a series hybrid powertrain arrangement. A Wankel engine used only in a generator setup has packaging and weight distribution advantages, maximizing interior passenger and luggage space when used in a vehicle. The engine/generator may be at one end of the vehicle with the electric driving motors at the other connected only by thin light cables. In 2010 Audi revealed a prototype series-hybrid electric car, the A1 e-tron, that incorporated a small 250 cc Wankel engine running at 5,000 rpm recharging the car's batteries as needed, and providing electricity directly to the electric driving motor. In 2010 FEV Inc revealed that in their prototype electric version of the Fiat 500 a Wankel engine would also be used as a range extender. Valmet Automotive of Finland in 2013 revealed a prototype car incorporating a Wankel powered series-hybrid powertrain car named the EVA, utilizing an engine manufactured by the German company Wankel SuperTec. The UK Aixro dealer offers a Range Extender based in the 294 cc per chamber Kart engine.

Mazda of Japan ceased production of direct drive Wankel engines in their model range in 2012, leaving the motor industry world-wide with no production cars using the engine. The company is continuing development of the next generation of their Wankel engines, the SkyActiv-R with a new rear wheel drive sports car model announced in October 2015 although with a launch date of 2020. Mazda states that the SkyActiv-R solves the three key issues with previous rotary engines: fuel economy, emissions and reliability. Mr Takashi Yamanouchi, the global CEO of Mazda stated, "The rotary engine has very good dynamic performance, but it's not so good on economy when you accelerate and decelerate. However, with a range extender you can use a rotary engine at a constant 2,000rpm, at its most efficient. It's compact, too." No Wankel engine in this arrangement, as yet has made it into production vehicles or planes. However, in November 2013 Mazda announced a series-hybrid prototype car to the motoring press, the Mazda2 EV using a Wankel engine as a range extender. The engine is a tiny, almost inaudible, single-rotor 330cc unit generating 30 bhp at 4,500rpm maintaining a continuous electric output of 20 kW. The engine is located under the rear luggage floor.

Other Uses

UEL UAV-741 Wankel engine for a UAV

Small Wankel engines are being found increasingly in other applications, such as go-karts, personal water craft and auxiliary power units for aircraft. Kawasaki patented also a mixture cooled rotary engine (US patent 3991722). Yanmar Diesel and Dolmar-Sachs had a rotary engine chain saw (SAE paper 760642) and outboard boat engines, and the French Outils Wolf, a Wankel rotary engine powered lawnmower (Rotondor), with the rotor in a horizontal position and no seals in the down side, for production costs savings. The Graupner/O.S. 49-PI is a 1.27 hp (947 W) 5 cc Wankel engine for model airplane use which has been in production essentially unchanged since 1970; even with a large muffler, the entire package weighs only 380 grams (13.4 ounces).

The simplicity of the Wankel makes it well-suited for mini, micro, and micro-mini engine designs. The Microelectromechanical systems (MEMS) Rotary Engine Lab at the University of California, Berkeley, has previously undertaken research towards the development of Wankel engines of down to 1 mm in diameter with displacements less than 0.1 cc. Materials include silicon and motive power includes compressed air. The goal of such research was to eventually develop an internal combustion engine with the ability to deliver 100 milliwatts of electrical power; with the engine itself serving as the rotor of the generator, with magnets built into the engine rotor itself. Development of the miniature Wankel engine stopped at UC Berkeley at the end of the DARPA contract. Miniature Wankel engines struggled to maintain compression due to sealing problems, similar to problems observed in the large scale versions. In addition, miniature engines suffer from an adverse surface to volume ratio causing excess heat losses; the relatively large surface area of the combustion chamber walls transfers away what little heat is generated in the small combustion volume resulting in quenching and low efficiency.

Ingersoll-Rand built the largest ever Wankel engine which was available between 1975 and 1985 producing 1,100 hp (820 kW) with two rotors. A one rotor version was available producing 550 hp (410 kW). The displacement per rotor was 41 liters with each rotor approximately one meter in diameter. The engine was derived from a previous, unsuccessful Curtiss-Wright design, which failed because of a well-known problem with all internal combustion engines: the fixed speed at which the flame front travels limits the distance combustion can travel from the point of ignition in a given time, and thereby limiting the maximum size of the cylinder or rotor chamber which can be used. This problem was solved by limiting the engine speed to only 1200 rpm and the use of natural gas as fuel; this was particularly well chosen, since one of the major uses of the engine was to drive compressors on natural gas pipelines.

Yanmar Diesel of Japan produced some small, charge-cooled rotor rotary engines for uses such as chainsaws and outboard engines, some of their contributions are the LDR (rotor recess in the leading edge of combustion chamber) engines having better exhaust emissions profiles, and that reed-valve controlled intake ports improved part-load and low rpm performance.

In 1971 and 1972, Arctic Cat produced snowmobiles powered by Sachs KM 914 303 cc and KC-24 294 cc Wankel Engine made in Germany.

In the early 1970s Johnson and other brands sold Snowmobiles powered by 35 or 45 HP Wankel engines designed and built by OMC.

Aixro of Germany produces and sells a 294 cc per chamber charge-cooled rotor and liquid-cooled housings kart engines, other makers are: Wankel AG, Cubewano, Rotron, Precision Technology USA.

The American M1 Abrams tank uses an auxiliary rotary engine power unit. The engines was developed by the TARDEC US Army lab. The engine has a high power density 330 cc rotary engine modified to operate with various fuels such as, high octane military grade jet fuel. The entire fleet of Abrams tanks were retrofitted with the rotary units.

Non-internal Combustion

In addition for use as an internal combustion engine, the basic Wankel design has also been used for gas compressors, and superchargers for internal combustion engines, but in these cases, although the design still offers advantages in reliability, the basic advantages of the Wankel in size and weight over the four-stroke internal combustion engine are irrelevant. In a design using a Wankel supercharger on a Wankel engine, the supercharger is twice the size of the engine.

The Wankel design is used in the seat belt pre-tensioner system of some Mercedes-Benz and Volkswagen cars. When the deceleration sensors sense a potential crash, small explosive cartridges are triggered electrically and the resulting pressurized gas feeds into tiny Wankel engines which rotate to take up the slack in the seat belt systems, anchoring the driver and passengers firmly in the seat before a collision.

Atkinson Cycle

The Atkinson-cycle engine is a type of internal combustion engine invented by James Atkinson in 1882. The Atkinson cycle is designed to provide efficiency at the expense of power density, or total power extracted per unit of displacement per rotation. A modern variation of this approach is used in some modern automobile engines. While originally seen exclusively in hybrid electric applications such as the Toyota Prius, some nonhybrid automobiles now feature engines that can run in the Atkinson cycle as a part-time operating regimen, giving good economy while running in Atkinson cycle, and conventional power density when running as a normal-cycle engine. Mazda Skyactiv models offering this capability include the Mazda 3 and MX-5.

Design

Atkinson produced three different designs that had a short compression stroke and a longer expansion stroke. The first Atkinson-cycle engine was named the "Differential Engine". This engine used opposed pistons. The second, and most well-known design, was called the "Cycle Engine", which used an over-center arm to create four piston strokes in one revolution of the crankshaft. Reciprocating engine allowed the intake, compression, power, and exhaust strokes of the four-stroke cycle to occur in a single turn of the crankshaft and was designed to avoid infringing certain patents covering Otto-cycle engines. Atkinson's third and final engine was called the "Utilite Engine" and it operated much like any two-stroke engine. The common thread throughout Atkinson's designs is the engines have an expansion stroke that is longer than compression stroke and by this method the engine can achieve greater thermal efficiency than a traditional piston engine. Atkinson's engines were produced by the British Gas Engine Company and also licensed to other overseas manufacturers. Many modern engines now use unconventional valve timing to produce the effect of a shorter compression stroke/longer power stroke, Miller applied this technique to the four-stroke engine, so it is sometimes referred as the Atkinson/Miller cycle, US patent 2817322 dated Dec 24, 1957. In 1888, Charon filed a French patent and displayed an engine at the Paris Exhibition in 1889 . The Charon gas engine (four- stroke) also used a similar cycle to

Miller but without a supercharger. It is referred to as the "Charon cycle" Modern engine designers are realizing the potential fuel-efficiency improvements the Atkinson-type cycle can provide.

Atkinson "Differential Engine"

The first implementation of the Atkinson cycle was in 1882; unlike later versions, it was arranged as an opposed piston engine, the Atkinson differential engine. In this, a single crankshaft was connected to two opposed pistons through a toggle jointed linkage that had a nonlinearity; for half a revolution, one piston remained almost stationary while the other approached it and returned, and then for the next half revolution, the pistons changed over which piston was almost stationary and which piston approached and returned. Thus, in each revolution, one piston provided a compression stroke and a power stroke, and then the other piston provided an exhaust stroke and a charging stroke. As the power piston remained withdrawn during exhaust and charging, it was practical to provide exhaust and charging using valves behind a port that was covered during the compression stroke and the power stroke, and so the valves did not need to resist high pressure and could be of the simpler sort used in many steam engines, or even reed valves.

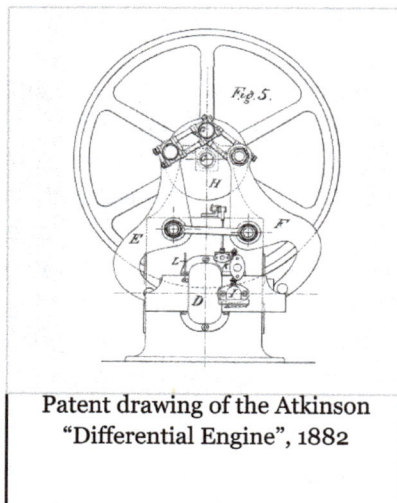

Patent drawing of the Atkinson
"Differential Engine", 1882

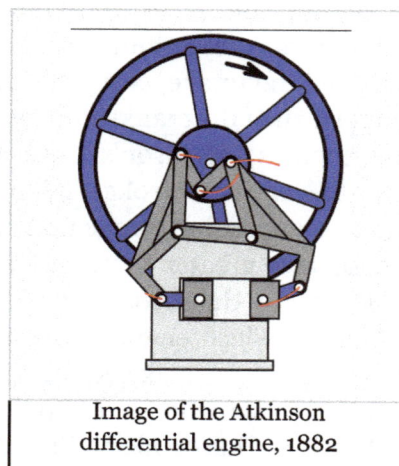

Image of the Atkinson
differential engine, 1882

Atkinson "Cycle Engine"

The next engine designed by Atkinson in 1887 was named the "Cycle Engine" This engine used poppet valves, a cam and an over-center arm to produce four piston strokes for every revolution of the crankshaft. The intake and compression strokes were significantly shorter than the expansion and exhaust strokes. The "Cycle" engines were produced and sold for several years by the British Engine Company. Atkinson also licensed production to other manufactures. Sizes ranged from a few up to 100 horsepower.

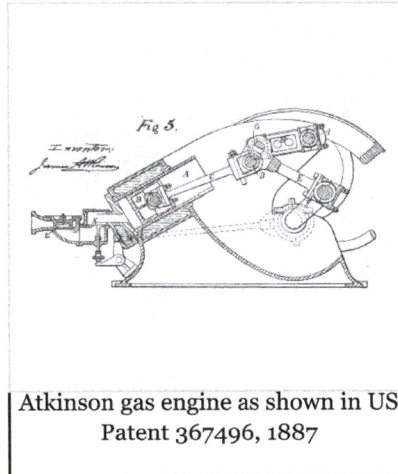

Atkinson gas engine as shown in US Patent 367496, 1887

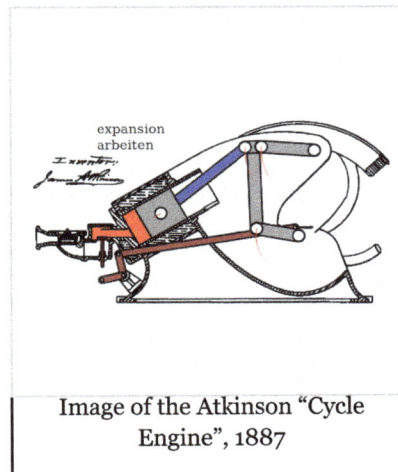

Image of the Atkinson "Cycle Engine", 1887

Atkinson "Utilite Engine"

Atkinson's third design was named the "Utilite Engine". Atkinson realized an improvement was needed to make his cycle more applicable to a high-speed engine. The final engine produced by the British Gas Engine company was the Utilite Engine. With this design, Atkinson was able to make a more conventional engine yet preserve the efficiency of having a short compression stroke and a long expansion stroke in a rather ingenious way. It operates much like a standard two-stroke except that the exhaust port is located at about the middle of the stroke. During the expansion stroke, a valve (which remains closed until the piston reaches the end of the stroke) prevents pressure from escaping as the piston moves past the exhaust port. Once the valve is opened, it remains open as the piston heads back toward compression, letting fresh air charge the cylinder and exhaust escape until the port is covered. A rich fuel/air mixture is injected by a small piston pump after the start of compression. This resulted in a two-stroke engine with a short compression and longer expansion stroke. The Utilite Engines were tested and found to be even more efficient than Atkinson's previous designs. Very few Utilite Engines were produced and none is known to survive. The British patent is from 1892, #2492. No US patent for the Utilite Engine is known.

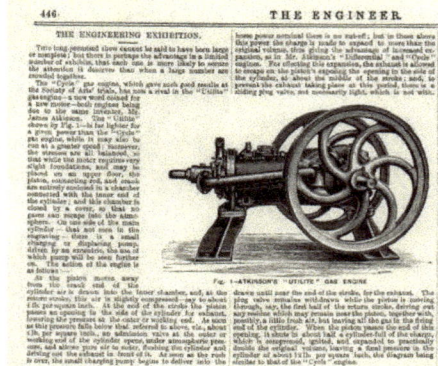

Atkinson Utilite Engine

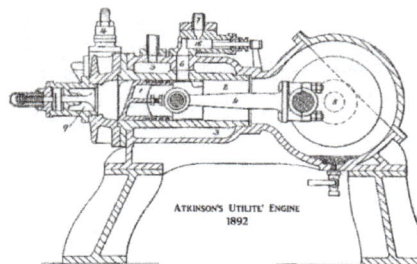

Atkinson's Utilite engine 1892

Ideal Thermodynamic Cycle

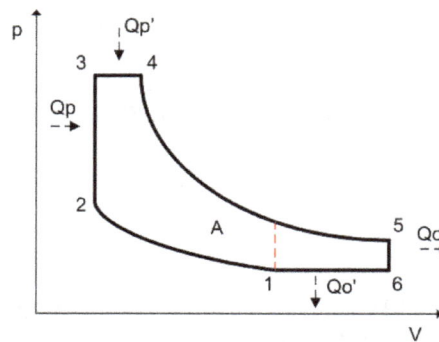

Atkinson gas cycle

The ideal Atkinson cycle consists of:

- 1–2 Isentropic, or reversible, adiabatic compression
- 2–3 Isochoric heating (Qp)

- 3–4 Isobaric heating (Qp')

- 4–5 Isentropic expansion

- 5–6 Isochoric cooling (Qo)

- 6–1 Isobaric cooling (Qo')

Modern Atkinson-cycle Engines

A small engine with Atkinson-style linkages between the piston and flywheel. Modern Atkinson-cycle engines do away with this complex energy path.

Recently, the term "Atkinson cycle" has been used to describe a modified Otto-cycle engine in which the intake valve is held open longer than normal to allow a reverse flow of intake air into the intake manifold, this simulated Atkinson cycle is most notably used in the widespread Toyota 1NZ-FXE engine. The effective compression ratio is reduced (for a time the air is escaping the cylinder freely rather than being compressed), but the expansion ratio is unchanged. This means the compression ratio is smaller than the expansion ratio. Heat gained from burning fuel increases the pressure, thereby forcing the piston to move, expanding the air volume beyond the volume when compression began. The goal of the modern Atkinson cycle is to allow the pressure in the combustion chamber at the end of the power stroke to be equal to atmospheric pressure; when this occurs, all the available energy has been obtained from the combustion process. For any given portion of air, the greater expansion ratio allows more energy to be converted from heat to useful mechanical energy, meaning the engine is more efficient.

The disadvantage of the four-stroke Atkinson-cycle engine versus the more common Otto-cycle engine is reduced power density. Due to a smaller portion of the compression stroke being devoted to compressing the intake air, an Atkinson-cycle engine does not take in as much air as would a similarly designed and sized Otto-cycle engine. Four-stroke engines of this type that use the same type of intake valve motion but with a supercharger to make up for the loss of power density are known as Miller-cycle engines.

Rotary Atkinson-cycle Engine

The Atkinson cycle can be used in a rotary engine. In this configuration, an increase in both power and efficiency can be achieved when compared to the Otto cycle. This type of engine retains the one power phase per revolution, together with the different compression and expansion volumes of the original Atkinson cycle. Exhaust gases are expelled from the engine by compressed-air scavenging. This modification of the Atkinson cycle allows the use of alternative fuels such as diesel and hydrogen. Disadvantages of this design include the requirement that rotor tips seal very tightly on the outer housing wall and the mechanical losses suffered through friction between rapidly oscillating parts of irregular shape.

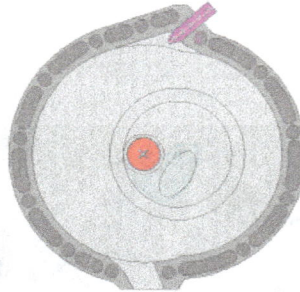

Rotary Atkinson-cycle engine

Vehicles using Atkinson-cycle Engines

2004 Toyota Prius hybrid

2010 Ford Fusion Hybrid (North America)

While a modified Otto-cycle piston engine using the Atkinson cycle provides good fuel efficiency, it is at the expense of a lower power-per-displacement as compared to a traditional four-stroke engine. If demand for more power is intermittent, the power of the engine can be supplemented by an electric motor during times when more power is needed. This forms the basis of an Atkinson cycle-based hybrid electric drivetrain. These electric motors can be used independently of, or in combination with, the Atkinson-cycle engine, to provide the most efficient means of producing the desired power. This drive-train first entered production in late 1997 in the Japanese-market Toyota Prius.

At this writing, many production full hybrid-electric vehicles use Atkinson-cycle theories:

- Chevrolet Volt
- Chrysler Pacifica (front-wheel drive) plug-in hybrid model minivan
- Ford C-Max (front-wheel drive / US market) hybrid and plug-in hybrid models
- Ford Escape/Mercury Mariner/Mazda Tribute electric (front- and four-wheel drive) with a compression ratio of 12.4:1
- Ford Fusion Hybrid/Mercury Milan Hybrid/Lincoln MKZ Hybrid electric (front-wheel drive) with a compression ratio of 12.3:1
- Honda Accord Plug-in Hybrid
- Honda Accord Hybrid (front-wheel drive)
- Hyundai Sonata Hybrid (front-wheel drive)
- Hyundai Elantra Atkinson-cycle models
- Hyundai Grandeur hybrid (front-wheel drive)
- Hyundai Ioniq hybrid, plug-in hybrid (front-wheel drive)
- Infiniti M35h hybrid (rear-wheel drive)
- Kia Niro hybrid (front-wheel drive)
- Kia Optima Hybrid Kia K5 hybrid 500h (front-wheel drive) with a compression ratio of 13:1
- Kia Cadenza Hybrid Kia K7 hybrid 700h (front-wheel drive)
- Lexus CT 200h (front-wheel drive)
- Lexus ES 300h (front-wheel drive)
- Lexus GS 450h hybrid electric (rear-wheel drive) with a compression ratio of 13:1
- Lexus GS F (rear-wheel drive)
- Lexus RC F (rear-wheel drive)
- Lexus HS 250h (front-wheel drive)
- Lexus IS 200t (2016)
- Lexus NX hybrid electric (four-wheel drive)
- Lexus RX 450h hybrid electric (four-wheel drive)
- Mazda Mazda6 (2014 and newer in North America, 2013 rest of world)
- Mercedes ML450 Hybrid (four-wheel drive) electric
- Mercedes S400 Blue Hybrid (rear-wheel drive) electric
- Toyota Camry Hybrid electric (front-wheel drive) with a compression ratio of 12.5:1
- Toyota Highlander Hybrid (2011 and newer)
- Toyota Prius hybrid electric (front-wheel drive) with a (purely geometric) compression ra-

tio of 13.0:1

- Toyota Yaris Hybrid (front-wheel drive) with a compression ratio of 13.4:1
- Toyota Auris Hybrid (front-wheel drive)
- Toyota Tacoma V6 (beginning in 2015 for the 2016 model year)

Summary of the Patent

The 1887 patent (US 367496) describes the mechanical linkages necessary to obtain all four strokes of the four-stroke cycle for a gas engine within one revolution of the crankshaft. There is also a reference to an 1886 Atkinson patent (US 336505) which describes an opposed-piston gas engine. The British patent for the "Utilite'" is from 1892 (#2492).

Pistonless Rotary Engine

A pistonless rotary engine is an internal combustion engine that does not use pistons in the way a reciprocating engine does, but instead uses one or more rotors, sometimes called rotary pistons. An example of a pistonless rotary engine is the Wankel engine.

The term rotary combustion engine has been suggested as an alternative name for these engines to distinguish them from early (generally up to the early 1920s) aircraft engines and motorcycle engines also known as rotary engines. However, both continue to be called rotary engines and only the context determines which type is meant.

Pistonless Rotary Engines

The basic concept of a (pistonless) rotary engine avoids the reciprocating motion of the piston with its inherent vibration and rotational-speed-related mechanical stress. As of 2006 the Wankel engine is the only successful pistonless rotary engine, but many similar concepts have been proposed and are under various stages of development. Examples of rotary engines include:

Production stage

- Beauchamp Tower's nineteenth century spherical steam engine (theoretically adaptable to use internal combustion)
- The Wankel engine
- The Atkinson cycle engine

Development stage

- The Baylin Engine
- The Engineair engine
- The Liquidpiston engine
- The Crankless engine

- The Hamilton Walker engine

- The Quasiturbine

- The Ramgen Integrated Supersonic Component Engine

- The Rand cam engine

- The RKM engine (RotationsKolbenMaschine)

- The Sarich orbital engine

- The Tri-Dyne Engine

- The Trochilic engine

- The Wave disk engine

- The Moto Turbine Radiale by Jean Claude Lefeuvre

- The Jonova engine

- The Renault-Rambler lobular rotor engine (Spanish pat nº 0313466)

Conceptual stage

- The Gerotor engine

- The Rotary Engine by Jose-Ignacio Martin-Artajo, SI

- The Jose Maria Bosch-Barata engines (Spanish pats nºs 0228187, 0254176 and 0407242)

Reciprocating Engine

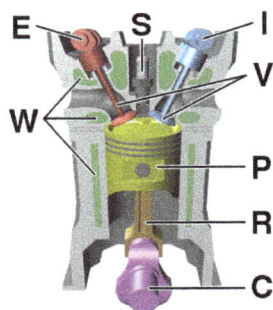

Internal combustion piston engine
Components of a typical, four stroke cycle, internal combustion piston engine.
E - Exhaust camshaft
I - Intake camshaft
S - Spark plug
V - Valves
P - Piston
R - Connecting rod
C - Crankshaft
W - Water jacket for coolant flow

A reciprocating engine, also often known as a piston engine, is a heat engine (usually, although there are also pneumatic and hydraulic reciprocating engines) that uses one or more reciprocating pistons to convert pressure into a rotating motion. This article describes the common features of all types. The main types are: the internal combustion engine, used extensively in motor vehicles; the steam engine, the mainstay of the Industrial Revolution; and the niche application Stirling engine. Internal Combustion engines are further classified in two ways: either a spark-ignition (SI) engine, where the spark plug initiates the combustion; or a compression-ignition (CI) engine, where the air within the cylinder is compressed, thus heating it, so that the heated air ignites fuel that is injected then or earlier.

Common Features in all Types

There may be one or more pistons. Each piston is inside a cylinder, into which a gas is introduced, either already under pressure (e.g. steam engine), or heated inside the cylinder either by ignition of a fuel air mixture (internal combustion engine) or by contact with a hot heat exchanger in the cylinder (Stirling engine). The hot gases expand, pushing the piston to the bottom of the cylinder. This position is also known as the Bottom Dead Center (BDC), or where the piston forms the largest volume in the cylinder. The piston is returned to the cylinder top (Top Dead Centre) (TDC) by a flywheel, the power from other pistons connected to the same shaft or (in a double acting cylinder) by the same process acting on the other side of the piston. This is where the piston forms the smallest volume in the cylinder. In most types the expanded or "exhausted" gases are removed from the cylinder by this stroke. The exception is the Stirling engine, which repeatedly heats and cools the same sealed quantity of gas. The stroke is simply the distance between the TDC and the BDC, or the greatest distance that the piston can travel in one direction.

In some designs the piston may be powered in both directions in the cylinder, in which case it is said to be double-acting.

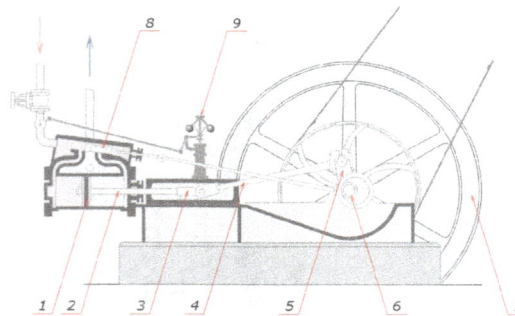

Steam piston engine
A labeled schematic diagram of a typical single-cylinder, simple expansion, double-acting high pressure steam engine.
Power takeoff from the engine is by way of a belt.
1 – Piston
2 – Piston rod
3 – Crosshead bearing
4 – Connecting rod
5 – Crank
6 – Eccentric valve motion
7 – Flywheel
8 – Sliding valve
9 – Centrifugal governor.

In most types, the linear movement of the piston is converted to a rotating movement via a connecting rod and a crankshaft or by a swashplate or other suitable mechanism. A flywheel is often used to ensure smooth rotation or to store energy to carry the engine through an un-powered part of the cycle. The more cylinders a reciprocating engine has, generally, the more vibration-free (smoothly) it can operate. The power of a reciprocating engine is proportional to the volume of the combined pistons' displacement.

A seal must be made between the sliding piston and the walls of the cylinder so that the high pressure gas above the piston does not leak past it and reduce the efficiency of the engine. This seal is usually provided by one or more piston rings. These are rings made of a hard metal, and are sprung into a circular groove in the piston head. The rings fit tightly in the groove and press against the cylinder wall to form a seal.

It is common to classify such engines by the number and alignment of cylinders and total volume of displacement of gas by the pistons moving in the cylinders usually measured in cubic centimetres (cm³ or cc) or litres (l) or (L) (US: liter). For example, for internal combustion engines, single and two-cylinder designs are common in smaller vehicles such as motorcycles, while automobiles typically have between four and eight, and locomotives, and ships may have a dozen cylinders or more. Cylinder capacities may range from 10 cm³ or less in model engines up to several thousand cubic centimetres in ships' engines.

The compression ratio affects the performance in most types of reciprocating engine. It is the ratio between the volume of the cylinder, when the piston is at the bottom of its stroke, and the volume when the piston is at the top of its stroke.

The bore/stroke ratio is the ratio of the diameter of the piston, or "bore", to the length of travel within the cylinder, or "stroke". If this is around 1 the engine is said to be "square", if it is greater than 1, i.e. the bore is larger than the stroke, it is "oversquare". If it is less than 1, i.e. the stroke is larger than the bore, it is "undersquare".

Cylinders may be aligned in line, in a V configuration, horizontally opposite each other, or radially around the crankshaft. Opposed-piston engines put two pistons working at opposite ends of the same cylinder and this has been extended into triangular arrangements such as the Napier Deltic. Some designs have set the cylinders in motion around the shaft, such as the Rotary engine.

In steam engines and internal combustion engines, valves are required to allow the entry and exit of gases at the correct times in the piston's cycle. These are worked by cams, eccentrics or cranks driven by the shaft of the engine. Early designs used the D slide valve but this has been largely superseded by Piston valve or Poppet valve designs. In steam engines the point in the piston cycle at which the steam inlet valve closes is called the cutoff and this can often be controlled to adjust the torque supplied by the engine and improve efficiency. In some steam engines, the action of the valves can be replaced by an oscillating cylinder.

Internal combustion engines operate through a sequence of strokes that admit and remove gases to and from the cylinder. These operations are repeated cyclically and an engine is said to be 2-stroke, 4-stroke or 6-stroke depending on the number of strokes it takes to complete a cycle.

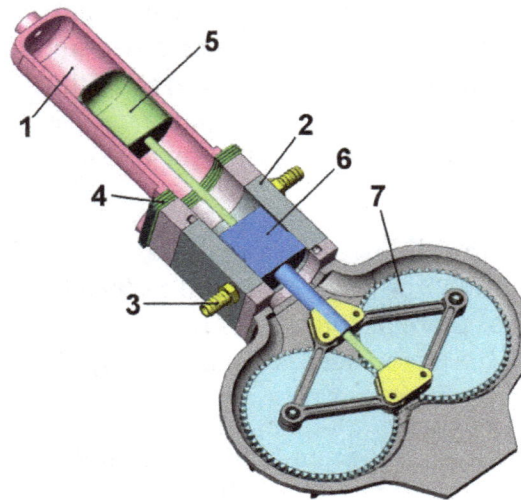

Stirling piston engine
Rhombic Drive – Beta Stirling Engine Design, showing the second displacer piston (green) within the cylinder, which shunts the working gas between the hot and cold ends, but produces no power itself.
Pink – Hot cylinder wall
Dark grey – Cold cylinder wall
Green – Displacer piston
Dark blue – Power piston
Light blue – Flywheels

In some steam engines, the cylinders may be of varying size with the smallest bore cylinder working the highest pressure steam. This is then fed through one or more, increasingly larger bore cylinders successively, to extract power from the steam at increasingly lower pressures. These engines are called Compound engines.

Aside from looking at the power that the engine can produce, the Mean Effective Pressure (MEP), can also be used in comparing the power output and performance of reciprocating engines of the same size. The mean effective pressure is the fictitious pressure which would produce the same amount of net work that was produced during the power stroke cycle. This is shown by:

W_{net} = MEP x Piston Area x Stroke = MEP x Displacement Volume and therefore: MEP = W_{net}/ Displacement Volume

Whichever engine with the larger value of MEP produces more net work per cycle and performs more efficiently.

History

An early known example of rotary to reciprocating motion can be found in a number of Roman saw mills (dating to the 3rd to 6th century AD) in which a crank and connecting rod mechanism converted the rotary motion of the waterwheel into the linear movement of the saw blades.

The reciprocating engine developed in Europe during the 18th century, first as the atmospheric engine then later as the steam engine. These were followed by the Stirling engine and internal combustion engine in the 19th century. Today the most common form of reciprocating engine is the internal combustion engine running on the combustion of petrol, diesel, Liquefied petroleum gas (LPG) or compressed natural gas (CNG) and used to power motor vehicles and engine power plants.

One notable reciprocating engine from the World War II Era was the 28-cylinder, 3,500 hp (2,600 kW) Pratt & Whitney R-4360 "Wasp Major" radial engine. It powered the last generation of large piston-engined planes before jet engines and turboprops took over from 1944 onward. It had a total engine capacity of 71.5 L (4,360 cu in), and a high power-to-weight ratio.

The largest reciprocating engine in production at present, but not the largest ever built, is the Wärtsilä-Sulzer RTA96-C turbocharged two-stroke diesel engine of 2006 built by Wärtsilä. It is used to power the largest modern container ships such as the Emma Mærsk. It is five stories high (13.5 m or 44 ft), 27 m (89 ft) long, and weighs over 2,300 metric tons (2,500 short tons) in its largest 14 cylinders version producing more than 84.42 MW (114,800 bhp). Each cylinder has a capacity of 1,820 L (64 cu ft), making a total capacity of 25,480 L (900 cu ft) for the largest versions.

Engine Capacity

For piston engines, an engine's capacity is the engine displacement, in other words the volume swept by all the pistons of an engine in a single movement. It is generally measured in litres (l) or cubic inches (c.i.d. or cu in or in^3) for larger engines, and cubic centimetres (abbreviated cc) for smaller engines. All else being equal, engines with greater capacities are more powerful and consumption of fuel increases accordingly, although power and fuel consumption are affected by many factors outside of engine displacement.

Other Modern Non-internal Combustion Types

Reciprocating engines that are powered by compressed air, steam or other hot gases are still used in some applications such as to drive many modern torpedoes or as pollution-free motive power. Most steam-driven applications use steam turbines, which are more efficient than piston engines.

The French-designed FlowAIR vehicles use compressed air stored in a cylinder to drive a reciprocating engine in a pollution-free urban vehicle.

Torpedoes may use a working gas produced by high test peroxide or Otto fuel II, which pressurise without combustion. The 230 kg (510 lb) Mark 46 torpedo, for example, can travel 11 km (6.8 mi) underwater at 74 km/h (46 mph) fuelled by Otto fuel without oxidant.

Reciprocating Quantum Heat Engine

Quantum heat engines are devices that generate power from heat that flows from a hot to a cold reservoir. The mechanism of operation of the engine can be described by the laws of quantum mechanics. Quantum refrigerators are devices that consume power with the purpose to pump heat from a cold to a hot reservoir.

In a reciprocating quantum heat engine the working medium is a quantum system such as spin systems or an harmonic oscillator. The Carnot cycle and Otto cycle are the ones most studied.

The quantum versions obey the laws of thermodynamics. In addition these models can justify the assumptions of endoreversible thermodynamics. A theoretical study has shown that it is possible and practical to build a reciprocating engine that is composed of a single oscillating atom. This is an area for future research and could have applications in nanotechnology.

Miscellaneous Engines

There are a large number of unusual varieties of piston engines that have various claimed advantages, many of which see little if any current use:

- Free-piston engine
- Swing-piston engine
- IRIS engine
- Bourke engine

Diesel Engine

Diesel generator on an oil tanker

The diesel engine (correctly known as a compression-ignition or CI engine) is an internal combustion engine in which ignition of the fuel that has been injected into the combustion chamber is caused by the high temperature which a gas achieves (i.e. the air) when greatly compressed (adiabatic compression). Diesel engines work by compressing only the air. This increases the air temperature inside the cylinder to such a high degree that it ignites atomised diesel fuel that is injected into the combustion chamber. This contrasts with spark-ignition engines such as a petrol engine (gasoline engine) or gas engine (using a gaseous fuel as opposed to petrol), which use a spark plug to ignite an air-fuel mixture. In compression-ignition engines, glow plugs (combustion chamber pre-warmers) may be used to aid starting in cold weather, or when the engine uses a lower compression-ratio, or both. The original compression-ignition engine operates on the "constant pressure" cycle of gradual combustion and produces no audible knock.

A diesel engine built by MAN AG in 1906

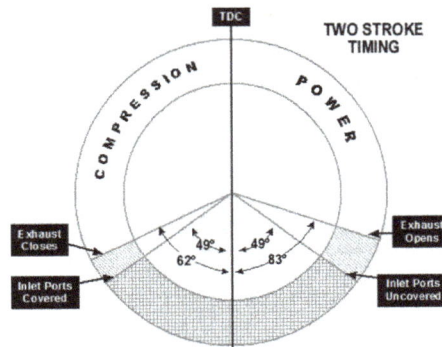

Detroit Diesel timing

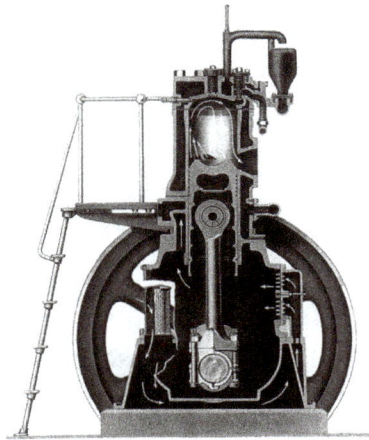

Fairbanks Morse model 32

The compression-ignition engine has the highest thermal efficiency (engine efficiency) of any practical internal or external combustion engine due to its very high expansion ratio and inherent lean burn which enables heat dissipation by the excess air. A small efficiency loss is also avoided compared to two-stroke non-direct-injection gasoline engines since unburnt fuel is not present at valve overlap and therefore no fuel goes directly from the intake/injection to the exhaust. Low-speed compression-ignition engines (as used in ships and other applications where overall engine weight is relatively unimportant) can have a thermal efficiency that exceeds 50%.

Compression-ignition engines are manufactured in two-stroke and four-stroke versions. They were originally used as a more efficient replacement for stationary steam engines. Since the 1910s they have been used in submarines and ships. Use in locomotives, trucks, heavy equipment and electricity generation plants followed later. In the 1930s, they slowly began to be used in a few automobiles. Since the 1970s, the use of compression-ignition engines in larger on-road and off-road vehicles in the US increased. According to the British Society of Motor Manufacturing and Traders, the EU average for compression-ignition cars accounts for 50% of the total sold, including 70% in France and 38% in the UK.

The world's largest compression-ignition engine is currently a Wärtsilä-Sulzer RTA96-C Common Rail marine compression-ignition, which produces a peak power output of 84.42 MW (113,210 hp) at 102 rpm.

History

Diesel's prototype engine

Diesels first experimental engine 1893

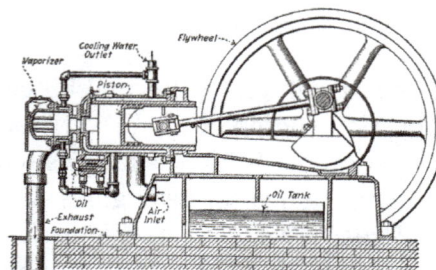

Hot bulb engine

The definition of a "Diesel" engine to many has become an engine that uses compression ignition. To some it may be an engine that uses heavy fuel oil. To others an engine that does not use spark ignition. However the original cycle proposed by Rudolph Diesel in 1892 was a constant temperature cycle (a cycle based on the Carnot theory) which would require much higher compression than what is needed for compression ignition. Diesel's idea was to compress the air so tightly that the temperature of the air would exceed that of combustion. In his 1892 US patent (granted in 1895) #542846 Diesel describes the compression required for his cycle:

> "pure atmospheric air is compressed, according to curve 1 2, to such a degree that, before ignition or combustion takes place, the highest pressure of the diagram and the highest temperature are obtained-that is to say, the temperature at which the subsequent combustion has to take place, not the burning or igniting point. To make this more clear, let it be assumed that the subsequent combustion shall take place at a temperature of 700°. Then in that case the initial pressure must be sixty-four atmospheres, or for 800° centigrade the pressure must be ninety atmospheres, and so on. Into the air thus compressed is then gradually introduced from the exterior finely divided fuel, which ignites on introduction, since the air is at a temperature far above the igniting-point of the fuel. The characteristic features of the cycle according to my present invention are therefore, increase of pressure and temperature up to the maximum, not by combustion, but prior to combustion by mechanical compression of air, and there upon the subsequent performance of work without increase of pressure and temperature by gradual combustion during a prescribed part of the stroke determined by the cut-oil".

In later years Diesel realized his original cycle would not work and he adopted the constant pressure cycle. Diesel describes the cycle in his 1895 patent application. Notice that there is no longer a mention of compression temperatures exceeding the temperature of combustion. Now all that is mentioned is the compression must be high enough for ignition.

> "1. In an internal-combustion engine, the combination of a cylinder and piston constructed and arranged to compress air to a degree producing a temperature above the igniting-point of the fuel, a supply for compressed air or gas; a fuel-supply; a distributing-valve for fuel, a passage from the air supply to the cylinder in communication with the fuel-distributing valve, an inlet to the cylinder in communication with the air-supply and with the fuel-valve, and a cut-oil, substantially as described."

History shows that the invention of the Diesel engine was not based solely on one man's idea but it was the culmination of many different ideas that would be developed over time.

In 1806 Claude and Nicéphore Niépce (brothers) developed the first known internal combustion engine and the first fuel injection system. The Pyréolophore fuel system used a blast of air provided by a bellows to atomize Lycopodium (a highly combustible fuel made from broad moss). Later coal dust mixed with resin became the fuel. Finally in 1816 they experimented with alcohol and white oil of petroleum (a fuel similar to kerosene). They discovered that the kerosene type fuel could be finely vaporized by passing it through a reed type device, this made the fuel highly combustible.

In 1874 George Brayton developed and patented a 2 stroke, oil fueled constant pressure engine "The Ready Motor". This engine used a metered pump to supply fuel to an injection device in which the oil was vaporize by air and burned as it entered the cylinder. These were some of the first practical internal combustion engines to supply motive power. Brayton's engines were installed in several boats, a rail car, 2 submarines and a bus. Early Diesel engines use a similar cycle.

Throughout the 1880s Brayton continued trying to improve his engines. In 1887 Brayton developed and patented a 4 stroke direct injection oil engine (US patent #432,114 of 1890, application filed in 1887) The fuel system used a variable quantity pump and liquid fuel high pressure spray type injection. The liquid was forced through a spring loaded relief type valve (injector) which caused the fuel to become divided into small droplets (vaporized). Injection was timed to occur at or near the peak of the compression stroke. A platinum igniter or ignitor provided the source of ignition. Brayton describes the invention as follows: "I have discovered that heavy oils can be mechanically converted into a finely-divided condition within a firing portion of the cylinder, or in a communicating firing chamber." Another part reads "I have for the first time, so far as my knowledge extends, regulated speed by variably controlling the direct discharge of liquid fuel into the combustion chamber or cylinder into a finely-divided condition highly favorable to immediate combustion". This was likely the first engine to use a lean burn system to regulate engine speed / output. In this manner the engine fired on every power stroke and speed / output was controlled solely by the quantity of fuel injected.

In 1890 Brayton developed and patented a 4 stroke air blast oil engine (US patent #432,260) The fuel system delivered a variable quantity of vaporized fuel to the center of the cylinder under pressure at or near the peak of the compression stroke. The ignition source was an igniter made from platinum wire. A variable quantity injection pump provided the fuel to an injector where it was mixed with air as it entered the cylinder. A small crank driven compressor provided the source for air. This engine also used the lean burn system.

Brayton died in 1893 but would be credited with the invention of the constant pressure Brayton cycle.

In 1885, the English inventor Herbert Akroyd Stuart began investigating the possibility of using paraffin oil (very similar to modern-day diesel) for an engine, which unlike petrol would be difficult to vaporise in a carburettor as its volatility is not sufficient to allow this.

The hot bulb engines, first prototyped in 1886 and built from 1891 by Richard Hornsby and Sons, used a pressurized fuel injection system. The Hornsby-Akroyd oil engine engine used a comparatively low compression ratio, so that the temperature of the air compressed in the combustion chamber at the end of the compression stroke was not high enough to initiate combustion. Combustion instead took place in a separated combustion chamber, the "vaporizer" or "hot bulb" mounted on the cylinder head, into which fuel was sprayed. Self-ignition occurred from contact between the fuel-air mixture and the hot walls of the vaporizer. As the engine's load increased, so did the temperature of the bulb, causing the ignition period to advance; to counteract pre-ignition, water was dripped into the air intake.

In 1892, Akroyd Stuart patented a water-jacketed vaporiser to allow compression ratios to be increased. In the same year, Thomas Henry Barton at Hornsbys built a working high-compression version for experimental purposes, whereby the vaporiser was replaced with a cylinder head, therefore not relying on air being preheated, but by combustion through higher compression ratios. It ran for six hours—the first time automatic ignition was produced by compression alone. This was five years before Rudolf Diesel built his well-known high-compression prototype engine in 1897.

Herbert Akroyd Stuart was a pioneer in developing compression ignition, Rudolf Diesel however, was subsequently credited with the compression ignition engine innovation. Higher compression and thermal efficiency is what distinguishes Diesel's patent of 3,500 kilopascals (508 psi).

In 1892 Diesel received patents in Germany, Switzerland, the United Kingdom and the United States for "Method of and Apparatus for Converting Heat into Work". In 1893 he described a "slow-combustion engine" that first compressed air thereby raising its temperature above the igniting-point of the fuel, then gradually introducing fuel while letting the mixture expand "against resistance sufficiently to prevent an essential increase of temperature and pressure", then cutting off fuel and "expanding without transfer of heat". In 1894 and 1895 he filed patents and addenda in various countries for his Diesel engine; the first patents were issued in Spain (No. 16,654), France (No. 243,531) and Belgium (No. 113,139) in December 1894, and in Germany (No. 86,633) in 1895 and the United States (No. 608,845) in 1898. He operated his first successful engine in 1897.

On February 17, 1894, the redesigned engine ran for 88 revolutions - one minute; with this news, Maschinefabrik Augsburg's stock rose by 30%, indicative of the tremendous anticipated demands for a more efficient engine. In 1896, Rudolph's rushed to have a prototype running, in order to maintain the patent. The first engine ready for testing was built on December 31, 1896; a much different engine than the one they had started with. In 1897, between deal signing, and brainstorming episodes they succeed, the engine runs; 16.93 kW with an efficiency of 16.6%, he is granted the patent. By 1898, Diesel had become a millionaire. His engines were used to power pipelines, electric and water plants, automobiles and trucks, and marine craft. They were soon to be used in mines, oil fields, factories, and transoceanic shipping.

Timeline

1800's.

- 1806 The Pyréolophore uses the first fuel injection system.
- 1874 George Brayton's constant pressure "Ready Motor" uses a metered fuel pump and burns oil fuel inside the cylinder.
- 1886: Herbert Akroyd Stuart builds a prototype hot bulb engine.
- 1887 George Brayton builds an engine that uses a spring loaded injector and solid metered injection system (lean burn combustion).
- 1890 George Brayton builds an "Air Blast" injection engine with a lean burn system.
- 1891: Herbert Akroyd Stuart patents an internal combustion engine that uses a "hot bulb" and pressurized fuel injection.
- 1892: February 23, Rudolf Diesel obtained a patent (RP 67207) titled "Arbeitsverfahren und Ausführungsart für Verbrennungsmaschinen" (Working Methods and Techniques for Internal Combustion Engines).

- 1892: Akroyd Stuart builds his first working Diesel engine.

- 1893: Diesel's essay titled Theory and Construction of a Rational Motor appeared.

- 1893: August 10, Diesel built his first prototype in Augsburg, This engine never ran under its own power.

- 1894 Diesel's second prototype runs for the first time.

- 1895 Diesel applies for a second patent US Patent # 608845

- 1896 Blackstone & Co, a Stamford farm implement they built lamp start oil engines.

- 1897: Adolphus Busch licenses rights to the Diesel Engine for the US and Canada.

- 1897 After 4 years Diesel's prototype engine is running and finally ready for efficiency testing and production.

- 1898: Diesel licensed his engine to Branobel, a Russian oil company interested in an engine that could consume non-distilled oil. Branobel's engineers spent four years designing a ship-mounted engine.

- 1899: Diesel licensed his engine to builders Krupp and Sulzer, who quickly became major manufacturers.

1900s

- 1902: Until 1910 MAN produced 82 copies of the stationary diesel engine.

- 1903: Two first diesel-powered ships were launched, both for river and canal operations: La Petite-Pierre in France, powered by Dyckhoff-built diesels, and Vandal tanker in Russia, powered by Swedish-built diesels with an electrical transmission.

- 1904: The French built the first diesel submarine, the Z.

- 1905: Four diesel engine turbochargers and intercoolers were manufactured by Büchl (CH), as well as a scroll-type supercharger from Creux (F) company.

- 1908: Prosper L'Orange and Deutz developed a precisely controlled injection pump with a needle injection nozzle.

- 1909: The prechamber with a hemispherical combustion chamber was developed by Prosper L'Orange with Benz.

1910s

- 1910: The Norwegian research ship Fram was a sailing ship fitted with an auxiliary diesel engine, and was thus the first ocean-going ship with a diesel engine.

- 1912: The Danish built the first ocean-going ship exclusively powered by a diesel engine, MS Selandia. The first locomotive with a diesel engine also appeared.

- 1913: US Navy submarines used NELSECO units. Rudolf Diesel died mysteriously when he

crossed the English Channel on the SS Dresden.

- 1914: German U-boats were powered by MAN diesels.

- 1919: Prosper L'Orange obtained a patent on a prechamber insert and made a needle injection nozzle. First diesel engine from Cummins.

1920s

One of the eight-cylinder 3200 I.H.P. Harland and Wolff—Burmeister & Wain Diesel engines installed in the motorship Glenapp. This was the highest powered Diesel engine yet (1920) installed in a ship. Note man standing lower right for size comparison.

- 1921: Prosper L'Orange built a continuous variable output injection pump.

- 1922: The first vehicle with a (pre-chamber) diesel engine was Agricultural Tractor Type 6 of the Benz Söhne agricultural tractor OE Benz Sendling.

- 1923: The first truck with pre-chamber diesel engine made by MAN and Benz. Daimler-Motoren-Gesellschaft testing the first air-injection diesel-engined truck.

- 1924: The introduction on the truck market of the diesel engine by commercial truck manufacturers in the IAA. Fairbanks-Morse starts building diesel engines.

- 1924-1925 Fairbanks Morse introduced the 2 stroke Y-VA and Model 32. It was the first cold start diesel manufactured by Fairbanks and would be come an icon of American industrial power.

- 1927: First truck injection pump and injection nozzles of Bosch. First passenger car prototype of Stoewer.

1930s

- 1930s: Caterpillar started building diesels for their tractors.

- 1930: First US diesel-power passenger car (Cummins powered Packard) built in Columbus, Indiana (US).

- 1930: Beardmore Tornado diesel engines power the British airship R101.

- 1932: Introduction of the strongest diesel truck in the world by MAN with 160 hp (120 kW).

- 1933: First European passenger cars with diesel engines (Citroën Rosalie); Citroën used an engine of the English diesel pioneer Sir Harry Ricardo. The car did not go into production due to legal restrictions on the use of diesel engines.

- 1934: First turbo diesel engine for a railway train by Maybach. First streamlined, stainless steel passenger train in the US, the Pioneer Zephyr, using a Winton engine.

- 1934: First tank equipped with diesel engine, the Polish 7TP.

- 1934–35: Junkers Motorenwerke in Germany started production of the Jumo aviation diesel engine family, the most famous of these being the Jumo 205, of which over 900 examples were produced by the outbreak of World War II.

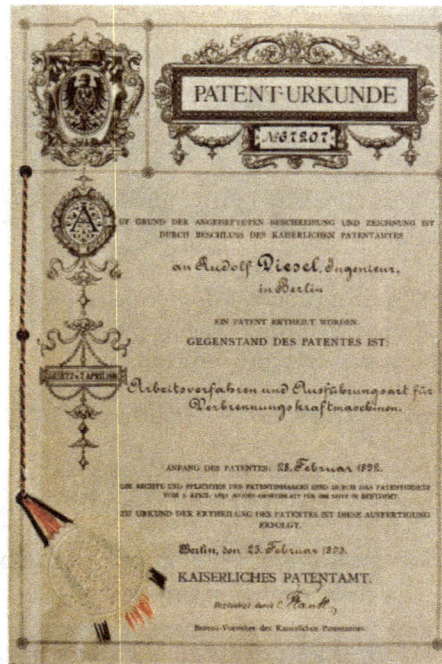

Rudolf Diesel's 1893 patent on his engine design

- 1936: Mercedes-Benz built the 260D diesel car. AT&SF inaugurated the diesel train Super Chief. The airship Hindenburg was powered by diesel engines. First series of passenger cars manufactured with diesel engine (Mercedes-Benz 260 D, Hanomag and Saurer). Daimler Benz airship diesel engine 602LOF6 for the LZ129 Hindenburg airship.

- 1937: The Soviet Union developed the Kharkiv model V-2 diesel engine, later used in the T-34 tanks, widely regarded as the best tank chassis of World War II.

- 1937: BMW 114 experimental airplane diesel engine development.

- 1938: General Motors forms the GM Diesel Division, later to become Detroit Diesel, and introduces the Series 71 inline high-speed medium-horsepower two stroke engine; GM's EMD subsidiary introduces the 567 two stroke medium-speed high-horsepower engine for locomotive, ship and stationary applications; These GM and EMD engines utilize GM's patented Unit injector.

- 1938: First turbo diesel engine of Saurer.

Fairbanks-Morse opposed piston diesel engines on the WWII submarine USS Pampanito (SS-383) (on display in San Francisco)

1940s

- 1942: Tatra started production of Tatra 111 with air-cooled V12 diesel engine.

- 1943–46: The common-rail (CRD) system was invented (and patented by) Clessie Cummins

- 1944: Development of air cooling for diesel engines by Klöckner Humboldt Deutz AG (KHD) for the production stage, and later also for Magirus Deutz.

- 1950s

- 1953: Turbo-diesel truck for Mercedes in small series.

- 1954: Turbo-diesel truck in mass production by Volvo. First diesel engine with an overhead cam shaft of Daimler Benz.

- 1958 EMD introduces turbocharging for its 567 series of medium speed, high horsepower locomotive, stationary and marine engines. Every subsequent engine (645 and 710) would incorporate this turbocharger.

1960s

- 1960: The diesel drive displaced steam turbines and coal fired steam engines.

- 1962–65: A diesel compression braking system, eventually to be manufactured by Jacobs (of drill chuck fame) and nicknamed the "Jake Brake", was invented and patented by Clessie Cummins.

- 1968: Peugeot introduced the first 204 small cars with a transversally mounted diesel engine and front-wheel drive.

1970s

- 1973: DAF produced an air-cooled diesel engine.

- 1976 February: Tested a diesel engine for the Volkswagen Golf passenger car. The Cummins Common Rail injection system was further developed by the ETH Zurich from 1976 to 1992.

- 1978: Mercedes-Benz produced the first passenger car with a turbo-diesel engine (Mercedes-Benz 300 SD). Oldsmobile introduced the first passenger car diesel engine produced by an American car company.

- 1979: Peugeot 604, the first turbo-diesel car to be sold in Europe.

1980s

- 1985: ATI Intercooler diesel engine from DAF. European Truck Common Rail system with the IFA truck type W50 introduced.

- 1986: BMW 524td, the world's first passenger car equipped with an electronically controlled injection pump (developed by Bosch). The same year, the Fiat Croma was the first passenger car in the world to have a direct injection (turbocharged) diesel engine.

- 1987: Most powerful production truck with a 460 hp (340 kW) MAN diesel engine.

- 1989: Audi 100, the first passenger car in the world with a turbocharged direct injection and electronic control diesel engine.

1990s

- 1991: European emission standards Euro 1 met with the truck diesel engine of Scania.

- 1993: Pump nozzle injection introduced in Volvo truck engines.

- 1994: Unit injector system by Bosch for diesel engines. Mercedes-Benz unveils the first automotive diesel engine with four valves per cylinder. Medium speed high horsepower locomotive, ship and stationary diesel engines have utilized four valves per cylinder since at least 1938.

- 1995: First successful use of common rail in a production vehicle, by Denso in Japan, Hino "Rising Ranger" truck.

- 1996: First diesel engine with direct injection and four valves per cylinder, used in the Opel Vectra.

- 1997: First common rail diesel engine in a passenger car, the Alfa Romeo 156.

- 1998: BMW made history by winning the 24 Hour Nürburgring race with the 320d, powered by a two-litre, four-cylinder diesel engine. The combination of high-performance with better fuel efficiency allowed the team to make fewer pit stops during the long endurance race. Volkswagen introduces three and four-cylinder turbodiesel engines, with Bosch-developed electronically controlled unit injectors. Smart presented the first common rail three-cylinder diesel engine used in a passenger car (the Smart City Coupé).

- 1999: Euro 3 of Scania and the first common rail truck diesel engine of Renault.

2000s

- 2002: A street-driven Dodge Dakota pickup with a 735 horsepower (548 kW) diesel engine built at Gale banks engineering hauls its own service trailer to the Bonneville Salt Flats and set an FIA land speed record as the world's fastest pickup truck with a one-way run of 222 mph (357 km/h) and a two-way average of 217 mph (349 km/h).

- 2003: Piezoelectric injector technology by Bosch, Siemens and Delphi.

- 2004: In Western Europe, the proportion of passenger cars with diesel engine exceeded 50%. Selective catalytic reduction (SCR) system in Mercedes, Euro 4 with EGR system and particle filters of MAN. Audi A8 3.0 TDI is the first production vehicle in the world with common rail injection and piezoelectric injectors.

- 2006: Audi R10 TDI won the 12 Hours of Sebring and defeated all other engine concepts. The same car won the 2006 24 Hours of Le Mans. Euro 5 for all Iveco trucks. JCB Diesel-max broke the FIA diesel land speed record from 1973, eventually setting the new record at over 350 mph (563 km/h).

- 2007: Lombardini develops a new 440 cc twin-cyinder common rail diesel engine, which two years later sees application in automotive use, in the Ligier microcars. At the time, this engine was considered to be the smallest twin-cyinder engine with a common rail system.

- 2008: Subaru introduced the first horizontally opposed diesel engine to be fitted to a passenger car. This is a Euro 5 compliant engine with an EGR system. SEAT wins the drivers' title and the manufacturers' title in the FIA World Touring Car Championship with the SEAT León TDI. The achievements are repeated in the following season.

- 2009: Volkswagen won the 2009 Dakar Rally held in Argentina and Chile. The first diesel to do so. Race Touareg 2 models finished first and second. The same year, Volvo is claimed the world's strongest truck with their FH16 700. An inline 6-cylinder, 16 L (976 cu in) 700 hp (522 kW) diesel engine producing 3150 Nm (2323.32 lb•ft) of torque and fully complying with Euro 5 emission standards.

2010s

- 2010: Mitsubishi developed and started mass production of its 4N13 1.8 L DOHC I4, the world's first passenger car diesel engine that features a variable valve timing system. Scania AB's V8 had the highest torque and power ratings of any truck engine, 730 hp (544 kW) and 3,500 N·m (2,581 ft·lb).

- 2011: Piaggio launches a twin-cyinder turbodiesel engine, with common rail injection, on its new range of microvans.

- 2012: Common rail systems working with pressures of 2,500 bar launched.

Operating Principle

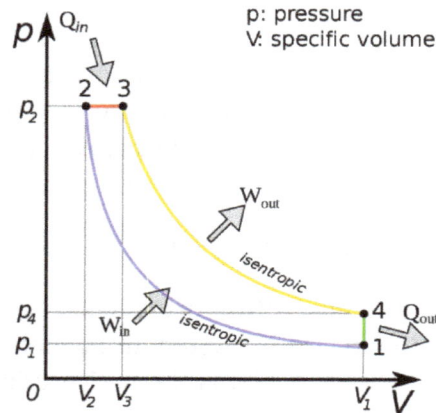

p-V Diagram for the Ideal Diesel cycle. The cycle follows the numbers 1–4 in clockwise direction. The horizontal axis is Volume of the cylinder. In the diesel cycle the combustion occurs at almost constant pressure. On this diagram the work that is generated for each cycle corresponds to the area within the loop.

Diesel engine model, left side

Diesel engine model, right side

The diesel internal combustion engine differs from the gasoline powered Otto cycle by using highly compressed hot air to ignite the fuel rather than using a spark plug (compression ignition rather than spark ignition).

In the true diesel engine, only air is initially introduced into the combustion chamber. The air is then compressed with a compression ratio typically between 15:1 and 23:1. This high compression causes the temperature of the air to rise. At about the top of the compression stroke, fuel is injected directly into the compressed air in the combustion chamber. This may be into a (typically toroidal) void in the top of the piston or a pre-chamber depending upon the design of the engine. The fuel injector ensures that the fuel is broken down into small droplets, and that the fuel is distributed evenly. The heat of the compressed air vaporizes fuel from the surface of the droplets. The vapour is then ignited by the heat from the compressed air in the combustion chamber, the droplets continue to vaporise from their surfaces and burn, getting smaller, until all the fuel in the droplets has been burnt. Combustion occurs at a substantially constant pressure during the initial part of the power stroke. The start of vaporisation causes a delay before ignition and the characteristic diesel knocking sound as the vapour reaches ignition temperature and causes an abrupt increase in pressure above the piston (not shown on the P-V indicator diagram). When combustion is complete the combustion gases expand as the piston descends further; the high pressure in the cylinder drives the piston downward, supplying power to the crankshaft.

As well as the high level of compression allowing combustion to take place without a separate ignition system, a high compression ratio greatly increases the engine's efficiency. Increasing the compression ratio in a spark-ignition engine where fuel and air are mixed before entry to the cylinder is limited by the need to prevent damaging pre-ignition. Since only air is compressed in a diesel engine, and fuel is not introduced into the cylinder until shortly before top dead centre (TDC), premature detonation is not a problem and compression ratios are much higher.

The p–V diagam is a simplified and idealised representation of the events involved in a Diesel engine cycle, arranged to illustrate the similarity with a Carnot cycle. Starting at 1, the piston is at bottom dead centre and both valves are closed at the start of the compression stroke; the cylinder contains air at atmospheric pressure. Between 1 and 2 the air is compressed adiabatically—that is without heat transfer to or from the environment—by the rising piston. (This is only approximately true since there will be some heat exchange with the cylinder walls.) During this compression, the volume is reduced, the pressure and temperature both rise. At or slightly before 2 (TDC) fuel is injected and burns in the compressed hot air. Chemical energy is released and this constitutes an injection of thermal energy (heat) into the compressed gas. Combustion and heating occur between 2 and 3. In this interval the pressure remains constant since the piston descends, and the volume increases; the temperature rises as a consequence of the energy of combustion. At 3 fuel injection and combustion are complete, and the cylinder contains gas at a higher temperature than at 2. Between 3 and 4 this hot gas expands, again approximately adiabatically. Work is done on the system to which the engine is connected. During this expansion phase the volume of the gas rises, and its temperature and pressure both fall. At 4 the exhaust valve opens, and the pressure falls abruptly to atmospheric (approximately). This is unresisted expansion and no useful work is done by it. Ideally the adiabatic expansion should continue, extending the line 3–4 to the right until the pressure falls to that of the surrounding air, but the loss of efficiency caused by this unresisted expansion is justified by the practical difficulties involved in recovering it (the engine would have to be much larger). After the opening of the exhaust valve, the exhaust stroke follows, but this (and the following induction stroke) are not shown on the diagram. If shown, they would be represented by a low-pressure loop at the bottom of the diagram. At 1 it is assumed that the exhaust and induction strokes have been completed, and the cylinder is again filled with air. The piston-cylinder system absorbs energy between 1 and 2—this is the work needed to compress the air in the cylinder, and is provided by mechanical kinetic energy stored in the flywheel of the engine. Work output is done by the piston-cylinder combination between 2 and 4. The difference between these two increments of work is the indicated work output per cycle, and is represented by the area enclosed by the p–V loop. The adiabatic expansion is in a higher pressure range than that of the compression because the gas in the cylinder is hotter during expansion than during compression. It is for this reason that the loop has a finite area, and the net output of work during a cycle is positive.

Major Advantages

Diesel engines have several advantages over other internal combustion engines:

- They burn less fuel than a petrol engine performing the same work, due to the engine's higher temperature of combustion and greater expansion ratio. Gasoline engines are typically 30% efficient while diesel engines can convert over 45% of the fuel energy into mechanical energy.

- They have no high voltage electrical ignition system, resulting in high reliability and easy adaptation to damp environments. The absence of coils, spark plug wires, etc., also eliminates a source of radio frequency emissions which can interfere with navigation and communication equipment, which is especially important in marine and aircraft applications, and for preventing interference with radio telescopes.

- The longevity of a diesel engine is generally about twice that of a petrol engine due to the increased strength of parts used. Diesel fuel has better lubrication properties than petrol as well. Indeed, in unit injectors, the fuel is employed for three distinct purposes: injector lubrication, injector cooling and injection for combustion.

Bus powered by biodiesel

Diesel fuel is distilled directly from petroleum. Distillation yields some gasoline, but the yield would be inadequate without catalytic reforming, which is a more costly process.

- Although diesel fuel will burn in open air using a wick, it does not release a large amount of flammable vapor which could lead to an explosion. The low vapor pressure of diesel is especially advantageous in marine applications, where the accumulation of explosive fuel-air mixtures is a particular hazard. For the same reason, diesel engines are immune to vapor lock.

- For any given partial load the fuel efficiency (mass burned per energy produced) of a diesel engine remains nearly constant, as opposed to petrol and turbine engines which use proportionally more fuel with partial power outputs.

- They generate less waste heat in cooling and exhaust.

- Diesel engines can accept super- or turbo-charging pressure without any natural limit, constrained only by the strength of engine components. This is unlike petrol engines, which inevitably suffer detonation at higher pressure.

- The carbon monoxide content of the exhaust is minimal.

- Biodiesel is an easily synthesized, non-petroleum-based fuel (through transesterification) which can run directly in many diesel engines, while gasoline engines either need adaptation to run synthetic fuels or else use them as an additive to gasoline (e.g., ethanol added to gasohol).

Early Fuel Injection Systems

Diesel's original engine injected fuel with the assistance of compressed air, which atomized the fuel and forced it into the engine through a nozzle (a similar principle to an aerosol spray). The nozzle opening was closed by a pin valve lifted by the camshaft to initiate the fuel injection before top dead centre (TDC). This is called an air-blast injection. Driving the compressor used some power but the efficiency and net power output was more than any other combustion engine at that time.

Diesel engines in service today raise the fuel to extreme pressures by mechanical pumps and deliver it to the combustion chamber by pressure-activated injectors without compressed air. With direct injected diesels, injectors spray fuel through 4 to 12 small orifices in its nozzle. The early air injection diesels always had a superior combustion without the sharp increase in pressure during combustion. Research is now being performed and patents are being taken out to again use some form of air injection to reduce the nitrogen oxides and pollution, reverting to Diesel's original implementation with its superior combustion and possibly quieter operation. In all major aspects, the modern diesel engine holds true to Rudolf Diesel's original design, that of igniting fuel by compression at an extremely high pressure within the cylinder. With much higher pressures and high technology injectors, present-day diesel engines use the so-called solid injection system applied by Herbert Akroyd Stuart for his hot bulb engine. The indirect injection engine could be considered the latest development of these low speed hot bulb ignition engines.

Fuel Delivery

Over the years many different injection methods have been used. These can be described as:

- Air blast, where the fuel is blown into the cylinder by a blast of air.

- Solid fuel / hydraulic injection where the fuel is pushed through a spring loaded valve / injector to produce a combustable mist.

- Mechanical Unit injector where the injector is directly operated by a cam and fuel quantity is controlled by a rack or lever .

- Mechanical Electronic Unit Injector where the injector is operated by a cam and fuel quantity is controlled electronically.

- Common Rail Mechanical Injection, Fuel is at high pressure in a common rail and controlled by mechanical means.

- Common Rail electronic injection, Fuel is at high pressure in a common rail and controlled electronically.

Diesel engines are also produced with two significantly different injection locations. "Direct" and "Indirect". Indirect injected engines place the injector in a pre-combustion chamber in the head which due to thermal losses generally require a "glow plug" to start and very high compression ratio, usually in the range of 21:1 to 23:1 ratio. Direct injected engines use a generally donut shaped combustion chamber void on the top of the piston. Thermal efficiency losses are significantly lower in DI engines which facilitates a much lower compression ratio generally between 14:1 and 20:1 but most DI engines are closer to 17:1. The direct injected process is significantly more internally violent and thus requires careful design, and more robust construction. The lower compression ratio also creates challenges for emissions due to partial burn. Turbocharging is particularly suited to DI engines since the low compression ratio facilitates meaningful forced induction, and the increase in airflow allows capturing additional fuel efficiency not only from more complete combustion, but also from lowering parasitic efficiency losses when properly operated, by widening both power and efficiency curves. The violent combustion process of direct injection also creates more noise, but modern designs using "split shot" injectors or similar multi shot processes have dramatically amended this issue by firing a small charge of fuel before the main delivery which pre-charges the combustion chamber for a less abrupt and in most cases slightly cleaner burn.

A vital component of all diesel engines is a mechanical or electronic governor which regulates the idling speed and maximum speed of the engine by controlling the rate of fuel delivery. Unlike Otto-cycle engines, incoming air is not throttled and a diesel engine without a governor cannot have a stable idling speed and can easily overspeed, resulting in its destruction. Mechanically governed fuel injection systems are driven by the engine's gear train. These systems use a combination of springs and weights to control fuel delivery relative to both load and speed. Modern electronically controlled diesel engines control fuel delivery by use of an electronic control module (ECM) or electronic control unit (ECU). The ECM/ECU receives an engine speed signal, as well as other operating parameters such as intake manifold pressure and fuel temperature, from a sensor and controls the amount of fuel and start of injection timing through actuators to maximise power and efficiency and minimise emissions. Controlling the timing of the start of injection of fuel into the cylinder is a key to minimizing emissions, and maximizing fuel economy (efficiency), of the engine. The timing is measured in degrees of crank angle of the piston before top dead centre. For example, if the ECM/ECU initiates fuel injection when the piston is 10° before TDC, the start of injection, or timing, is said to be 10° BTDC. Optimal timing will depend on the engine design as well as its speed and load, and is usually 4° BTDC in 1,350–6,000 HP, net, "medium speed" locomotive, marine and stationary diesel engines.

Advancing the start of injection (injecting before the piston reaches to its SOI-TDC) results in higher in-cylinder pressure and temperature, and higher efficiency, but also results in increased engine noise due to faster cylinder pressure rise and increased oxides of nitrogen (NO_x) formation due to higher combustion temperatures. Delaying start of injection causes incomplete combustion, reduced fuel efficiency and an increase in exhaust smoke, containing a considerable amount of particulate matter and unburned hydrocarbons.

Mechanical and Electronic Injection

Many configurations of fuel injection have been used over the course of the 20th century.

Most present-day diesel engines use a mechanical single plunger high-pressure fuel pump driven by the engine crankshaft. For each engine cylinder, the corresponding plunger in the fuel pump measures out the correct amount of fuel and determines the timing of each injection. These engines use injectors that are very precise spring-loaded valves that open and close at a specific fuel pressure. Separate high-pressure fuel lines connect the fuel pump with each cylinder. Fuel volume for each single combustion is controlled by a slanted groove in the plunger which rotates only a few degrees releasing the pressure and is controlled by a mechanical governor, consisting of weights rotating at engine speed constrained by springs and a lever. The injectors are held open by the fuel pressure. On high-speed engines the plunger pumps are together in one unit. The length of fuel lines from the pump to each injector is normally the same for each cylinder in order to obtain the same pressure delay.

A cheaper configuration on high-speed engines with fewer than six cylinders is to use an axial-piston distributor pump, consisting of one rotating pump plunger delivering fuel to a valve and line for each cylinder (functionally analogous to points and distributor cap on an Otto engine).

Many modern systems have a single fuel pump which supplies fuel constantly at high pressure with a common rail (single fuel line common) to each injector. Each injector has a solenoid operated by an electronic control unit, resulting in more accurate control of injector opening times that depend on other control conditions, such as engine speed and loading, and providing better engine performance and fuel economy.

Both mechanical and electronic injection systems can be used in either direct or indirect injection configurations.

Two-stroke diesel engines with mechanical injection pumps can be inadvertently run in reverse, albeit in a very inefficient manner, possibly damaging the engine. Large ship two-stroke diesels are designed to run in either direction, obviating the need for a gearbox.

Indirect Injection

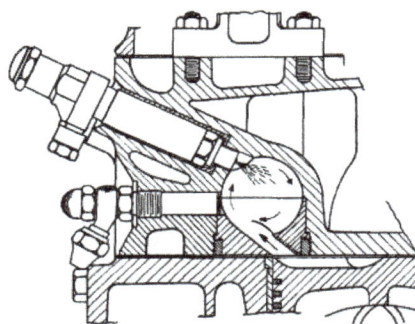

Ricardo Comet indirect injection chamber

An indirect Diesel injection system (IDI) engine delivers fuel into a small chamber called a swirl chamber, pre combustion chamber, pre chamber or ante-chamber, which is connected to the cylinder by a narrow air passage. Generally the goal of the pre camber is to create increased turbulence for better air / fuel mixing. This system also allows for a smoother, quieter running engine, and because fuel mixing is assisted by turbulence, injector pressures can be lower. Most IDI systems tend to use a single orifice injector. The pre-chamber has the disadvantage of lowering efficiency due to increased heat loss to the engine's cooling system, restricting the combustion burn, thus reducing the efficiency by 5–10%.. IDI engines are also more difficult to start and usually require the use of glow plugs. IDI engines may be cheaper to build but generally require a higher compression ratio than the DI counterpart. IDI also makes it easier to produce smooth, quieter running engines with a simple mechanical injection system since exact injection timing is not as critical. Most modern automotive engines are DI which have the benefits of greater efficiency, easier starting, however IDI engines can still be found in the many ATV and small Diesel applications.

Direct Injection

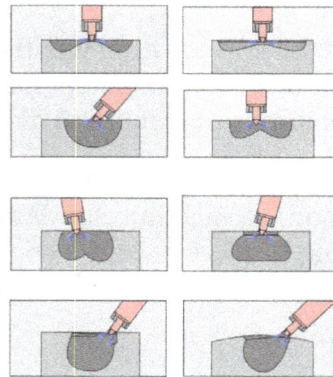

Different types of piston bowls

Direct injection diesel engines inject fuel directly into the cylinder. Usually there is a combustion cup in the top of the piston where the fuel is sprayed. Many different methods of injection can be used.

Electronic control of the fuel injection transformed the direct injection engine by allowing much greater control over the combustion.

Unit Direct Injection

Unit direct injection also injects fuel directly into the cylinder of the engine. In this system the injector and the pump are combined into one unit positioned over each cylinder controlled by the camshaft. Each cylinder has its own unit eliminating the high-pressure fuel lines, achieving a more consistent injection. This type of injection system, also developed by Bosch, is used by Volkswagen AG in cars (where it is called a Pumpe-Düse-System—literally pump-nozzle system) and by Mercedes-Benz ("PLD") and most major diesel engine manufacturers in large commercial engines (MAN SE, CAT, Cummins, Detroit Diesel, Electro-Motive Diesel, Volvo). With recent advancements, the pump pressure has been raised to 2,400 bars (240 MPa; 35,000 psi), allowing injection parameters similar to common rail systems.

Common Rail Direct Injection

"Common Rail" injection was first used in production by Atlas Imperial Diesel in the 1920s. The rail pressure was kept at a steady 2,000 - 4,000 psi. In the injectors a needle was mechanically lifted off of the seat to create the injection event. Modern common rail systems use very high-pressures. In these systems an engine driven pump pressurizes fuel at up to 2,500 bar (250 MPa; 36,000 psi), in a "common rail". The common rail is a tube that supplies each computer-controlled injector containing a precision-machined nozzle and a plunger driven by a solenoid or piezoelectric actuator.

Cold Weather Problems

Starting

In cold weather, high speed diesel engines can be difficult to start because the mass of the cylinder block and cylinder head absorb the heat of compression, preventing ignition due to the higher surface-to-volume ratio. Pre-chambered engines make use of small electric heaters inside the pre-chambers called glowplugs, while direct-injected engines have these glowplugs in the combustion chamber.

Many engines use resistive heaters in the intake manifold to warm the inlet air for starting, or until the engine reaches operating temperature. Engine block heaters (electric resistive heaters in the engine block) connected to the utility grid are used in cold climates when an engine is turned off for extended periods (more than an hour), to reduce startup time and engine wear. Block heaters are also used for emergency power standby Diesel-powered generators which must rapidly pick up load on a power failure. In the past, a wider variety of cold-start methods were used. Some engines, such as Detroit Diesel engines used[when?] a system to introduce small amounts of ether into the inlet manifold to start combustion. Others used a mixed system, with a resistive heater burning methanol. An impromptu method, particularly on out-of-tune engines, is to manually spray an aerosol can of ether-based engine starter fluid into the intake air stream (usually through the intake air filter assembly).

Gelling

Diesel fuel is also prone to waxing or gelling in cold weather; both are terms for the solidification of diesel oil into a partially crystalline state. The crystals build up in the fuel line (especially in fuel filters), eventually starving the engine of fuel and causing it to stop running. Low-output electric heaters in fuel tanks and around fuel lines are used to solve this problem. Also, most engines have a spill return system, by which any excess fuel from the injector pump and injectors is returned to the fuel tank. Once the engine has warmed, returning warm fuel prevents waxing in the tank.

Due to improvements in fuel technology with additives, waxing rarely occurs in all but the coldest weather when a mix of diesel and kerosene may be used to run a vehicle. Gas stations in regions with a cold climate are required to offer winterized diesel in the cold seasons that allow operation below a specific Cold Filter Plugging Point. In Europe these diesel characteristics are described in the EN 590 standard.

Supercharging and Turbocharging

Many diesels are now turbocharged and some are both turbo charged and supercharged. A turbocharged engine can produce more power than a naturally aspirated engine of the same configuration. A supercharger is powered mechanically by the engine's crankshaft, while a turbocharger is powered by the engine exhaust. Turbocharging can improve the fuel economy of diesel engines by recovering waste heat from the exhaust, increasing the excess air factor, and increasing the ratio of engine output to friction losses.

A two-stroke engine does not have a discrete exhaust and intake stroke and thus is incapable of self-aspiration. Therefore, all two-stroke engines must be fitted with a blower or some form of compressor to charge the cylinders with air and assist in dispersing exhaust gases, a process referred to as scavenging. In some cases, the engine may also be fitted with a turbocharger, whose output is directed into the blower inlet.

A few designs employ a hybrid blower / turbocharger (a turbo-compressor system) for scavenging and charging the cylinders, which device is mechanically driven at cranking and low speeds to act as a blower, but which acts as a true turbocharger at higher speeds and loads. A hybrid turbocharger can revert to compressor mode during commands for large increases in engine output power.

As turbocharged or supercharged engines produce more power for a given engine size as compared to naturally aspirated engines, attention must be paid to the mechanical design of components, lubrication, and cooling to handle the power. Pistons are usually cooled with lubrication oil sprayed on the bottom of the piston. Large "Low speed" engines may use water, sea water, or oil supplied through telescoping pipes attached to the crosshead to cool the pistons.

Types

Size Groups

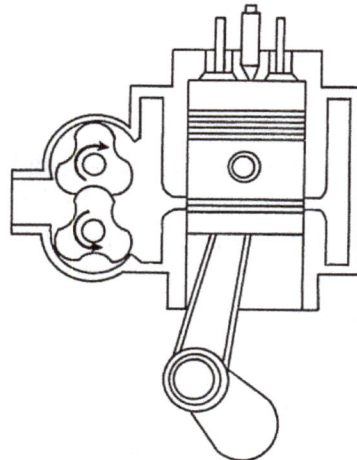

Two Cycle Diesel engine with Roots blower, typical of Detroit Diesel and some Electro-Motive Diesel Engines

There are three size groups of Diesel engines

- Small—under 188 kW (252 hp) output

- Medium

- Large

Basic Types

There are two basic types of Diesel Engines

- Four stroke cycle
- Two stroke cycle

Early Engines

In 1897, when the first Diesel engine was completed Adolphus Busch traveled to Cologne and negotiated exclusive right to produce the Diesel engine in the US and Canada. In his examination of the engine, it was noted that the Diesel at that time operated at thermodynamic efficiencies of 27%, while a typical expansion steam engine would operate at about 7-10%.

In the early decades of the 20th century, when large diesel engines were first being used, the engines took a form similar to the compound steam engines common at the time, with the piston being connected to the connecting rod by a crosshead bearing. Following steam engine practice some manufacturers made double-acting two-stroke and four-stroke diesel engines to increase power output, with combustion taking place on both sides of the piston, with two sets of valve gear and fuel injection. While it produced large amounts of power, the double-acting diesel engine's main problem was producing a good seal where the piston rod passed through the bottom of the lower combustion chamber to the crosshead bearing, and no more were built. By the 1930s turbochargers were fitted to some engines. Crosshead bearings are still used to reduce the wear on the cylinders in large long-stroke main marine engines.

Modern High and Medium-speed Engines

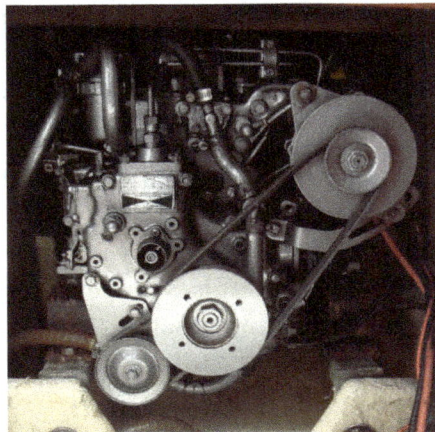

A Yanmar 2GM20 marine diesel engine, installed in a sailboat

As with petrol engines, there are two classes of diesel engines in current use: two-stroke and four-stroke. The four-stroke type is the "classic" version, tracing its lineage back to Rudolf Diesel's prototype. It is also the most commonly used form, being the preferred power source for many motor vehicles, especially buses and trucks. Much larger engines, such as used for railroad locomotion and marine propulsion, are often two-stroke units, offering a more favourable power-to-weight ratio, as well as better fuel economy. The most powerful engines in the world are two-stroke diesels of mammoth dimensions.

Two-stroke diesel engine operation is similar to that of petrol counterparts, except that fuel is not mixed with air before induction, and the crankcase does not take an active role in the cycle. The traditional two-stroke design relies upon a mechanically driven positive displacement blower to charge the cylinders with air before compression and ignition. The charging process also assists in expelling (scavenging) combustion gases remaining from the previous power stroke.

The archetype of the modern form of the two-stroke diesel is the (high-speed) Detroit Diesel Series 71 engine, designed by Charles F. "Boss" Kettering and his colleagues at General Motors Corporation in 1938, in which the blower pressurizes a chamber in the engine block that is often referred to as the "air box". The (very much larger medium-speed) Electro-Motive Diesel engine is used as the prime mover in EMD diesel-electric locomotive, marine and stationary applications, and was designed by the same team, and is built to the same principle. However, a significant improvement built into most later EMD engines is the mechanically assisted turbo-compressor, which provides charge air using mechanical assistance during starting (thereby obviating the necessity for Roots-blown scavenging), and provides charge air using an exhaust gas-driven turbine during normal operations—thereby providing true turbocharging and additionally increasing the engine's power output by at least fifty percent.

Three English Electric 7SRL diesel-alternator sets being installed at the Saateni Power Station, Zanzibar 1955

In a two-stroke diesel engine, as the cylinder's piston approaches the bottom dead centre exhaust ports or valves are opened relieving most of the excess pressure after which a passage between the air box and the cylinder is opened, permitting air flow into the cylinder. The air flow blows the remaining combustion gases from the cylinder—this is the scavenging process. As the piston passes through bottom centre and starts upward, the passage is closed and compression commences, culminating in fuel injection and ignition. Refer to two-stroke diesel engines for more detailed coverage of aspiration types and supercharging of two-stroke diesel engines.

Normally, the number of cylinders are used in multiples of two, although any number of cylinders can be used as long as the load on the crankshaft is counterbalanced to prevent excessive vibration. The inline-six-cylinder design is the most prolific in light- to medium-duty engines, though small V8 and larger inline-four displacement engines are also common. Small-capacity engines (generally considered to be those below five litres in capacity) are generally four- or six-cylinder types, with the four-cylinder being the most common type found in automotive uses. Five-cylinder diesel engines have also been produced, being a compromise between the smooth running of the six-cylinder and the space-efficient dimensions of the four-cylinder. Diesel engines for smaller plant machinery, boats, tractors, generators and pumps may be four, three or two-cylinder types, with the single-cylinder diesel engine remaining for light stationary work. Direct reversible two-stroke marine diesels need at least three cylinders for reliable restarting forwards and reverse, while four-stroke diesels need at least six cylinders.

The desire to improve the diesel engine's power-to-weight ratio produced several novel cylinder arrangements to extract more power from a given capacity. The uniflow opposed-piston engine uses two pistons in one cylinder with the combustion cavity in the middle and gas in- and outlets at the ends. This makes a comparatively light, powerful, swiftly running and economic engine suitable for use in aviation. An example is the Junkers Jumo 204/205. The Napier Deltic engine, with three cylinders arranged in a triangular formation, each containing two opposed pistons, the whole engine having three crankshafts, is one of the better known.

Gas Generator

Before 1950, Sulzer started experimenting with two-stroke engines with boost pressures as high as 6 atmospheres, in which all the output power was taken from an exhaust gas turbine. The two-stroke pistons directly drove air compressor pistons to make a positive displacement gas generator. Opposed pistons were connected by linkages instead of crankshafts. Several of these units could be connected to provide power gas to one large output turbine. The overall thermal efficiency was roughly twice that of a simple gas turbine. This system was derived from Raúl Pateras Pescara's work on free-piston engines in the 1930s.

Advantages and Disadvantages Versus Spark-Ignition Engines

Fuel Economy

The MAN S80ME-C7 low speed diesel engines use 155 grams (5.5 oz) of fuel per kWh for an overall energy conversion efficiency of 54.4%, which is the highest conversion of fuel into power by any single-cycle internal or external combustion engine (The efficiency of a combined cycle gas turbine system can exceed 60%.) Diesel engines are more efficient than gasoline (petrol) engines of the same power rating, resulting in lower fuel consumption. A common margin is 40% more miles per gallon for an efficient turbodiesel. For example, the current model Škoda Octavia, using Volkswagen Group engines, has a combined Euro rating of 6.2 L/100 km (46 mpg$_{-imp}$; 38 mpg$_{-US}$) for the 102 bhp (76 kW) petrol engine and 4.4 L/100 km (64 mpg$_{-imp}$; 53 mpg$_{-US}$) for the 105 bhp (78 kW) diesel engine.

However, such a comparison does not take into account that diesel fuel is denser and contains about 15% more energy by volume. Although the calorific value of the fuel is slightly lower at 45.3 MJ/kg (megajoules per kilogram) than petrol at 45.8 MJ/kg, liquid diesel fuel is significantly denser than liquid petrol. This is significant because volume of fuel, in addition to mass, is an important consideration in mobile applications.

Adjusting the numbers to account for the energy density of diesel fuel, the overall energy efficiency is still about 20% greater for the diesel version.

While a higher compression ratio is helpful in raising efficiency, diesel engines are much more efficient than gasoline (petrol) engines when at low power and at engine idle. Unlike the petrol engine, diesels lack a butterfly valve (throttle) in the inlet system, which closes at idle. This creates parasitic loss and destruction of availability of the incoming air, reducing the efficiency of petrol engines at idle. In many applications, such as marine, agriculture, and railways, diesels are left idling and unattended for many hours, sometimes even days. These advantages are especially attractive in locomotives.

Even though diesel engines have a theoretical fuel efficiency of 75%, in practice it is lower. Engines in large diesel trucks, buses, and newer diesel cars can achieve peak efficiencies around 45%, and could reach 55% efficiency in the near future. However, average efficiency over a driving cycle is lower than peak efficiency. For example, it might be 37% for an engine with a peak efficiency of 44%.

Torque

Diesel engines produce more torque than petrol engines for a given displacement due to their higher compression ratio. Higher pressure in the cylinder and higher forces on the connecting rods and crankshaft require stronger, heavier components. Heavier rotating components prevent diesel engines from revving as high as petrol engines for a given displacement. Diesel engines generally have similar power and inferior power to weight ratios as compared to petrol engines. Petrol engines must be geared lower to get the same torque as a comparable diesel but since petrol engines rev higher both will have similar acceleration. An arbitrary amount of torque at the wheels can be gained by gearing any power source down sufficiently (including a hand crank). For example, a theoretical engine with a constant 200 ft.lbs of torque and a 3000 rpm rev limit has just as much power (a little over 114 hp) as another theoretical engine with a constant maximum 100 ft.lbs of torque and a 6000 rpm rev limit. A (lossless) 2 to 1 reduction gear on the second engine will output a constant maximum 200 ft.lbs of torque at a maximum of 3000 rpm, with no change in power. Comparing engines based on (maximum) torque is just as useful as comparing them based on (maximum) rpm.

Power

Conditions in the diesel engine differ from the spark-ignition engine due to the different thermodynamic cycle. In addition the power and engine speed are directly controlled by the fuel supply, rather than by controlling the air supply as in an otto cycle engine.

The average diesel engine has a poorer power-to-weight ratio than the petrol engine. This is because the diesel must operate at lower engine speeds due to the need for heavier, stronger parts to resist the operating pressure caused by the high compression ratio of the engine, which increases the forces on the parts due to inertial forces. In addition, diesels are often built with stronger parts to give them longer lives and better reliability, important considerations in industrial applications. Diesel engines usually have longer stroke lengths chiefly to facilitate achieving the necessary compression ratios. As a result, piston and connecting rods are heavier and more force must be transmitted through the connecting rods and crankshaft to change the momentum of the piston. This is another reason that a diesel engine must be stronger for the same power output as a petrol engine.

Yet it is this characteristic that has allowed some enthusiasts to acquire significant power increases with turbocharged engines by making fairly simple and inexpensive modifications. A petrol engine of similar size cannot put out a comparable power increase without extensive alterations because the stock components cannot withstand the higher stresses placed upon them. Since a diesel engine is already built to withstand higher levels of stress, it makes an ideal candidate for performance tuning at little expense. However, it should be said that any modification that raises the amount of fuel and air put through a diesel engine will increase its operating temperature, which will reduce its life and increase service requirements. These are issues with newer, lighter, high-performance diesel engines which are not "overbuilt" to the degree of older engines and they are being pushed to provide greater power in smaller engines.

Emissions

Since the diesel engine uses less fuel than the petrol engine per unit distance, the diesel produces less carbon dioxide (CO_2) per unit distance. Recent advances in production and changes in the political climate have increased the availability and awareness of biodiesel, an alternative to petroleum-derived diesel fuel with a much lower net-sum emission of CO_2, due to the absorption of CO_2 by plants used to produce the fuel. However, the use of waste vegetable oil, sawmill waste from managed forests in Finland, and advances in the production of vegetable oil from algae demonstrate great promise in providing feed stocks for sustainable biodiesel that are not in competition with food production.

When a diesel engine runs at low power, there is enough oxygen present to burn the fuel—diesel engines only make significant amounts of carbon monoxide when running under a load.

Diesel fuel is injected just before the power stroke. As a result, the fuel cannot burn completely unless it has a sufficient amount of oxygen. This can result in incomplete combustion and black smoke in the exhaust if more fuel is injected than there is air available for the combustion process. Modern engines with electronic fuel delivery can adjust the timing and amount of fuel delivered, and so operate with less waste of fuel. In a mechanical system fuel timing system, the injection and duration must be set to be efficient at the anticipated operating rpm and load, and so the settings are less than ideal when the engine is running at any other RPM. The electronic injection can "sense" engine revs, load, even boost and temperature, and continuously alter the timing to match the given situation. In the petrol engine, air and fuel are mixed for the entire compression stroke, ensuring complete mixing even at higher engine speeds.

Diesel exhaust is well known for its characteristic smell, but this smell in recent years has become much less due to use of low sulfur fuel.

Diesel exhaust has been found to contain a long list of toxic air contaminants. Among these pollutants, fine particle pollution is an important as a cause of diesel's harmful health effects. However, when diesel engines burn their fuel with high oxygen levels, this results in high combustion temperatures and higher efficiency, and these particles tend to burn, but the amount of NOx pollution tends to increase.

NOx pollution can be reduced with diesel exhaust fluid, which is injected into the exhaust stream, and catalytically destroys the NOx chemical species. Exhaust gas recirculation which works by recirculating a portion of an engine's exhaust gas back to the engine cylinders also has very positive effects on NOx emissions.

Noise

The distinctive noise of a diesel engine is variably called diesel clatter, diesel nailing, or diesel knock. Diesel clatter is caused largely by the diesel combustion process; the sudden ignition of the diesel fuel when injected into the combustion chamber causes a pressure wave. Engine designers can reduce diesel clatter through: indirect injection; pilot or pre-injection; injection timing; injection rate; compression ratio; turbo boost; and exhaust gas recirculation (EGR). Common rail diesel injection systems permit multiple injection events as an aid to noise reduction. Diesel fuels with a higher octane rating modify the combustion process and reduce diesel clatter. CN (Cetane number) can be raised by distilling higher quality crude oil, by catalyzing a higher quality product or by using a cetane improving additive.

A combination of improved mechanical technology such as multi-stage injectors which fire a short "pilot charge" of fuel into the cylinder to initiate combustion before delivering the main fuel charge, higher injection pressures that have improved the atomisation of fuel into smaller droplets, and electronic control (which can adjust the timing and length of the injection process to optimise it for all speeds and temperatures), have partially mitigated these problems in the latest generation of common-rail designs, while improving engine efficiency.

Reliability

For most industrial or nautical applications, reliability is considered more important than light weight and high power.

The lack of an electrical ignition system greatly improves the reliability. The high durability of a diesel engine is also due to its overbuilt nature. Diesel fuel is a better lubricant than petrol and thus, it is less harmful to the oil film on piston rings and cylinder bores as occurs in petro powered engines; it is routine for diesel engines to cover 400,000 km (250,000 mi) or more without a rebuild.

Due to the greater compression ratio and the increased weight of the stronger components, starting a diesel engine is harder than starting a gasoline engine of similar design and displacement. More torque from the starter motor is required to push the engine through the compression cycle when starting compared to a petrol engine. This can cause difficulty when starting in winter time if using conventional automotive batteries because of the lower current available.

Either an electrical starter or an air-start system is used to start the engine turning. On large engines, pre-lubrication and slow turning of an engine, as well as heating, are required to minimise the amount of engine damage during initial start-up and running. Some smaller military diesels can be started with an explosive cartridge, called a Coffman starter, which provides the extra power required to get the machine turning. In the past, Caterpillar and John Deere used a small petrol pony engine in their tractors to start the primary diesel engine. The pony engine heated the diesel to aid in ignition and used a small clutch and transmission to spin up the diesel engine. Even more unusual was an International Harvester design in which the diesel engine had its own carburetor and ignition system, and started on petrol. Once warmed, the operator moved two levers to switch the engine to diesel operation, and work could begin. These engines had very complex cylinder heads, with their own petrol combustion chambers, and were vulnerable to expensive damage if special care was not taken (especially in letting the engine cool before turning it off).

Cylinder Cavitation and Erosion Damage

One phenomenon that can affect water-cooled diesel engines is cylinder cavitation and erosion. This is due to a phenomenon in high-compression engines where the ignition of the fuel in the cylinder causes a high-frequency vibration that causes bubbles to form in the coolant in contact with the cylinder. When these tiny bubbles collapse, coolant impacts the cylinder wall, over time causing small holes to form in the cylinder wall. This damage is mitigated in some engines with coatings, or with a coolant additive specifically designed to prevent cavitation and erosion damage. Engines damaged in this way will require the affected cylinder to be repaired (where possible) or will be rendered unusable.

Quality and Variety of Fuels

Petrol/gasoline engines are limited in the variety and quality of the fuels they can burn. Older petrol engines fitted with a carburetor required a volatile fuel that would vaporise easily to create the necessary air-fuel ratio for combustion. Because both air and fuel are admitted to the cylinder, if the compression ratio of the engine is too high or the fuel too volatile (with too low an octane rating), the fuel will ignite under compression, as in a diesel engine, before the piston reaches the top of its stroke. This pre-ignition causes a power loss and over time major damage to the piston and cylinder. The need for a fuel that is volatile enough to vaporise but not too volatile (to avoid pre-ignition) means that petrol engines will only run on a narrow range of fuels. There has been some success at dual-fuel engines that use petrol and ethanol, petrol and propane, and petrol and methane.

In diesel engines, a mechanical injector system vaporizes the fuel directly into the combustion chamber or a pre-combustion chamber (as opposed to a Venturi jet in a carburetor, or a fuel injector in a fuel injection system vaporising fuel into the intake manifold or intake runners as in a petrol engine). This forced vaporisation means that less-volatile fuels can be used. More crucially, because only air is inducted into the cylinder in a diesel engine, the compression ratio can be much higher as there is no risk of pre-ignition provided the injection process is accurately timed. This means that cylinder temperatures are much higher in a diesel engine than a petrol engine, allowing less volatile fuels to be used.

Diesel fuel is a form of light fuel oil, very similar to kerosene (paraffin), but diesel engines, especially older or simple designs that lack precision electronic injection systems, can run on a wide variety of other fuels. Some of the most common alternatives are Jet A-1 type jet fuel or vegetable oil from a very wide variety of plants. Some engines can be run on vegetable oil without modification, and most others require fairly basic alterations. Biodiesel is a pure diesel-like fuel refined from vegetable oil and can be used in nearly all diesel engines. Requirements for fuels to be used in diesel engines are the ability of the fuel to flow along the fuel lines, the ability of the fuel to lubricate the injector pump and injectors adequately, and its ignition qualities (ignition delay, cetane number). Inline mechanical injector pumps generally tolerate poor-quality or bio-fuels better than distributor-type pumps. Also, indirect injection engines generally run more satisfactorily on bio-fuels than direct injection engines. This is partly because an indirect injection engine has a much greater 'swirl' effect, improving vaporisation and combustion of fuel, and because (in the case of vegetable oil-type fuels) lipid depositions can condense on the cylinder walls of a direct-injection engine if combustion temperatures are too low (such as starting the engine from cold).

It is often reported that Diesel designed his engine to run on peanut oil, but this is false. Patent number 608845 describes his engine as being designed to run on pulverulent solid fuel (coal dust). Diesel stated in his published papers, "at the Paris Exhibition in 1900 (Exposition Universelle) there was shown by the Otto Company a small diesel engine, which, at the request of the French Government ran on Arachide (earth-nut or peanut) oil, and worked so smoothly that only a few people were aware of it. The engine was constructed for using mineral oil, and was then worked on vegetable oil without any alterations being made. The French Government at the time thought of testing the applicability to power production of the Arachide, or earth-nut, which grows in considerable quantities in their African colonies, and can easily be cultivated there." Diesel himself later conducted related tests and appeared supportive of the idea.

Most large marine diesels run on heavy fuel oil (sometimes called "bunker oil"), which is a thick, viscous and almost flameproof fuel which is very safe to store and cheap to buy in bulk as it is a waste product from the petroleum refining industry. The fuel must not only be pre-heated, but must be kept heated during handling and storage in order to maintain its pumpability. This is usually accomplished by steam tracing on fuel lines and steam coils in fuel oil tanks. The fuel is then preheated to over 100C before entering the engine in order to attain the proper viscosity for atomisation.

Fuel and Fluid Characteristics

Diesel engines can operate on a variety of different fuels, depending on configuration, though the eponymous diesel fuel derived from crude oil is most common. The engines can work with the full spectrum of crude oil distillates, from natural gas, alcohols, petrol, wood gas to the fuel oils from diesel oil to residual fuels. Many automotive diesel engines would run on 100% biodiesel without any modifications.

The type of fuel used is selected to meet a combination of service requirements, and fuel costs. Good-quality diesel fuel can be synthesised from vegetable oil and alcohol. Diesel fuel can be made from coal or other carbon base using the Fischer–Tropsch process. Biodiesel is growing in popularity since it can frequently be used in unmodified engines, though production remains limited. Recently, biodiesel from coconut, which can produce a very promising coco methyl ester (CME), has characteristics which enhance lubricity and combustion giving a regular diesel engine without any modification more power, less particulate matter or black smoke, and smoother engine performance. The Philippines pioneers in the research on Coconut based CME with the help of German and American scientists. Petroleum-derived diesel is often called petrodiesel if there is need to distinguish the source of the fuel.

Pure plant oils are increasingly being used as a fuel for cars, trucks and remote combined heat and power generation especially in Germany where hundreds of decentralised small- and medium-sized oil presses cold press oilseed, mainly rapeseed, for fuel. There is a Deutsches Institut für Normung fuel standard for rapeseed oil fuel.

Residual fuels are the "dregs" of the distillation process and are a thicker, heavier oil, or oil with higher viscosity, which are so thick that they are not readily pumpable unless heated. Residual fuel oils are cheaper than clean, refined diesel oil, although they are dirtier. Their main considerations are for use in ships and very large generation sets, due to the cost of the large volume of fuel consumed, frequently amounting to many tonnes per hour. The poorly refined biofuels straight vegetable oil (SVO) and waste vegetable oil (WVO) can fall into this category, but can be viable fuels on non-common rail or TDI PD diesels with the simple conversion of fuel heating to 80 to 100 degrees Celsius to reduce viscosity, and adequate filtration to OEM standards. Engines using these heavy oils have to start and shut down on standard diesel fuel, as these fuels will not flow through fuel lines at low temperatures. Moving beyond that, use of low-grade fuels can lead to serious maintenance problems because of their high sulphur and lower lubrication properties. Most diesel engines that power ships like supertankers are built so that the engine can safely use low-grade fuels due to their separate cylinder and crankcase lubrication.

Normal diesel fuel is more difficult to ignite and slower in developing fire than petrol because of its higher flash point, but once burning, a diesel fire can be fierce.

Fuel contaminants such as dirt and water are often more problematic in diesel engines than in petrol engines. Water can cause serious damage, due to corrosion, to the injection pump and injectors; and dirt, even very fine particulate matter, can damage the injection pumps due to the close tolerances that the pumps are machined to. All diesel engines will have a fuel filter (usually much finer than a filter on a petrol engine), and a water trap. The water trap (which is sometimes part of the fuel filter) often has a float connected to a warning light, which warns when there is too much water in the trap, and must be drained before damage to the engine can result. The fuel filter must be replaced much more often on a diesel engine than on a petrol engine, changing the fuel filter every 2–4 oil changes is not uncommon for some vehicles.

Safety

Fuel Flammability

Diesel fuel is less flammable than petrol, leading to a lower risk of fire caused by fuel in a vehicle equipped with a diesel engine.

In yachts, diesel engines are often used because the petrol (gasoline) that fuels spark-ignition engines releases combustible vapors which can lead to an explosion if it accumulates in a confined space such as the bottom of a vessel. Ventilation systems are mandatory on petrol-powered vessels.

The United States Army and NATO use only diesel engines and turbines because of fire hazard. Although neither gasoline nor diesel is explosive in liquid form, both can create an explosive air/vapor mix under the right conditions. However, diesel fuel is less prone due to its lower vapor pressure, which is an indication of evaporation rate. The Material Safety Data Sheet for ultra-low sulfur diesel fuel indicates a vapor explosion hazard for diesel indoors, outdoors, or in sewers.

US Army gasoline-engined tanks during World War II were nicknamed Ronsons, because of their greater likelihood of catching fire when damaged by enemy fire, although tank fires were usually caused by detonation of the ammunition rather than fuel, while diesel tanks such as the Soviet T-34 were less prone to catching fire.

Maintenance Hazards

Fuel injection introduces potential hazards in engine maintenance due to the high fuel pressures used. Residual pressure can remain in the fuel lines long after an injection-equipped engine has been shut down. This residual pressure must be relieved, and if it is done so by external bleed-off, the fuel must be safely contained. If a high-pressure diesel fuel injector is removed from its seat and operated in open air, there is a risk to the operator of injury by hypodermic jet-injection, even with only 100 pounds per square inch (690 kPa) pressure. The first known such injury occurred in 1937 during a diesel engine maintenance operation.

Cancer

Diesel exhaust has been classified as an IARC Group 1 carcinogen. It causes lung cancer and is associated with an increased risk for bladder cancer.

Applications

The characteristics of diesel have different advantages for different applications.

Passenger Cars

Diesel engines have long been popular in bigger cars and have been used in smaller cars such as superminis in Europe since the 1980s. They were popular in larger cars earlier, as the weight and cost penalties were less noticeable. Diesel engines tend to be more economical at regular driving speeds and are much better at city speeds. Their reliability and life-span tend to be better (as detailed). Some 40 percent or more of all cars sold in Europe are diesel-powered where they are considered a low CO_2 option. Mercedes-Benz in conjunction with Robert Bosch GmbH produced diesel-powered passenger cars starting in 1936 and very large numbers are used all over the world (often as "Grande Taxis" in the Third World). Diesel-powered passenger cars are very popular in India too, since the price of diesel fuel there is lower as compared to petrol. As a result, predominantly petrol-powered car manufacturers including the Japanese car manufacturers produce and market diesel-powered cars in India. Diesel-powered cars also dominate the Indian taxi industry.

Railroad Rolling Stock

Diesel engines have eclipsed steam engines as the prime mover on all non-electrified railroads in the industrialized world. The first diesel locomotives appeared in the early 20th century, and diesel multiple units soon after. While electric locomotives have replaced the diesel locomotive for some passenger traffic in Europe and Asia, diesel is still today very popular for cargo-hauling freight trains and on tracks where electrification is not feasible. Most modern diesel locomotives are actually diesel-electric locomotives: the diesel engine is used to power an electric generator that in turn powers electric traction motors with no mechanical connection between diesel engine and traction. After 2000, environmental requirements has caused higher development cost for engines, and it has become common for passenger multiple units to use engines and automatic mechanical gearboxes made for trucks. Up to four such combinations might be used to achieve enough power in a train.

Other Transport Uses

Larger transport applications (trucks, buses, etc.) also benefit from the Diesel's reliability and high torque output. Diesel displaced paraffin (or tractor vaporising oil, TVO) in most parts of the world by the end of the 1950s with the US following some 20 years later.

- Aircraft
- Marine
- Motorcycles

In merchant ships and boats, the same advantages apply with the relative safety of Diesel fuel an additional benefit. The German pocket battleships were the largest Diesel warships, but the German torpedo-boats known as E-boats (Schnellboot) of the Second World War were also Diesel craft. Conventional submarines have used them since before World War I, relying on the almost total absence of carbon monoxide in the exhaust. American World War II Diesel-electric submarines operated on two-stroke cycle, as opposed to the four-stroke cycle that other navies used.

Non-road Diesel Engines

Non-road diesel engines include mobile equipment and vehicles that are not used on the public roadways such as construction equipment and agricultural tractors.

Military Fuel Standardisation

NATO has a single vehicle fuel policy and has selected diesel for this purpose. The use of a single fuel simplifies wartime logistics. NATO and the United States Marine Corps have even been developing a diesel military motorcycle based on a Kawasaki off road motorcycle the KLR 650, with a purpose designed naturally aspirated direct injection diesel at Cranfield University in England, to be produced in the US, because motorcycles were the last remaining gasoline-powered vehicle in their inventory. Before this, a few civilian motorcycles had been built using adapted stationary diesel engines, but the weight and cost disadvantages generally outweighed the efficiency gains.

Non-transport Uses

A 1944 V12 2,300 kW power plant undergoing testing & restoration

Diesel engines are also used to power permanent, portable, and backup generators, irrigation pumps, corn grinders, and coffee de-pulpers.

Engine Speeds

Within the diesel engine industry, engines are often categorized by their rotational speeds into three unofficial groups:

- High-speed engines (> 1,000 rpm),

- Medium-speed engines (300–1,000 rpm), and

- Slow-speed engines (< 300 rpm).

High- and medium-speed engines are predominantly four-stroke engines; except for the Detroit Diesel two-stroke range. Medium-speed engines are physically larger than high-speed engines and can burn lower-grade (slower-burning) fuel than high-speed engines. Slow-speed engines are predominantly large two-stroke crosshead engines, hence very different from high- and medium-speed engines. Due to the lower rotational speed of slow- and medium-speed engines, there is more time for combustion during the power stroke of the cycle, allowing the use of slower-burning fuels than high-speed engines.

High-speed Engines

High-speed (approximately 1,000 rpm and greater) engines are used to power trucks (lorries), buses, tractors, cars, yachts, compressors, pumps and small electrical generators. As of 2008, most high-speed engines have direct injection. Many modern engines, particularly in on-highway applications, have common rail direct injection, which is cleaner burning.

Medium-speed Engines

Medium-speed engines are used in large electrical generators, ship propulsion and mechanical drive applications such as large compressors or pumps. Medium speed diesel engines operate on either diesel fuel or heavy fuel oil by direct injection in the same manner as low-speed engines.

Engines used in electrical generators run at approximately 300 to 1000 rpm and are optimized to run at a set synchronous speed depending on the generation frequency (50 or 60 hertz) and provide a rapid response to load changes. Typical synchronous speeds for modern medium-speed engines are 500/514 rpm (50/60 Hz), 600 rpm (both 50 and 60 Hz), 720/750 rpm, and 900/1000 rpm.

As of 2009, the largest medium-speed engines in current production have outputs up to approximately 20 MW (27,000 hp) and are supplied by companies like MAN B&W, Wärtsilä, and Rolls-Royce (who acquired Ulstein Bergen Diesel in 1999). Most medium-speed engines produced are four-stroke machines, however there are some two-stroke medium-speed engines such as by EMD (Electro-Motive Diesel), and the Fairbanks Morse OP (Opposed-piston engine) type.

Typical cylinder bore size for medium-speed engines ranges from 20 cm to 50 cm, and engine configurations typically are offered ranging from in-line 4-cylinder units to V-configuration 20-cylinder units. Most larger medium-speed engines are started with compressed air direct on pistons, using an air distributor, as opposed to a pneumatic starting motor acting on the flywheel, which tends to be used for smaller engines. There is no definitive engine size cut-off point for this.

It should also be noted that most major manufacturers of medium-speed engines make natural gas-fueled versions of their diesel engines, which in fact operate on the Otto cycle, and require spark ignition, typically provided with a spark plug. There are also dual (diesel/natural gas/coal gas) fuel versions of medium and low speed diesel engines using a lean fuel air mixture and a small injection of diesel fuel (so-called "pilot fuel") for ignition. In case of a gas supply failure or maximum power demand these engines will instantly switch back to full diesel fuel operation.

Low-speed Engines

The MAN B&W 5S50MC 5-cylinder, 2-stroke, low-speed marine diesel engine. This particular engine is found aboard a 29,000 tonne chemical carrier.

Also known as slow-speed, or traditionally oil engines, the largest diesel engines are primarily used to power ships, although there are a few land-based power generation units as well. These extremely large two-stroke engines have power outputs up to approximately 85 MW (114,000 hp), operate in the range from approximately 60 to 200 rpm and are up to 15 m (50 ft) tall, and can weigh over 2,000 short tons (1,800 t). They typically use direct injection running on cheap low-grade heavy fuel, also known as bunker C fuel, which requires heating in the ship for tanking and before injection due to the fuel's high viscosity. Often, the waste heat recovery steam boilers attached to the engine exhaust ducting generate the heat required for fuel heating. Provided the heavy fuel system is kept warm and circulating, engines can be started and stopped on heavy fuel.

Large and medium marine engines are started with compressed air directly applied to the pistons. Air is applied to cylinders to start the engine forwards or backwards because they are normally directly connected to the propeller without clutch or gearbox, and to provide reverse propulsion either the engine must be run backwards or the ship will use an adjustable propeller. At least three cylinders are required with two-stroke engines and at least six cylinders with four-stroke engines to provide torque every 120 degrees.

Companies such as MAN B&W Diesel, and Wärtsilä design such large low-speed engines. They are unusually narrow and tall due to the addition of a crosshead bearing. As of 2007, the 14-cylinder Wärtsilä-Sulzer 14RTFLEX96-C turbocharged two-stroke diesel engine built by Wärtsilä licensee Doosan in Korea is the most powerful diesel engine put into service, with a cylinder bore of 960 mm (37.8 in) delivering 114,800 hp (85.6 MW). It was put into service in September 2006, aboard what was then the world's largest container ship Emma Maersk which belongs to the A.P. Moller-Maersk Group. Typical bore size for low-speed engines ranges from approximately 35 to 98 cm (14 to 39 in). As of 2008, all produced low-speed engines with crosshead bearings are in-line configurations; no Vee versions have been produced.

Low-speed diesel engines (as used in ships and other applications where overall engine weight is relatively unimportant) often have a thermal efficiency which exceeds 50%.

Current and Future Developments

As of 2008, many common rail and unit injection systems already employ new injectors using stacked piezoelectric wafers in lieu of a solenoid, giving finer control of the injection event.

Variable geometry turbochargers have flexible vanes, which move and let more air into the engine depending on load. This technology increases both performance and fuel economy. Boost lag is reduced as turbo impeller inertia is compensated for.

Accelerometer pilot control (APC) uses an accelerometer to provide feedback on the engine's level of noise and vibration and thus instruct the ECU to inject the minimum amount of fuel that will produce quiet combustion and still provide the required power (especially while idling).

The next generation of common rail diesels is expected to use variable injection geometry, which allows the amount of fuel injected to be varied over a wider range, and variable valve timing similar to that of petrol engines. Particularly in the United States, coming tougher emissions regulations present a considerable challenge to diesel engine manufacturers. Ford's HyTrans Project has developed a system which starts the ignition in 400 ms, saving a significant amount of fuel on city routes, and there are other methods to achieve even more efficient combustion, such as homogeneous charge compression ignition, being studied.

Japanese and Swedish vehicle manufacturers are also developing diesel engines that run on dimethyl ether (DME).

Some recent diesel engine models utilize a copper alloy heat exchanger technology (CuproBraze) to take advantage of benefits in terms of thermal performance, heat transfer efficiency, strength/durability, corrosion resistance, and reduced emissions from higher operating temperatures.

Low heat rejection engines

A special class of experimental prototype internal combustion piston engines has been developed over several decades with the goal of improving efficiency by reducing heat loss. These engines are variously called adiabatic engines; due to better approximation of adiabatic expansion; low heat rejection engines, or high temperature engines. They are generally piston engines with combustion chamber parts lined with ceramic thermal barrier coatings. Some make use of pistons and other parts made of titanium which has a low thermal conductivity and density. Some designs are able to eliminate the use of a cooling system and associated parasitic losses altogether. Developing lubricants able to withstand the higher temperatures involved has been a major barrier to commercialization.

Variable Valve Timing

In internal combustion engines, variable valve timing (VVT) is the process of altering the timing of a valve lift event, and is often used to improve performance, fuel economy or emissions. It is increasingly being used in combination with variable valve lift systems. There are many ways in which this can be achieved, ranging from mechanical devices to electro-hydraulic and camless systems. Increasingly strict emissions regulations are causing many automotive manufacturers to use VVT systems.

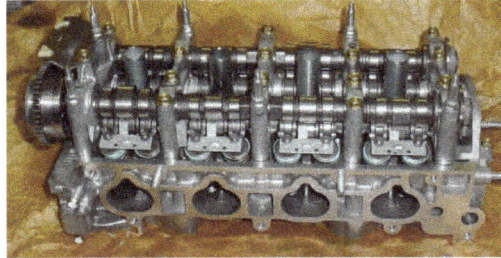

Cylinder head of Honda K20Z3. This engine uses continuously variable timing for the inlet valves

Two-stroke engines use a power valve system to get similar results to VVT.

Background Theory

The valves within an internal combustion engine are used to control the flow of the intake and exhaust gases into and out of the combustion chamber. The timing, duration and lift of these valve events has a significant impact on engine performance. Without variable valve timing or variable valve lift, the valve timing must be the same for all engine speeds and conditions, therefore compromises are necessary. An engine equipped with a variable valve timing actuation system is freed from this constraint, allowing performance to be improved over the engine operating range.

Piston engines normally use valves which are driven by camshafts. The cams open the valves (lift) for a certain amount of time (duration) during each intake and exhaust cycle. The timing of the valve opening and closing is also important. The camshaft is driven by the crankshaft through timing belts, gears or chains.

An engine requires large amounts of air when operating at high speeds. However, the intake valves may close before enough air has entered each combustion chamber, reducing performance. On the other hand, if the camshaft keeps the valves open for longer periods of time, as with a racing cam, problems start to occur at the lower engine speeds. This will cause unburnt fuel to exit the engine since the valves are still open. This leads to lower engine performance and increased emissions.

Continuous Versus Discrete

Early variable valve timing systems used discrete (stepped adjustment). For example, one timing would be used below 3500 rpm and another used above 3500 rpm.

More advanced "continuous variable valve timing" systems offer continuous (infinite) adjustment of the valve timing. Therefore, the timing can be optimized to suit all engine speeds and conditions.

Cam Phasing Versus Variable Duration

The simplest form of VVT is cam-phasing, where the angle of a camshaft is rotated forwards or backwards (relative to the crankshaft). Thus the valves open and close earlier or later; however, the camshaft lift and duration cannot be altered with a cam-phasing system.

Achieving variable duration on a VVT system requires a more complex system, such as multiple cam profiles or oscillating cams.

Typical Effect of Timing Adjustments

Late intake valve closing (LIVC) The first variation of continuous variable valve timing involves holding the intake valve open slightly longer than a traditional engine. This results in the piston actually pushing air out of the cylinder and back into the intake manifold during the compression stroke. The air which is expelled fills the manifold with higher pressure, and on subsequent intake strokes the air which is taken in is at a higher pressure. Late intake valve closing has been shown to reduce pumping losses by 40% during partial load conditions, and to decrease nitric oxide (NOx) emissions by 24%. Peak engine torque showed only a 1% decline, and hydrocarbon emissions were unchanged.

Early intake valve closing (EIVC) Another way to decrease the pumping losses associated with low engine speed, high vacuum conditions is by closing the intake valve earlier than normal. This involves closing the intake valve midway through the intake stroke. Air/fuel demands are so low at low-load conditions and the work required to fill the cylinder is relatively high, so Early intake valve closing greatly reduces pumping losses. Studies have shown early intake valve closing reduces pumping losses by 40%, and increases fuel economy by 7%. It also reduced nitric oxide emissions by 24% at partial load conditions. A possible downside to early intake valve closing is that it significantly lowers the temperature of the combustion chamber, which can increase hydrocarbon emissions.

Early intake valve opening Early intake valve opening is another variation that has significant potential to reduce emissions. In a traditional engine, a process called valve overlap is used to aid in controlling the cylinder temperature. By opening the intake valve early, some of the inert/combusted exhaust gas will back flow out of the cylinder, via the intake valve, where it cools momentarily in the intake manifold. This inert gas then fills the cylinder in the subsequent intake stroke, which aids in controlling the temperature of the cylinder and nitric oxide emissions. It also improves volumetric efficiency, because there is less exhaust gas to be expelled on the exhaust stroke.

Early/late exhaust valve closing Early and late exhaust valve closing can also reduce emissions. Traditionally, the exhaust valve opens, and exhaust gas is pushed out of the cylinder and into the exhaust manifold by the piston as it travels upward. By manipulating the timing of the exhaust valve, engineers can control how much exhaust gas is left in the cylinder. By holding the exhaust valve open slightly longer, the cylinder is emptied more and ready to be filled with a bigger air/fuel charge on the intake stroke. By closing the valve slightly early, more exhaust gas remains in the cylinder which increases fuel efficiency. This allows for more efficient operation under all conditions.

Challenges

The main factor preventing this technology from wide use in production automobiles is the ability to produce a cost effective means of controlling the valve timing under the conditions internal to an engine. An engine operating at 3000 revolutions per minute will rotate the camshaft 25 times per second, so the valve timing events have to occur at precise times to offer performance benefits. Electromagnetic and pneumatic camless valve actuators offer the greatest control of precise valve timing, but, in 2016, are not cost effective for production vehicles.

History

Steam Engines

The history of the search for a method of variable valve opening duration goes back to the age of steam engines when the valve opening duration was referred to as "steam cut-off". The Stephenson valve gear, as used on early steam locomotives, supported variable cutoff, that is, changes to the time at which the admission of steam to the cylinders is cut off during the power stroke.

Early approaches to variable cutoff coupled variations in admission cutoff with variations in exhaust cutoff. Admission and exhaust cutoff were decoupled with the development of the Corliss valve. These were widely used in constant speed variable load stationary engines, with admission cutoff, and therefore torque, mechanically controlled by a centrifugal governor and trip valves.

As poppet valves came into use, a simplified valve gear using a camshaft came into use. With such engines, variable cutoff could be achieved with variable profile cams that were shifted along the camshaft by the governor.

Aircraft

An early experimental 200 hp Clerget V-8 from the 1910s used a sliding camshaft to change the valve timing. Some versions of the Bristol Jupiter radial engine of the early 1920s incorporated variable valve timing gear, mainly to vary the inlet valve timing in connection with higher compression ratios. The Lycoming R-7755 engine had a Variable Valve Timing system consisting of two cams that can be selected by the pilot. One for take off, pursuit and escape, the other for economical cruising.

Automotive

The desirability of being able to vary the valve opening duration to match an engine's rotational speed first became apparent in the 1920s when maximum allowable RPM limits were generally starting to rise. Until about this time an engine's idle RPM and its operating RPM were very similar, meaning that there was little need for variable valve duration. It was in the 1920s that the first patents for variable duration valve opening started appearing – for example United States patent U.S. Patent 1,527,456.

In 1958 Porsche made application for a German Patent, also applied for and published as British Patent GB861369 in 1959. The Porsche patent used an oscillating cam to increase the valve lift and duration. The desmodromic cam driven via a push/pull rod from an eccentric shaft or swashplate. It is unknown if any working prototype was ever made.

Fiat was the first auto manufacturer to patent a functional automotive variable valve timing system which included variable lift. Developed by Giovanni Torazza in the late 1960s, the system used hydraulic pressure to vary the fulcrum of the cam followers (US Patent 3,641,988). The hydraulic pressure changed according to engine speed and intake pressure. The typical opening variation was 37%.

Alfa Romeo was the first manufacturer to use a variable valve timing system in production cars (US Patent 4,231,330). The fuel injected models of the 1980 Alfa Romeo Spider 2000 had a mechanical VVT system. The system was engineered by Ing Giampaolo Garcea in the 1970s.

In 1989, Honda released the VTEC system. While the earlier Nissan NVCS alters the phasing of the camshaft, VTEC switches to a separate cam profile at high engine speeds to improve peak power. The first VTEC engine Honda produced was the B16A which was installed in the Integra, CRX, and Civic hatchback available in Japan and Europe.

In 1992, Porsche first introduced VarioCam, which was the first system to provide continuous adjustment (all previous systems used discrete adjustment). The system was released in the Porsche 968 and operated on the intake valves only.

Marine

Variable valve timing has begun to trickle down to marine engines. Volvo Penta's VVT marine engine uses a cam phaser, controlled by the ECM, continuously varies advance or retardation of camshaft timing.

Diesel

In 2007, Caterpillar developed the C13 and C15 Acert engines which used VVT technology to reduce NOx emissions, to avoid the use of EGR after 2002 EPA requirements.

In 2010, Mitsubishi developed and started mass production of its 4N13 1.8 L DOHC I4, the world's first passenger car diesel engine that features a variable valve timing system.

Automotive Nomenclature

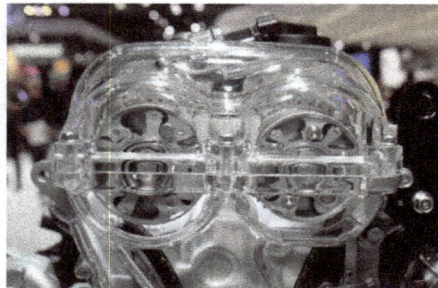

Hydraulic vane-type phasers on a cut-out model of Hyundai T-GDI engine

Manufacturers use many different names to describe their implementation of the various types of variable valve timing systems. These names include:

- AVCS (Subaru)
- AVLS (Subaru)
- CPS (Proton)

- CVTCS (Nissan, Infiniti)

- CVVT (Alfa Romeo, Citroën, Geely, Hyundai, Iran Khodro, Kia, Peugeot, Renault, Volvo)

- DCVCP - dual continuous variable cam phasing (General Motors)

- DVVT (Daihatsu) (Perodua)

- MIVEC (Mitsubishi)

- MultiAir (Fiat)

- N-VCT (Nissan)

- S-VT (Mazda)

- Ti-VCT (Ford)

- VANOS (BMW)

- VarioCam (Porsche)

- VCT (Ford, Yamaha)

- VTEC (Honda, Acura)

- VVC (MG Rover)

- VVL (Nissan)

- Valvelift (Audi)

- VVEL (Nissan, Infiniti)

- VVT (Chrysler, General Motors, Proton, Suzuki, Volkswagen Group)

- VVT-i (Toyota, Lexus)

- VTVT (Hyundai, Kia)

Methods for Implementing Variable Valve Control (VVC)

Cam Switching

This method uses two cam profiles, with an actuator to swap between the profiles (usually at a specific engine speed). Cam switching can also provide variable valve lift and variable duration, however the adjustment is discrete rather than continuous.

The first production use of this system was Honda's VTEC system. VTEC changes hydraulic pressure to actuate a pin that locks the high lift, high duration rocker arm to an adjacent low lift, low duration rocker arm(s).

Cam Phasing

Many production VVT systems are the cam phasing type, using a device known as a variator. This allows continuous adjustment of the cam timing (although many early systems only used discrete adjustment), however the duration and lift cannot be adjusted.

Oscillating Cam

These designs use an oscillating or rocking motion in a part cam lobe, which acts on a follower. This follower then opens and closes the valve. Some oscillating cam systems use a conventional cam lobe, while others use an eccentric cam lobe and a connecting rod. The principle is similar to steam engines, where the amount of steam entering the cylinder was regulated by the steam "cut-off" point.

The advantage of this design is that adjustment of lift and duration is continuous. However, in these systems, lift is proportional to duration, so lift and duration cannot be separately adjusted.

The BMW (valvetronic), Nissan (VVEL), and Toyota (valvematic) oscillating cam systems act on the intake valves only.

Eccentric Cam Drive

Eccentric cam drive systems operates through an eccentric disc mechanism which slows and speeds up the angular speed of the cam lobe during its rotation. Arranging the lobe to slow during its open period is equivalent to lengthening its duration.

The advantage of this system is that duration can be varied independent of lift (however this system does not vary lift). The drawback is two eccentric drives and controllers are needed for each cylinder (one for the intake valves and one for the exhaust valves), which increases complexity and cost.

MG Rover is the only manufacturer that has released engines using this system.

Three-dimensional Cam Lobe

This system consists of a cam lobe that varies along its length (similar to a cone shape). One end of the cam lobe has a short duration/reduced lift profile, and the other end has a longer duration/ greater lift profile. In between, the lobe provides a smooth transition between these two profiles. By shifting area of the cam lobe which is in contact with the follower, the lift and duration can be continuously altered. This is achieved by moving the camshaft axially (sliding it across the engine) so a stationary follower is exposed to a varying lobe profile to produce different amounts of lift and duration. The downside to this arrangement is that the cam and follower profiles must be carefully designed to minimise contact stress (due to the varying profile).

Ferrari is commonly associated with this system, however it is unknown whether any production models to date have used this system.

Two Shaft Combined Cam Lobe Profile

This system is not known to be used in any production engines.

It consists of two (closely spaced) parallel camshafts, with a pivoting follower that spans both camshafts and is acted on by two lobes simultaneously. Each camshaft has a phasing mechanism which allows its angular position relative to the engine's crankshaft to be adjusted. One lobe controls the opening of a valve and the other controls the closing of the same valve, therefore variable duration is achieved through the spacing of these two events.

The drawbacks to this design include:

- At long duration settings, one lobe may be starting to reduce its lift as the other is still increasing. This has the effect of lessening the overall lift and possibly causing dynamic problems. One company claims to have solved the uneven rate of opening of the valve problem to some extent thus allowing long duration at full lift.

- Size of the system, due to the parallel shafts, the larger followers etc.

Coaxial Two Shaft Combined Cam Lobe Profile

This system is not known to be used in any production engines.

The operating principle is that the one follower spans the pair of closely spaced lobes. Up to the angular limit of the nose radius the follower 'sees' the combined surface of the two lobes as a continuous, smooth surface. When the lobes are exactly aligned the duration is at a minimum (and equal to that of each lobe alone) and when at the extreme extent of their misalignment the duration is at a maximum. The basic limitation of the scheme is that only a duration variation equal to that of the lobe nose true radius (in camshaft degrees or double this value in crankshaft degrees) is possible. In practice this type of variable cam has a maximum range of duration variation of about forty crankshaft degrees.

This is the principle behind what seems to be the very first variable cam suggestion appearing in the USPTO patent files in 1925 (1527456). The "Clemson camshaft" is of this type.

Helical Camshaft

Also known as "Combined two shaft coaxial combined profile with helical movement", this system is not known to be used in any production engines.

It has a similar principle to the previous type, and can use the same base duration lobe profile. However instead of rotation in a single plane, the adjustment is both axial and rotational giving a helical or three-dimensional aspect to its movement. This movement overcomes the restricted duration range in the previous type. The duration range is theoretically unlimited but typically would be of the order of one hundred crankshaft degrees, which is sufficient to cover most situations.

The cam is reportedly difficult and expensive to produce, requiring very accurate helical machining and careful assembly.

Camless Engines

Engine designs which do not rely on a camshaft to operate the valves have greater flexibility in achieving variable valve timing and variable valve lift. However, there has not been a production camless engine released for road vehicles as yet.

Types of camless engines include:

- electro-mechanical (using electromagnets)
- hydraulic
- stepper motors
- pneumatic

Engine Control Unit

An ECU from a 1996 Chevrolet Beretta.

An engine control unit (ECU) is a type of electronic control unit that controls a series of actuators on an internal combustion engine to ensure optimal engine performance. It does this by reading values from a multitude of sensors within the engine bay, interpreting the data using multidimensional performance maps (called lookup tables), and adjusting the engine actuators accordingly. Before ECUs, air-fuel mixture, ignition timing, and idle speed were mechanically set and dynamically controlled by mechanical and pneumatic means.

Working of ECU

Control of Air/Fuel Ratio

Most modern engines use some type of fuel injection to deliver fuel to the cylinders. The ECU determines the amount of fuel to inject based on a number of sensor readings. Oxygen sensors tell the ECU whether the engine is running rich (too much fuel/too little oxygen) or running lean (too much oxygen/too little fuel) as compared to ideal conditions (known as stoichiometric). The throttle position sensors tell the ECU how far the throttle plate is opened when you press the accelerator. The mass air flow sensor measures the amount of air flowing into the engine through the throttle plate. The engine coolant temperature sensor measures whether the engine is warmed up or cool. (If the engine is still cool, additional fuel will be injected.)

Air/fuel mixture control of carburetors with computers is designed with a similar principle, but a mixture control solenoid or stopper motor is incorporated in the float bowl of the carburetor.

Control of Idle Speed

Most engine systems have idle speed control built into the ECU. The engine RPM is monitored by the crankshaft position sensor which plays a primary role in the engine timing functions for fuel injection, spark events, and valve timing. Idle speed is controlled by a programmable throttle stop or an idle air bypass control stepper motor. Early carburetor-based systems used a programmable throttle stop using a bidirectional DC motor. Early Throttle body injection (TBI) systems used an idle air control stepper motor. Effective idle speed control must anticipate the engine load at idle.

A full authority throttle control system may be used to control idle speed, provide cruise control functions and top speed limitation.

Control of Variable Valve Timing

Some engines have Variable Valve Timing. In such an engine, the ECU controls the time in the engine cycle at which the valves open. The valves are usually opened sooner at higher speed than at lower speed. This can optimize the flow of air into the cylinder, increasing power and fuel economy.

Electronic Valve Control

Experimental engines have been made and tested that have no camshaft, but have full electronic control of the intake and exhaust valve opening, valve closing and area of the valve opening. Such engines can be started and run without a starter motor for certain multi-cylinder engines equipped with precision timed electronic ignition and fuel injection. Such a static-start engine would provide the efficiency and pollution-reduction improvements of a mild hybrid-electric drive, but without the expense and complexity of an oversized starter motor.

The first production engine of this type was invented (in 2002) and introduced (in 2009) by Italian automaker Fiat in the Alfa Romeo MiTo. Their Multiair engines use electronic valve control which dramatically improve torque and horsepower, while reducing fuel consumption as much as 15%. Basically, the valves are opened by hydraulic pumps, which are operated by the ECU. The valves can open several times per intake stroke, based on engine load. The ECU then decides how much fuel should be injected to optimize combustion.

At steady load conditions, the valve opens, fuel is injected, and the valve closes. Under a sudden increase in throttle, the valve opens in the same intake stroke and a greater amount of fuel is injected. This allows immediate acceleration. For the next stroke, the ECU calculates engine load at the new, higher RPM, and decides how to open the valve: early or late, wide-open or half-open. The optimal opening and timing are always reached and combustion is as precise as possible. This, of course, is impossible with a normal camshaft, which opens the valve for the whole intake period, and always to full lift.

The elimination of cams, lifters, rockers, and timing set reduces not only weight and bulk, but also friction. A significant portion of the power that an engine actually produces is used up just driving the valve train, compressing all those valve springs thousands of times a minute.

Once more fully developed, electronic valve operation will yield even more benefits. Cylinder deactivation, for instance, could be made much more fuel efficient if the intake valve could be opened on every downstroke and the exhaust valve opened on every upstroke of the deactivated cylinder or "dead hole". Another even more significant advancement will be the elimination of the conventional throttle. When a car is run at part throttle, this interruption in the airflow causes excess vacuum, which causes the engine to use up valuable energy acting as a vacuum pump. BMW attempted to get around this on their V-10 powered M5, which had individual throttle butterflies for each cylinder, placed just before the intake valves. With electronic valve operation, it will be possible to control engine speed by regulating valve lift. At part throttle, when less air and gas are needed, the valve lift would not be as great. Full throttle is achieved when the gas pedal is depressed, sending an electronic signal to the ECU, which in turn regulates the lift of each valve event, and opens it all the way up.

Programmable E.C.U.s

A special category of E.C.U.s are those which are programmable. These units do not have a fixed behavior and can be reprogrammed by the user.

Programmable E.C.U.s are required where significant aftermarket modifications have been made to a vehicle's engine. Examples include adding or changing of a turbocharger, adding or changing of an intercooler, changing of the exhaust system or a conversion to run on alternative fuel. As a consequence of these changes, the old ECU may not provide appropriate control for the new configuration. In these situations, a programmable ECU can be wired in. These can be programmed/mapped with a laptop connected using a serial or USB cable, while the engine is running.

The programmable ECU may control the amount of fuel to be injected into each cylinder. This varies depending on the engine's RPM and the position of the accelerator pedal (or the manifold air pressure). The engine tuner can adjust this by bringing up a spreadsheet-like page on the laptop where each cell represents an intersection between a specific RPM value and an accelerator pedal position (or the throttle position, as it is called). In this cell a number corresponding to the amount of fuel to be injected is entered. This spreadsheet is often referred to as a fuel table or fuel map.

By modifying these values while monitoring the exhausts using a wide band lambda probe to see if the engine runs rich or lean, the tuner can find the optimal amount of fuel to inject to the engine at every different combination of RPM and throttle position. This process is often carried out at a dynamometer, giving the tuner a controlled environment to work in. An engine dynamometer gives a more precise calibration for racing applications. Tuners often utilize a chassis dynamometer for street and other high performance applications.

Other parameters that are often mappable are:

- Ignition Timing: Defines at what point in the engine cycle the spark plug should fire for each cylinder. Modern systems allow for individual trim on each cylinder for per-cylinder optimization of the ignition timing.

- Rev. limit: Defines the maximum RPM that the engine is allowed to reach. After this fuel and/or ignition is cut. Some vehicles have a "soft" cut-off before the "hard" cut-off. This "soft cut" generally functions by retarding ignition timing to reduce power output and thereby slow the acceleration rate just before the "hard cut" is hit.

- Water temperature correction: Allows for additional fuel to be added when the engine is cold, such as in a winter cold-start scenario or when the engine is dangerously hot, to allow for additional cylinder cooling (though not in a very efficient manner, as an emergency only).

- Transient fueling: Tells the E.C.U. to add a specific amount of fuel when throttle is applied. This is referred to as "acceleration enrichment".

- Low fuel pressure modifier: Tells the ECU to increase the injector fire time to compensate for an increase or loss of fuel pressure.

- Closed loop lambda: Lets the E.C.U. monitor a permanently installed lambda probe and modify the fueling to achieve the targeted air/fuel ratio desired. This is often the stoichiometric (ideal) air fuel ratio, which on traditional petrol (gasoline) powered vehicles this air:fuel ratio is 14.7:1. This can also be a much richer ratio for when the engine is under high load, or possibly a leaner ratio for when the engine is operating under low load cruise conditions for maximum fuel efficiency.

Some of the more advanced standalone/race E.C.U.s include functionality such as launch control, operating as a rev limiter while the car is at the starting line to keep the engine revs in a 'sweet spot', waiting for the clutch to be released to launch the car as quickly and efficiently as possible. Other examples of advanced functions are:

- Waste gate control: Controls the behavior of a turbocharger's waste gate, controlling boost. This can be mapped to command a specific duty cycle on the valve, or can use a P.I.D. based closed-loop control algorithm.

- Staged injection: Allows for an additional injector per cylinder, used to get a finer fuel injection control and atomization over a wide R.P.M. range. An example being the use of small injectors for smooth idle and low load conditions, and a second, larger set of injectors that are 'staged in' at higher loads, such as when the turbo boost climbs above a set point.

- Variable cam timing: Allows for control variable intake and exhaust cams (V.V..T), mapping the exact advance/retard curve positioning the camshafts for maximum benefit at all load/rpm positions in the map. This functionality is often used to optimize power output at high load/R.P.M.s, and to maximize fuel efficiency and emissions as lower loads/R.P.M.s.

- Gear control: Tells the ECU to cut ignition during (sequential gearbox) up shifts or blip the throttle during downshifts.

- Anti-lag: Is an option which is provided by racing E.C.U.s only for turbocharged vehicles. When it is on, it changes the ignition timing to late, providing a fast charge of the turbocharger. When anti-lag is on, gunshot sounds and flames come from the exhaust, indicating extreme temperatures and pressures.

A race ECU is often equipped with a data logger recording all sensors for later analysis using special software in a PC. This can be useful to track down engine stalls, misfires or other undesired behaviors during a race by downloading the log data and looking for anomalies after the event. The data logger usually has a capacity between 0.5 and 16 megabytes.

In order to communicate with the driver, a race ECU can often be connected to a "data stack", which is a simple dash board presenting the driver with the current RPM, speed and other basic engine data. These race stacks, which are almost always digital, talk to the ECU using one of several proprietary protocols running over RS232 or CANbus, connecting to the DLC connector (Data Link Connector) usually located on the underside of the dash, inline with the steering wheel.

History

Early Designs

One of the earliest attempts to use such a unitized and automated device to manage multiple engine control functions simultaneously was the "Kommandogerät" created by BMW in 1939, for their 801 14-cylinder aviation radial engine. This device replaced the 6 controls used to initiate hard acceleration with one control in the 801 series-equipped aircraft. However, it had some problems: it would surge the engine, making close formation flying of the Fw 190 (Focke-Wulf Fw 190 Wurger), a single-engine single-seat German fighter aircraft, somewhat difficult, and at first it switched supercharger gears harshly and at random, which could throw the aircraft into an extremely dangerous stall.

Hybrid Digital Designs

Hybrid digital/analog designs were popular in the mid-1980s. This used analog techniques to measure and process input parameters from the engine, then used a lookup table stored in a digital ROM chip to yield precomputed output values. Later systems compute these outputs dynamically. The ROM type of system is amenable to tuning if one knows the system well. The disadvantage of such systems is that the precomputed values are only optimal for an idealised, new engine. As the engine wears, the system is less able to compensate than a CPU based system.

Modern E.C.U.s

Modern E.C.U.s use a microprocessor which can process the inputs from the engine sensors in real-time. An electronic control unit contains the hardware and software (firmware). The hardware consists of electronic components on a printed circuit board (P.C.B.), ceramic substrate or a thin laminate substrate. The main component on this circuit board is a micro controller chip (CPU). The software is stored in the microcontroller or other chips on the P.C.B., typically in EPROMs or flash memory so the C.P.U. can be re-programmed by uploading updated code or replacing chips. This is also referred to as an (electronic) Engine Management System (E.M.S.).

Sophisticated engine management systems receive inputs from other sources, and control other parts of the engine; for instance, some variable valve timing systems are electronically controlled, and turbocharger waste gates can also be managed. They also may communicate with transmission control units or directly interface electronically controlled automatic transmissions, traction control systems, and the like. The Controller Area Network or CAN bus automotive network is often used to achieve communication between these devices.

Modern E.C.U.s sometimes include features such as cruise control, transmission control, anti-skid brake control, and anti-theft control, etc.

General Motors'(GM) first E.C.U.s had a small application of hybrid digital E.C.U.s as a pilot program in 1979, but by 1980, all active programs were using microprocessor based systems. Due to the large ramp up of volume of E.C.U.s that were produced to meet the Clean Air Act requirements for 1981, only one ECU model could be built for the 1981 model year. The high volume E.C.U. that was installed in G.M. vehicles from the first high volume year, 1981, onward was a modern microprocessor based system. GM moved rapidly to replace carburation with fuel injection as the preferred method of fuel delivery for vehicles it manufactured. This process first saw fruition in 1980 with fuel injected Cadillac engines, followed by the Pontiac 2.5L I4 "Iron Duke" and the Chevrolet 5.7L V8 L83 "Cross-Fire" engine powering the Chevrolet Corvette in 1982. The 1990 Cadillac Brougham powered by the Oldsmobile 5.0L V8 LV2 engine was the last carbureted passenger car manufactured for sale in the North American market (a 1992 Volkswagen Beetle model powered by a carbureted engine was available for purchase in Mexico but not offered for sale in the United States or Canada) and by 1991 GM was the last of the major US and Japanese automakers to abandon carburetion and manufacture all of its passenger cars exclusively with fuel injected engines. In 1988 Delco (GM's electronics division), had produced more than 28,000 E.C.U.s per day, making it the world's largest producer of on-board digital control computers at the time.

Other Applications

Such systems are used for many internal combustion engines in other applications. In aeronautical applications, the systems are known as "FADECs" (Full Authority Digital Engine Controls). This kind of electronic control is less common in piston-engined light fixed-wing aircraft and helicopters than in automobiles. This is due to the common configuration of a carbureted engine with a magneto ignition system that does not require electrical power generated by an alternator to run, which is considered a safety advantage.

References

- Pulkrabek, Willard W. (1997). Engineering Fundamentals of the Internal Combustion Engine. Prentice Hall. p. 2. ISBN 9780135708545.

- Yamaguchi, Jack K (1985). The New Mazda RX-7 and Mazda Rotary Engine Sports Cars. New York: St. Martin's Press. ISBN 0-312-69456-3.

- Jan P Norbye: "Rivals to the Wankel", Popular Science, Jan 1967; 'The Wankel Engine. Design, development, applications'; Chilton, 1972. ISBN 0-8019-5591-2

- B Lawton: 'The Turbocharged Diesel Wankel Engine', C68/78, of: 'Institution of Mechanical Engineers Conference Publications. 1978-2, Turbocharging and Turbochargers, ISBN 0 85298 395 6, pp 151–160.

- Gingery, Vincent. Building the Atkinson Differential Engine. David J. Gingery Publishing, LLC. ISBN 1878087231.

- Ransome-Wallis, Patrick (2001). Illustrated Encyclopedia of World Railway Locomotives. Courier Dover Publications. p. 28. ISBN 0-486-41247-4.

- Icons of Invention: The Makers of the Modern World from Gutenberg to Gates. ABC-CLIO. 20. ISBN 9780313347436. Retrieved 2013-02-07. Check date values in: |date= (help)

- Zhao, Hua (2010). Advanced Direct Injection Combustion Engine Technologies and Development: Diesel Engines. Woodhead Publishing Limited. p. 8. ISBN 9781845697457.

- The Biodiesel Handbook, Chapter 2—The History of Vegetable Oil Based Diesel Fuels, by Gerhard Knothe, ISBN 978-1-893997-79-0

- Lumley, John L. (1999). Engines - An Introduction. Cambridge UK: Cambridge University Press. pp. 63–64. ISBN 0-521-64277-9.

- Gunston, Bill (1989). World Encyclopedia of Aero Engines. Cambridge, England: Patrick Stephens Limited. p. 26. ISBN 1-85260-163-9.

Diverse Aspects of Automotive Engines

The aspects of automotive engines are vehicle emissions control, automotive electronics, automatic transmission, g-force, fuel efficiency and revolutions per minute. The study of reducing the emissions produced by motor vehicles is known as vehicle emissions control whereas automotive electronics is the system that is used in road vehicles. The major components of automotive engines are discussed in this chapter.

Vehicle Emissions Control

Vehicle emissions control is the study of reducing the motor vehicle emissions—emissions produced by motor vehicles, especially internal combustion engines.

Types of Emissions

Emissions of many air pollutants have been shown to have variety of negative effects on public health and the natural environment. Emissions that are principal pollutants of concern include:

- Hydrocarbons (HC) - A class of burned or partially burned fuel, hydrocarbons are toxins. Hydrocarbons are a major contributor to smog, which can be a major problem in urban areas. Prolonged exposure to hydrocarbons contributes to asthma, liver disease, lung disease, and cancer. Regulations governing hydrocarbons vary according to type of engine and jurisdiction; in some cases, "non-methane hydrocarbons" are regulated, while in other cases, "total hydrocarbons" are regulated. Technology for one application (to meet a non-methane hydrocarbon standard) may not be suitable for use in an application that has to meet a total hydrocarbon standard. Methane is not directly toxic, but is more difficult to break down in a catalytic converter, so in effect a "non-methane hydrocarbon" regulation can be considered easier to meet. Since methane is a greenhouse gas, interest is rising in how to eliminate emissions of it. HC emissions can come not only from a vehicle's engine buts also directly from the fuel tank and lines, 24 hours a day, even when the engine is off; the complex system of fuel vent lines and a charcoal canister is meant to collect and contain fuel vapors and route them either back to the fuel tank or, after the engine is started and warmed up, into the air intake to be burned in the engine.

- Carbon monoxide (CO) - A product of incomplete combustion, inhaled carbon monoxide reduces the blood's ability to carry oxygen; overexposure (carbon monoxide poisoning) may be fatal. (Carbon monoxide persistently binds to hemoglobin, the oxygen-carrying chemical in red blood cells, where oxygen (O_2) would temporarily bind; the bonding of CO excludes O_2 and also reduces the ability of the hemoglobin to release already-bound oxygen, on both counts rendering the red blood cells ineffective. Recovery is by the slow release of bound CO and the body's production of new hemoglobin—a healing process—so full recovery from moderate to severe [but nonfatal] CO poisoning takes hours or days. Removing a person from a CO-poisoned atmosphere to fresh air stops the injury but does not yield prompt recovery, unlike the case where a person is removing from an asphyxiating atmosphere [i.e. one deficient in oxygen]. Toxic effects delayed by days are also common.)

- NO_x - Generated when nitrogen in the air reacts with oxygen at the high temperature and pressure inside the engine. NO_x is a precursor to smog and acid rain. NO_x is the sum of NO and NO_2. NO_2 is extremely reactive. NO_x production is increased when an engine runs at its most efficient (i.e. hottest) operating point, so there tends to be a natural tradeoff between efficiency and control of NO_x emissions.

- Particulate matter – Soot or smoke made up of particles in the micrometre size range: Particulate matter causes negative health effects, including but not limited to respiratory disease and cancer. Very fine particulate matter has been linked to cardiovascular disease.

- Sulfur oxide (SO_x) - A general term for oxides of sulfur, which are emitted from motor vehicles burning fuel containing sulfur. Reducing the level of fuel sulfur reduces the level of Sulfur oxide emitted from the tailpipe.

- Volatile organic compounds (VOCs) - Organic compounds which typically have a boiling point less than or equal to 250 °C; for example chlorofluorocarbons (CFCs) and formaldehyde. Volatile organic compounds are a subsection of Hydrocarbons that are mentioned separately because of their dangers to public health.

History

Throughout the 1950s and 1960s, various federal, state and local governments in the United States conducted studies into the numerous sources of air pollution. These studies ultimately attributed a significant portion of air pollution to the automobile, and concluded air pollution is not bounded by local political boundaries. At that time, such minimal emission control regulations as existed in the U.S. were promulgated at the municipal or, occasionally, the state level. The ineffective local regulations were gradually supplanted by more comprehensive state and federal regulations. By 1967 the State of California created the California Air Resources Board, and in 1970, the federal United States Environmental Protection Agency (EPA) was established. Both agencies, as well as other state agencies, now create and enforce emission regulations for automobiles in the United States. Similar agencies and regulations were contemporaneously developed and implemented in Canada, Western Europe, Australia, and Japan.

The first effort at controlling pollution from automobiles was the PCV (positive crankcase ventilation) system. This draws crankcase fumes heavy in unburned hydrocarbons — a precursor to photochemical smog — into the engine's intake tract so they are burned rather than released unburned from the crankcase into the atmosphere. Positive crankcase ventilation was first installed on a widespread basis by law on all new 1961-model cars first sold in California. The following year, New York required it. By 1964, most new cars sold in the U.S. were so equipped, and PCV quickly became standard equipment on all vehicles worldwide.

The first legislated exhaust (tailpipe) emission standards were promulgated by the State of

California for 1966 model year for cars sold in that state, followed by the United States as a whole in model year 1968. Also in 1966, the first emission test cycle was enacted in the State of California measuring tailpipe emissions in PPM (parts per million). The standards were progressively tightened year by year, as mandated by the EPA.

By the 1974 model year, the emission standards had tightened such that the de-tuning techniques used to meet them were seriously reducing engine efficiency and thus increasing fuel usage. The new emission standards for 1975 model year, as well as the increase in fuel usage, forced the invention of the catalytic converter for after-treatment of the exhaust gas. This was not possible with existing leaded gasoline, because the lead residue contaminated the platinum catalyst. In 1972, General Motors proposed to the American Petroleum Institute the elimination of leaded fuels for 1975 and later model year cars. The production and distribution of unleaded fuel was a major challenge, but it was completed successfully in time for the 1975 model year cars. All modern cars are now equipped with catalytic converters and leaded fuel is nearly impossible to buy in most First World countries.

Regulatory Agencies

The agencies charged with regulating exhaust emissions vary from jurisdiction to jurisdiction, even in the same country. For example, in the United States, overall responsibility belongs to the EPA, but due to special requirements of the State of California, emissions in California are regulated by the Air Resources Board. In Texas, the Texas Railroad Commission is responsible for regulating emissions from LPG-fueled rich burn engines (but not gasoline-fueled rich burn engines).

North America

- California Air Resources Board - California, United States (most sources)
- Environment Canada - Canada (most sources)
- Environmental Protection Agency - United States (most sources)
- Texas Railroad Commission - Texas, United States (LPG-fueled engines only)
- Transport Canada - Canada (trains and ships)

Europe

The European Union has control over regulation of emissions in EU member states; however, many member states have their own government bodies to enforce and implement these regulations in their respective countries. In short, the EU forms the policy (by setting limits such as the European emission standard) and the member states decide how to best implement it in their own country.

United Kingdom

In the United Kingdom, matters concerning environmental policy are what is known as "devolved powers" which means, each of the constituent countries deals with it separately through their own government bodies set up to deal with environmental issues in their respective country:

- Environment Agency - England and Wales

- Scottish Environment Protection Agency (SEPA) - Scotland

- Department of the Environment - Northern Ireland

However, many UK-wide policies are handled by the Department of the Environment Food and Rural Affairs (DEFRA) and they are still subject to EU regulations.

Emissions Control

Engine efficiency has been steadily improved with improved engine design, more precise ignition timing and electronic ignition, more precise fuel metering, and computerized engine management.

Advances in engine and vehicle technology continually reduce the toxicity of exhaust leaving the engine, but these alone have generally been proved insufficient to meet emissions goals. Therefore, technologies to detoxify the exhaust are an essential part of emissions control.

Air Injection

One of the first-developed exhaust emission control systems is secondary air injection. Originally, this system was used to inject air into the engine's exhaust ports to provide oxygen so unburned and partially burned hydrocarbons in the exhaust would finish burning. Air injection is now used to support the catalytic converter's oxidation reaction, and to reduce emissions when an engine is started from cold. After a cold start, an engine needs an air-fuel mixture richer than what it needs at operating temperature, and the catalytic converter does not function efficiently until it has reached its own operating temperature. The air injected upstream of the converter supports combustion in the exhaust headpipe, which speeds catalyst warmup and reduces the amount of unburned hydrocarbon emitted from the tailpipe.

Exhaust Gas Recirculation

In the United States and Canada, many engines in 1973 and newer vehicles (1972 and newer in California) have a system that routes a metered amount of exhaust into the intake tract under particular operating conditions. Exhaust neither burns nor supports combustion, so it dilutes the air/fuel charge to reduce peak combustion chamber temperatures. This, in turn, reduces the formation of NO_x.

Catalytic Converter

The catalytic converter is a device placed in the exhaust pipe, which converts hydrocarbons, carbon monoxide, and NO_x into less harmful gases by using a combination of platinum, palladium and rhodium as catalysts.

There are two types of catalytic converter, a two-way and a three-way converter. Two-way converters were common until the 1980s, when three-way converters replaced them on most automobile engines.

Evaporative Emissions Control

EVAP may also refer to Evaporation.

Evaporative emissions are the result of gasoline vapors escaping from the vehicle's fuel system. Since 1971, all U.S. vehicles have had fully sealed fuel systems that do not vent directly to the atmosphere; mandates for systems of this type appeared contemporaneously in other jurisdictions. In a typical system, vapors from the fuel tank and carburetor bowl vent (on carbureted vehicles) are ducted to canisters containing activated carbon. The vapors are adsorbed within the canister, and during certain engine operational modes fresh air is drawn through the canister, pulling the vapor into the engine, where it burns.

Remote Sensing Emission Testing - Field Studies

Some US states are also using a technology developed by Dr. Donald H. Stedman of the University of Denver, which uses infra-red and ultraviolet light to detect emissions while vehicles pass by on public roads, thus eliminating the need for owners to go to a test center. Stedman's invisible light flash detection of exhaust gases is commonly used in metropolitan areas, is offered by the US-Swedish company OPUS Inspection and becoming more broadly known in Europe.

Use of Emission Test Data

Emission test results from individual vehicles are in many cases compiled to evaluate the emissions performance of various classes of vehicles, the efficacy of the testing program and of various other emission-related regulations (such as changes to fuel formulations) and to model the effects of auto emissions on public health and the environment. For example, the Environmental Working Group used California ASM emissions data to create an "Auto Asthma Index" that rates vehicle models according to emissions of hydrocarbons and nitrogen oxides, chemical precursors to photochemical smog.

Automotive Electronics

Automotive electronics are any electrically-generated systems used in road vehicles, such as: carputers, telematics, in-car entertainment systems, etc..

Automotive electronics originated from the need to control engines. The first electronic pieces were used to control engine functions and were referred to as engine control units (ECU). As electronic controls began to be used for more automotive applications, the acronym ECU took on the more general meaning of "electronic control unit", and then specific ECU's were developed. Now, ECU's are modular. Two types include engine control modules (ECM) or transmission control modules (TCM).

A modern car may have up to 100 ECU's and a commercial vehicle up to 40.

Automotive electronics or automotive embedded systems are distributed systems, and according to different domains in the automotive field, they can be classified into:

1. Engine electronics

2. Transmission electronics

3. Chassis electronics

4. Active safety

5. Driver assistance

6. Passenger comfort

7. Entertainment systems

Engine Electronics

One of the most demanding electronic parts of an automobile is the engine control unit. Engine controls demand one of the highest real time deadlines, as the engine itself is a very fast and complex part of the automobile. Of all the electronics in any car the computing power of the engine control unit is the highest, typically a 32-bit processor.

It controls such things as:

In a diesel engine:

- Fuel injection rate

- Emission control, NOx control

- Regeneration of oxidation catalytic converter

- Turbocharger control

- Cooling system control

- Throttle control

In a gasoline engine:

- Lambda control

- OBD (On-Board Diagnostics)

- Cooling system control

- Ignition system control

- Lubrication system control (only a few have electronic control)

- Fuel injection rate control

- Throttle control

Many more engine parameters are actively monitored and controlled in real-time. There are about 20 to 50 that measure pressure, temperature, flow, engine speed, oxygen level and NOx level plus other parameters at different points within the engine. All these sensor signals are sent to the

ECU, which has the logic circuits to do the actual controlling. The ECU output is connected to different actuators for the throttle valve, EGR valve, rack (in VGTs), fuel injector (using a pulse-width modulated signal), dosing injector and more. There are about 20 to 30 actuators in all.

Transmission Electronics

These control the transmission system, mainly the shifting of the gears for better shift comfort and to lower torque interrupt while shifting. Automatic transmissions use controls for their operation, and also many semi-automatic transmissions having a fully automatic clutch or a semi-auto clutch (declutching only). The engine control unit and the transmission control exchange messages, sensor signals and control signals for their operation.

Chassis Electronics

The chassis system has lot of sub-systems which monitor various parameters and are actively controlled:

- ABS - Anti-lock Braking System
- TCS – Traction Control System
- EBD – Electronic Brake Distribution
- ESP – Electronic Stability Program

Passive Safety

These systems are always ready to act when there is a collision in progress or to prevent it when it senses a dangerous situation:

- Air bags
- Hill descent control
- Emergency brake assist system

Driver Assistance

- Lane assist system
- Speed assist system
- Blind spot detection
- Park assist system
- Adaptive cruise control system

Passenger Comfort

- Automatic climate control
- Electronic seat adjustment with memory

- Automatic wipers

- Automatic headlamps - adjusts beam automatically

- Automatic cooling - temperature adjustment

Infotainment Systems

- Navigation system

- Vehicle audio

- Information access

All of the above systems forms an infotainment system. Developmental methods for these systems vary according to each manufacturer. Different tools are used for both hardware and software development.

Functional Safety Requirements

In order to minimize the risk of dangerous failures, safety related electronic systems have to be developed following the applicable product liability requirements. Disregard for, or inadequate application of these standards can lead to not only personal injuries, but also severe legal and economic consequences such as product cancellations or recalls.

The IEC 61508 standard, generally applicable to electrical/electronic/programmable safety-related products, is only partially adequate for automotive-development requirements. Consequently, for the automotive industry, this standard is replaced by the existing ISO 26262, currently released as a Final Draft International Standard (FDIS). ISO/DIS 26262 describes the entire product life-cycle of safety related electrical/electronic systems for road vehicles. It has been published as an international standard in its final version in November 2011. The implementation of this new standard will result in modifications and various innovations in the automobile electronics development process, as it covers the complete product life-cycle from the concept phase until its decommissioning.

Automatic Transmission

An 8-gear automatic transmission

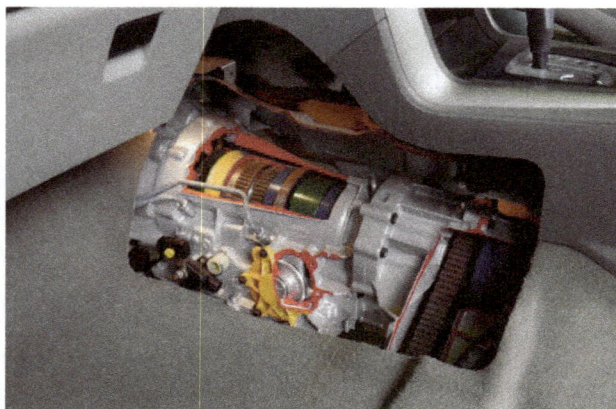

Cutaway showing the typical positioning of an automatic transmission from the interior of an automobile

An automatic transmission, also called auto, self-shifting transmission, n-speed automatic (where n is its number of forward gear ratios), or AT, is a type of motor vehicle transmission that can automatically change gear ratios as the vehicle moves, freeing the driver from having to shift gears manually. Like other transmission systems on vehicles, it allows an internal combustion engine, best suited to run at a relatively high rotational speed, to provide a range of speed and torque outputs necessary for vehicular travel. The number of forward gear ratios is often expressed for manual transmissions as well (e.g., 6-speed manual).

The most popular form found in automobiles is the hydraulic automatic transmission. Similar but larger devices are also used for heavy-duty commercial and industrial vehicles and equipment. This system uses a fluid coupling in place of a friction clutch, and accomplishes gear changes by hydraulically locking and unlocking a system of planetary gears. These systems have a defined set of gear ranges, often with a parking pawl that locks the output shaft of the transmission to keep the vehicle from rolling either forward or backward. Some machines with limited speed ranges or fixed engine speeds, such as some forklifts and lawn mowers, only use a torque converter to provide a variable gearing of the engine to the wheels.

Besides the traditional hydraulic automatic transmissions, there are also other types of automated transmissions, such as a continuously variable transmission (CVT) and semi-automatic transmissions, that free the driver from having to shift gears manually, by using the transmission's computer to change gear, if for example the driver were redlining the engine. Despite superficial similarity to other transmissions, traditional automatic transmissions differ significantly in internal operation and driver's feel from semi-automatics and CVTs. In contrast to conventional automatic transmissions, a CVT uses a belt or other torque transmission scheme to allow an "infinite" number of gear ratios instead of a fixed number of gear ratios. A semi-automatic retains a clutch like a manual transmission, but controls the clutch through electrohydraulic means. The ability to shift gears manually, often via paddle shifters, can also be found on certain automated transmissions (manumatics such as Tiptronic), semi-automatics (BMW SMG, VW Group DSG), and CVTs (such as Lineartronic).

The obvious advantage of an automatic transmission to the driver is the lack of a clutch pedal and manual shift pattern in normal driving. This allows the driver to operate the car with as few as two limbs (possibly using assist devices to position controls within reach of usable limbs), allowing amputees and other disabled individuals to drive. The lack of manual shifting also reduces the attention and workload required inside the cabin, such as monitoring the tachometer and taking a hand off the wheel to move the shifter, allowing the driver to ideally keep both hands on the wheel at all times and to focus more on the road. Control of the car at low speeds is often easier with an automatic than a manual, due to a side effect of the clutchless fluid-coupling design called "creep" that causes the car to want to move while in a driving gear, even at idle. The primary disadvantage of the most popular hydraulic designs is reduced mechanical efficiency of the power transfer between engine and drivetrain, due to the fluid coupling connecting the engine to the gearbox. This can result in lower power/torque ratings for automatics compared to manuals with the same engine specs, as well as reduced fuel efficiency in city driving as the engine must maintain idle against the resistance of the fluid coupling. Advances in transmission and coupler design have narrowed this gap considerably, but clutch-based transmissions (manual or semi-automatic) are still preferred in sport-tuned trim levels of various production cars, as well as in many auto racing leagues.

The automatic transmission was invented in 1921 by Alfred Horner Munro of Regina, Saskatchewan, Canada, and patented under Canadian patent CA 235757 in 1923. (Munro obtained UK patent GB215669 215,669 for his invention in 1924 and US patent 1,613,525 on 4 January 1927). Being a steam engineer, Munro designed his device to use compressed air rather than hydraulic fluid, and so it lacked power and never found commercial application. The first automatic transmission using hydraulic fluid may have been developed in 1932 by two Brazilian engineers, José Braz Araripe and Fernando Lehly Lemos; subsequently the prototype and plans were sold to General Motors who introduced it in the 1940 Oldsmobile as the "Hydra-Matic" transmission. They were incorporated into GM-built tanks during World War II and, after the war, GM marketed them as being "battle-tested." However, a Wall Street Journal article credits ZF Friedrichshafen with the invention, occurring shortly after World War I. ZF's origins were in manufacturing gears for airship engines beginning in 1915; the company was founded by Ferdinand von Zeppelin.

History

Modern automatic transmissions can trace their origins to an early "horseless carriage" gearbox that was developed in 1904 by the Sturtevant brothers of Boston, Massachusetts. This unit had two forward speeds, the ratio change being brought about by flyweights that were driven by the engine. At higher engine speeds, high gear was engaged. As the vehicle slowed down and engine RPM decreased, the gearbox would shift back to low. Unfortunately, the metallurgy of the time wasn't up to the task, and owing to the abruptness of the gear change, the transmission would often fail without warning.

The next significant phase in the automatic transmission's development occurred in 1908 with the introduction of Henry Ford's remarkable Model T. The Model T, in addition to being cheap and reliable by the standards of the day, featured a simple, two speed plus reverse planetary transmission whose operation was manually controlled by the driver using pedals. The pedals actuated the transmission's friction elements (bands and clutches) to select the desired gear. In some respects, this type of transmission was less demanding of the driver's skills than the contemporary, unsynchronized manual transmission, but still required that the driver know when to make a shift, as well as how to get the car off to a smooth start.

In 1934, both REO and General Motors developed semi-automatic transmissions that were less difficult to operate than a fully manual unit. These designs, however, continued to use a clutch to engage the engine with the transmission. The General Motors unit, dubbed the "Automatic Safety Transmission," was notable in that it employed a power-shifting planetary gearbox that was hydraulically controlled and was sensitive to road speed, anticipating future development.

Parallel to the development in the 1930s of an automatically shifting gearbox was Chrysler's work on adapting the fluid coupling to automotive use. Invented early in the 20th century, the fluid coupling was the answer to the question of how to avoid stalling the engine when the vehicle was stopped with the transmission in gear. Chrysler itself never used the fluid coupling with any of its automatic transmissions, but did use it in conjunction with a hybrid manual transmission called "Fluid Drive" (the similar Hy-Drive used a torque converter). These developments in automatic gearbox and fluid coupling technology eventually culminated in the introduction in 1939 of the General Motors Hydra-Matic, the world's first mass-produced automatic transmission.

Available as an option on 1940 Oldsmobiles and later Cadillacs, the Hydra-Matic combined a fluid coupling with three hydraulically controlled planetary gearsets to produce four forward speeds plus reverse. The transmission was sensitive to engine throttle position and road speed, producing fully automatic up- and down-shifting that varied according to operating conditions.

The Hydra-Matic was subsequently adopted by Cadillac and Pontiac, and was sold to various other automakers, including Bentley, Hudson, Kaiser, Nash, and Rolls-Royce. It also found use during World War II in some military vehicles. From 1950 to 1954, Lincoln cars were also available with the Hydra-Matic. Mercedes-Benz subsequently devised a four-speed fluid coupling transmission that was similar in principle to the Hydra-Matic, but of a different design.

Interestingly, the original Hydra-Matic incorporated two features which are widely emulated in today's transmissions. The Hydra-Matic's ratio spread through the four gears produced excellent "step-off" and acceleration in first, good spacing of intermediate gears, and the effect of an overdrive in fourth, by virtue of the low numerical rear axle ratio used in the vehicles of the time. In addition, in third and fourth gear, the fluid coupling only handled a portion of the engine's torque, resulting in a high degree of efficiency. In this respect, the transmission's behavior was similar to modern units incorporating a lock-up torque converter.

In 1956, GM introduced the "Jetaway" Hydra-Matic, which was different in design than the older model. Addressing the issue of shift quality, which was an ongoing problem with the original Hydra-Matic, the new transmission utilized two fluid couplings, the primary one that linked the transmission to the engine, and a secondary one that replaced the clutch assembly that controlled the forward gearset in the original. The result was much smoother shifting, especially from first to second gear, but with a loss in efficiency and an increase in complexity. Another innovation for this new style Hydra-Matic was the appearance of a Park position on the selector. The original Hydra-Matic, which continued in production until the mid-1960s, still used the reverse position for parking pawl engagement.

The first torque converter automatic, Buick's Dynaflow, was introduced for the 1948 model year.

It was followed by Packard's Ultramatic in mid-1949 and Chevrolet's Powerglide for the 1950 model year. Each of these transmissions had only two forward speeds, relying on the converter for additional torque multiplication. In the early 1950s, BorgWarner developed a series of three-speed torque converter automatics for American Motors, Ford Motor Company, Studebaker, and several other manufacturers in the US and other countries. Chrysler was late in developing its own true automatic, introducing the two-speed torque converter PowerFlite in 1953, and the three-speed TorqueFlite in 1956. The latter was the first to utilize the Simpson compound planetary gearset.

General Motors produced multiple-turbine torque converters from 1954 to 1961. These included the Twin-Turbine Dynaflow and the triple-turbine Turboglide transmissions. The shifting took place in the torque converter, rather than through pressure valves and changes in planetary gear connections. Each turbine was connected to the drive shaft through a different gear train. These phased from one ratio to another according to demand, rather than shifting. The Turboglide actually had two speed ratios in reverse, with one of the turbines rotating backwards.

By the late 1960s, most of the fluid-coupling four-speed and two-speed transmissions had disappeared in favor of three-speed units with torque converters. Also around this time, whale oil was removed from automatic transmission fluid. By the early 1980s, these were being supplemented and eventually replaced by overdrive-equipped transmissions providing four or more forward speeds. Many transmissions also adopted the lock-up torque converter (a mechanical clutch locking the torque converter pump and turbine together to eliminate slip at cruising speed) to improve fuel economy.

As computerized engine control units (ECUs) became more capable, much of the logic built into the transmission's valve body was offloaded to the ECU. Some manufacturers use a separate computer dedicated to the transmission called a transmission control unit (TCU), also known as the transmission control module (TCM), which share information with the engine management computer. In this case, solenoids turned on and off by the computer control shift patterns and gear ratios, rather than the spring-loaded valves in the valve body. This allows for more precise control of shift points, shift quality, lower shift times, and (on some newer cars) semi-automatic control, where the driver tells the computer when to shift. The result is an impressive combination of efficiency and smoothness. Some computers even identify the driver's style and adapt to best suit it.

ZF Friedrichshafen and BMW were responsible for introducing the first six-speed (the ZF 6HP26 in the 2002 BMW E65 7-Series). Mercedes-Benz's 7G-Tronic was the first seven-speed in 2003, with Toyota introducing an eight-speed in 2007 on the Lexus LS 460. Derived from the 7G-Tronic, Mercedes-Benz unveiled a semi-automatic transmission with the torque converter replaced with a wet multi clutch called the AMG SPEEDSHIFT MCT. The 2014 Jeep Cherokee has the world's first nine-speed automatic transmission for a passenger vehicle to market.

Parts and Operation

Hydraulic Automatic Transmissions

The predominant form of automatic transmission is hydraulically operated; using a fluid coupling or torque converter, and a set of planetary gearsets to provide a range of gear ratios.

A cutaway of an 8-speed ZF 8HP showing the major stages of a hydraulic automatic transmission: the torque converter (left), the planetary gearsets and clutch plates (center), as well as hydraulic and electronic controls (bottom).

Hydraulic automatic transmissions consist of three major components:

Torque Converter

A type of fluid coupling, hydraulically connecting the engine to the transmission. This takes the place of a friction clutch in a manual transmission. It transmits and decouples the engine power to the planetary gears, allowing the vehicle to come to stop with the engine still running without stalling.

A torque converter differs from a fluid coupling, in that it provides a variable amount of torque multiplication at low engine speeds, increasing breakaway acceleration. A fluid coupling works well when both the impeller and turbine are rotating at similar speeds, but it is very inefficient at initial acceleration, where rotational speeds are very different. This torque multiplication is accomplished with a third member in the coupling assembly known as the stator, which acts to modify the fluid flow depending on the relative rotational speeds of the impeller and turbine. The stator itself does not rotate, but its vanes are so shaped that when the impeller (which is driven by the engine) is rotating at a high speed and the turbine (which receives the transmitted power) is spinning at a low speed, the fluid flow hits the vanes of the turbine in a way that multiplies the torque being applied. This causes the turbine to begin spinning faster as the vehicle accelerates (ideally), and as the relative rotational speeds equalize, the torque multiplication diminishes. Once the impeller and turbine are rotating within 10% of each other's speed, the stator ceases to function and the torque converter acts as a simple fluid coupling.

Planetary Gears Train

Consisting of planetary gear sets as well as clutches and bands. These are the mechanical systems that provide the various gear ratios, altering the speed of rotation of the output shaft depending on which planetary gears are locked.

To effect gear changes, one of two types of clutches or bands are used to hold a particular member of the planetary gearset motionless, while allowing another member to rotate, thereby transmitting torque and producing gear reductions or overdrive ratios. These clutches are actuated by the valve body, their sequence controlled by the transmission's internal programming. Principally, a type of device known as a sprag or roller clutch is used for routine upshifts/downshifts. Operating much as a ratchet, it transmits torque only in one direction, free-wheeling or "overrunning" in the other. The advantage of this type of clutch is that it eliminates the sensitivity of timing a simultaneous clutch release/apply on two planetaries, simply "taking up" the drivetrain load when actuated, and releasing automatically when the next gear's sprag clutch assumes the torque transfer. The bands come into play for manually selected gears, such as low range or reverse, and operate on the planetary drum's circumference. Bands are not applied when drive/overdrive range is selected, the torque being transmitted by the sprag clutches instead. Bands are used for braking; the GM Turbo-Hydramatics incorporated this..

Hydraulic Controls

Uses special transmission fluid sent under pressure by an oil pump to control various clutches and bands modifying the speed of the output depending on the vehicle's running condition.

The pump is typically a gear pump mounted between the torque converter and the planetary gearset. It draws transmission fluid from a sump and pressurizes it, which is needed for transmission components to operate. The input for the pump is connected to the torque converter housing, which in turn is bolted to the engine's flexplate, so the pump provides pressure whenever the engine is running and there is enough transmission fluid, but the disadvantage is that when the engine is not running, no oil pressure is available to operate the main components of the transmission, and is thus impossible to push-start a vehicle equipped with an automatic transmission. Early automatic transmissions also had a rear pump for towing purposes, ensuring the lubrication of the rear-end components.

The governor is connected to the output shaft and regulates the hydraulic pressure depending on the vehicle speed. The engine load is monitored either by a throttle cable or a vacuum modulator. The valve body is the hydraulic control center that receives pressurized fluid from the main pump operated by the fluid coupling/torque converter. The pressure coming from this pump is regulated and used to run a network of spring-loaded valves, check balls and servo pistons. The valves use the pump pressure and the pressure from a centrifugal governor on the output side (as well as hydraulic signals from the range selector valves and the throttle valve or modulator) to control which ratio is selected on the gearset; as the vehicle and engine change speed, the difference between the pressures changes, causing different sets of valves to open and close. The hydraulic pressure controlled by these valves drives the various clutch and brake band actuators, thereby controlling the operation of the planetary gearset to select the optimum gear ratio for the current operating conditions. However, in many modern automatic transmissions, the valves are controlled by electro-mechanical servos which are controlled by the electronic engine control unit (ECU) or a separate transmission control unit (TCU, also known as transmission control module (TCM).

The hydraulic & lubricating oil, called automatic transmission fluid (ATF), provides lubrication, corrosion prevention, and a hydraulic medium to convey mechanical power (for the operation of the transmission). Primarily made from refined petroleum, and processed to provide properties that promote smooth power transmission and increase service life, the ATF is one of the few parts of the automatic transmission that needs routine service as the vehicle ages.

The multitude of parts, along with the complex design of the valve body, originally made hydraulic automatic transmissions much more complicated (and expensive) to build and repair than manual transmissions. In most cars (except US family, luxury, sport-utility vehicle, and minivan models) they have usually been extra-cost options for this reason. Mass manufacturing and decades of improvement have reduced this cost gap.

In some modern cars, computers use sensors on the engine to detect throttle position, vehicle speed, engine speed, engine load, etc. to control the exact shift point. The computer transmits the information via solenoids that redirect the fluid the appropriate clutch or servo to control shifting.

Continuously Variable Transmissions

A fundamentally different type of automatic transmission is the continuously variable transmission or CVT, which can smoothly and steplessly alter its gear ratio by varying the diameter of a pair of belt or chain-linked pulleys, wheels or cones. Some continuously variable transmissions use a hydrostatic drive — consisting of a variable displacement pump and a hydraulic motor — to transmit power without gears. Some early forms, such as the Hall system (which dates back to 1896), used a fixed displacement pump and a variable displacement motor, and were designed to provide robust variable transmission for early commercial heavy motor vehicles. CVT designs are usually as fuel efficient as manual transmissions in city driving, but early designs lose efficiency as engine speed increases.

A slightly different approach to CVT is the concept of toroidal CVT or infinitely variable transmission (IVT). These concepts provide zero and reverse gear ratios.

E-CVT

Some hybrid vehicles, notably those of Toyota, Lexus and Ford Motor Company, have an electronically controlled CVT (E-CVT). In this system, the transmission has fixed gears, but the ratio of wheel-speed to engine-speed can be continuously varied by controlling the speed of the third input to a differential using motor-generators.

Automatic Transmission Modes

Most automatic transmissions offer the driver a certain amount of manual control over the transmission's shifts.

Conventionally, in order to select the transmission operating mode, the driver moves a selection lever located either on the steering column or on the floor (as with a manual on the floor, except that automatic selectors on the floor do not move in the same type of pattern as manual levers do). In order to select modes, or to manually select specific gear ratios, the driver must push a button in (called the shift-lock button) or pull the handle (only on column mounted shifters) out. Some vehicles position selector buttons for each mode on the cockpit instead, freeing up space on the central console.

Vehicles conforming to US Government standards must have the modes ordered P-R-N-D-L (left to right, top to bottom, or clockwise). Previously, quadrant-selected automatic transmissions often used a P-N-D-L-R layout, or similar. Such a pattern led to a number of deaths and injuries owing to driver error causing unintentional gear selection, as well as the danger of having a selector (when worn) jump into reverse from low gear during engine braking maneuvers.

A floor selection lever in a 1992 Ford Escort showing the P-R-N-[D]-D-L modes as well as the shift lock button on the top of the lever

Depending on the model and make of the transmission, these controls can take several forms. However most include the following:

Park (P)

> This selection mechanically locks the output shaft of transmission, restricting the vehicle from moving in any direction. A parking pawl prevents the transmission from rotating, and therefore the vehicle from moving. However, it should be noted that the vehicle's non-driven wheels are still free to rotate, and the driven wheels may still rotate individually (because of the differential). For this reason, it is recommended to use the hand brake (parking brake) because this actually locks (in most cases) the wheels and prevents them from moving. It is typical of front-wheel-drive vehicles for the parking brake to lock the rear (non-driving) wheels, so use of both the parking brake and the transmission park lock provides the greatest security against unintended movement on slopes. This also increases the life of the transmission and the park pin mechanism, because parking on an incline with the transmission in park without the parking brake engaged will cause undue stress on the parking pin, and may even prevent the pin from releasing. A hand brake should also prevent the car from moving if a worn selector accidentally drops into reverse gear while idling.

> A car should be allowed to come to a complete stop before setting the transmission into park to prevent damage. Usually, Park (P) is one of only two selections in which the car's engine can be started, the other being Neutral (N). This is typically achieved via a normally open inhibitor switch (sometimes called a "neutral safety switch") wired in series with the starter motor engagement circuit, which is closed when P or N is selected, completing the circuit (when the key is turned to the start position). In many modern cars and trucks, the driver must have the foot brake applied before the transmission can be taken out of park. The Park position is omitted on buses/coaches (and some road tractors) with automatic transmission (on which a parking pawl is not practical), which must instead be placed in neutral with the air-operated parking brakes set.

Reverse (R)

This engages reverse gear within the transmission, permitting the vehicle to be driven backward, and operates a switch to turn on the white backup lights for improved visibility (the switch may also activate a beeper on delivery trucks or other large vehicles to audibly warn other drivers and nearby pedestrians of the driver's reverse movement). To select reverse in most transmissions, the driver must come to a complete stop, depress the shift-lock button (or move the shift lever toward the driver in a column shifter, or move the shifter sideways along a notched channel in a console shifter) and select reverse. The driver should avoid engaging reverse while the vehicle is moving forwards, and likewise avoid engaging any forward gear while travelling backwards. On transmissions with a torque converter, doing so at very low speed (walking pace) is not harmful, but causes unnecessary wear on clutches and bands, and a sudden deceleration that not only is uncomfortable, but also uncontrollable since the brakes and the throttle contribute in the same direction. This sudden acceleration, or jerk, can still be felt when engaging the gear at standstill, but the driver normally suppresses this by holding the brakes. Travelling slowly in the right direction while engaging the gear minimizes the jerk further, which is actually beneficial to the wearing parts of the transmission. Electronically controlled transmissions may behave differently, as engaging a gear at speed is essentially undefined behaviour. Some modern transmissions have a safety mechanism that will resist putting the car in reverse when the vehicle is moving forward; such a mechanism may consist of a solenoid-controlled physical barrier on either side of the reverse position, electronically engaged by a switch on the brake pedal, so that the brake pedal needs to be depressed in order to allow the selection of reverse. Some electronic transmissions prevent or delay engagement of reverse gear altogether while the car is moving.

Some shifters with a shift button allow the driver to freely move the shifter from R to N or D without actually depressing the button. However, the driver cannot shift back to R without depressing the shift button, to prevent accidental shifting which could damage the transmission, especially at high speeds.

Neutral / No gear (N)

This disengages all gear trains within the transmission, effectively disconnecting the transmission from the driven wheels, allowing the vehicle to coast freely under its own weight and gain momentum without the motive force from the engine. Coasting in idle down long grades (where law permits) should be avoided, though, as the transmission's lubrication pump is driven by non-idle engine RPMs. Similarly, emergency towing with an automatic transmission in neutral should be a last resort. Manufacturers understand emergency situations and list limitations of towing a vehicle in neutral (usually not to exceed 55 mph and 50 miles). This is the only other selection in which the vehicle's engine may be started.

Drive (D)

This position allows the transmission to engage the full range of available forward gear ratios, allowing the vehicle to move forward and accelerate through its range of gears. The number of gear ratios within the transmission depends on the model, but they initially ranged from three (predominant before the 1990s), to four and five speeds (losing popularity to six-speed autos). Six-speed automatic transmissions are probably the most common offering in cars and trucks from 2010 in carmakers as Toyota, GM and Ford. However, seven-speed automatics are becoming available in some high-performance production luxury cars (found in Mercedes 7G gearbox, Infiniti), as are eight-speed autos in models from 2006 introduced by Aisin Seiki Co. in Lexus, ZF and Hyundai Motor Company. From 2013 are available nine speeds transmissions produced by ZF and Mercedes 9G.

Overdrive ('D', 'OD', or a boxed [D] or the absence of an illuminated 'O/D OFF')

This mode is used in some transmissions to allow early computer-controlled transmissions to engage the automatic overdrive. In these transmissions, Drive (D) locks the automatic overdrive off, but is identical otherwise. OD (Overdrive) in these cars is engaged under steady speeds or low acceleration at approximately 35–45 mph (56–72 km/h). Under hard acceleration or below 35–45 mph (56–72 km/h), the transmission will automatically downshift. Other vehicles with this selector (for example light trucks) will not only disable up-shift to the overdrive gear, but keep the remaining available gears continuously engaged to the engine for use of compression braking. Drivers should verify the behaviour of this switch and consider the benefits of reduced friction brake use when city driving where speeds typically do not necessitate the overdrive gear.

Most automatic transmissions include some means of forcing a downshift (Throttle kickdown) into the lowest possible gear ratio if the throttle pedal is fully depressed. In many older designs, kickdown is accomplished by mechanically actuating a valve inside the transmission. Most modern designs use a solenoid-operated valve that is triggered by a switch on the throttle linkage or by the engine control unit (ECU) in response to an abrupt increase in engine power.

Mode selection allows the driver to choose between preset shifting programs. For example, Economy mode saves fuel by upshifting at lower engine speeds, while Sport mode (aka "Power" or "Performance") delays upshifting for maximum acceleration. Some transmission units also have Winter mode, where higher gear ratios are chosen to keep revs as low as possible while on slippery surfaces. The modes also change how the computer responds to throttle input.

Conventionally, automatic transmissions have selector positions that allow the driver to limit the maximum ratio that the transmission may engage. On older transmissions, this was accomplished by a mechanical lockout in the transmission valve body preventing an upshift until the lockout was disengaged; on computer-controlled transmissions, the same effect is accomplished by firmware. The transmission can still upshift and downshift automatically between the remaining ratios: for example, in the 3 range, a transmission could shift from first to second to third, but not into fourth or higher ratios. Some transmissions will still upshift automatically into the higher ratio if the engine reaches its maximum permissible speed in the selected range.

Third (3)

This mode limits the transmission to the first three gear ratios, or sometimes locks the transmission in third gear. This can be used to climb or going down hill. Some vehicles will automatically shift up out of third gear in this mode if a certain revolutions per minute

(RPM) range is reached in order to prevent engine damage. This gear is also recommended while towing a trailer.

Second (2 or S)

This mode limits the transmission to the first two gear ratios, or locks the transmission in second gear on Ford, Kia, and Honda models. This can be used to drive in adverse conditions such as snow and ice, as well as climbing or going down hills in winter. It is usually recommended to use second gear for starting on snow and ice, and use of this position enables this with an automatic transmission. Some vehicles will automatically shift up out of second gear in this mode if a certain RPM range is reached in order to prevent engine damage.

Although traditionally considered second gear, there are other names used. Chrysler models with a three-speed automatic since the late 1980s have called this gear 3 while using the traditional names for Drive and Low. Oldsmobile has called second gear as the 'Super' range — which was first used on their 4-speed Hydramatic transmissions, although the use of this term continued until the early 1980s when GM's Turbo Hydramatic automatic transmissions were standardized by all of their divisions years after the 4-speed Hydramatic was discontinued.

Some automatics, particularly those fitted to larger capacity or high torque engines, either when "2" is manually selected, or by engaging a winter mode, will start off in second gear instead of first, and then not shift into a higher gear until returned to "D." Also note that as with most American automatic transmissions, selecting "2" using the selection lever will not tell the transmission to be in only 2nd gear; rather, it will simply limit the transmission to 2nd gear after prolonging the duration of 1st gear through higher speeds than normal operation. The 2000–2002 Lincoln LS V8 (the five-speed automatic without manumatic capabilities, as opposed to the optional sport package w/ manu-matic 5-speed) started in 2nd gear during most starts both in winter and other seasons by selecting the "D5" transmission selection notch in the shiftgate (for fuel savings), whereas "D4" would always start in 1st gear. This is done to reduce torque multiplication when proceeding forward from a standstill in conditions where traction was limited — on snow- or ice-covered roads, for example.

First (1 or L [Low])

This mode locks the transmission in first gear only. In older vehicles, it will not change to any other gear range. Some vehicles will automatically shift up out of first gear in this mode if a certain RPM range is reached in order to prevent engine damage. This, like second, can be used during the winter season, for towing, or for downhill driving to increase the engine braking effect. The "Austin Mini" automatic transmission is different in this respect - This mode locks the transmission in first gear, but the gearbox has a freewheel on the overrun. Closing the throttle after acceleration results in the vehicle continuing at the same speed and only slowing down due to friction and wind resistance. During this time, the engine RPM will drop back to idle until the throttle is pressed again. What this means is that in "First", engine braking is not available and "2" is the lowest gear that should be used whilst descending hills.

One type of manumatic shifting system available on automatic transmissions are paddle shifters. The paddle depicted here is the upshift paddle in a 2013 Honda Accord, with the driver's hand on it. Manumatics and paddle shifters may control any type of automatic transmission, including the continuously variable transmission in the Accord.

Manual Controls

Some transmissions have a mode in which the driver has full control of ratios change (either by moving the selector, or through the use of buttons or paddles), completely overriding the automated function of the hydraulic controller. Such control is particularly useful in cornering, to avoid unwanted upshifts or downshifts that could compromise the vehicle's balance or traction. "Manumatic" shifters, first popularized by Porsche in the 1990s under the trade name Tiptronic, have become a popular option on sports cars and other performance vehicles. With the near-universal prevalence of electronically controlled transmissions, they are comparatively simple and inexpensive, requiring only software changes, and the provision of the actual manual controls for the driver. The amount of true manual control provided is highly variable: some systems will override the driver's selections under certain conditions, generally in the interest of preventing engine damage. Since these gearboxes also have a throttle kickdown switch, it is impossible to fully exploit the engine power at low to medium engine speeds.

Manufacturer-specific Modes

As well as the above modes there are other modes, dependent on the manufacturer and model. Some examples include:

D5

In Hondas and Acuras equipped with five-speed automatic transmissions, this mode is used commonly for highway use (as stated in the manual), and uses all five forward gear ratios.

D4

This mode is also found in Honda and Acura four or five-speed automatics, and only uses the first four gear ratios. According to the manual, it is used for stop-and-go traffic, such as city driving.

D3 or 3

This mode is found in Honda, Acura, Volkswagen and Pontiac four-speed automatics and only uses the first three gear ratios. According to the manual, it is used for stop-and-go traffic, such as city driving.

D2 and D1

These modes are found on older Ford transmissions (C6, etc.). In D1, all three gears are used, whereas in D2 the car starts in second gear and upshifts to third.

S or Sport

This is commonly described as Sport mode. It operates in an identical manner as "D" mode, except that the upshifts change much higher up the engine's rev range. This has the effect on maximising all the available engine output, and therefore enhances the performance of the vehicle, particularly during acceleration. This mode will also downchange much higher up the rev range compared to "D" mode, maximising the effects of engine braking. This mode will have a detrimental effect on fuel economy. Hyundai has a Norm/Power switch next to the gearshift for this purpose on the Tiburon.

Some early GMs equipped with HYDRA-MATIC transmissions used (S) to indicate Second gear, being the same as the 2 position on a Chrysler, shifting between only first and second gears. This would have been recommended for use on steep grades, or slippery roads like dirt, or ice, and limited to speeds under 40 mph. (L) was used in some early GMs to indicate (L)ow gear, being the same as the 2 position on a Chrysler, locking the transmission into first gear. This would have been recommended for use on steep grades, or slippery roads like dirt, or ice, and limited to speeds under 15 mph.

+ −, and M

This is for the Manual mode selection of gears in certain automatics, such as Porsche and Honda's Tiptronic and BMW and Kia's Steptronic. The M feature can also be found in vehicles such as the Dodge Magnum and Journey; Pontiac G6; Mazda3, Mazda6, and CX-7; Toyota Camry, Corolla, Fortuner, Previa and Innova; Kia Forte (K3/Cerato), Optima (K5), Cadenza (K7) and K9 (Quoris). Mitsubishi montero sport / pajero sport and some Audi models (Audi TT) do not have the M, and instead have the + and -, which is separated from the rest of the shift modes; the same is true for some Peugeot products like the Peugeot 206. Meanwhile, the driver can shift up and down at will by toggling the (console mounted) shift lever similar to a semi-automatic transmission. This mode may be engaged either through a selector/position or by actually changing the gears (e.g., tipping the gear-down paddles mounted near the driver's fingers on the steering wheel).

Winter (W)

In some Volvo, Mercedes-Benz, BMW and General Motors Europe models, a winter mode can be engaged so that second gear is selected instead of first when pulling away from stationary, to reduce the likelihood of loss of traction due to wheel spin on snow or ice. On GM cars, this was D2 in the 1950s, and is Second Gear Start after 1990. On Ford, Kia, and Honda automatics, this feature can be accessed by moving the gear selector to 2 to start, then taking your foot off the accelerator while selecting D once the car is moving.

Brake (B)

A mode selectable on some Toyota models, as well as electric cars from several manufacturers. It can be used to decelerate, or maintain speed going downhill, without using the conventional brakes. In non-hybrid cars, B mode selects a lower gear to increase engine braking. GM called this "HR" ("hill retarder") and "GR" ("grade retarder") in the 1950s. In hybrids such as the Toyota Prius, which have a fixed gear ratio, B mode slows the car in part by increasing engine air intake, which enhances engine braking. In electric cars such as the Nissan Leaf and Mitsubishi i-MiEV, B mode increases the level of regenerative braking when the accelerator pedal is released.

Some automatic transmissions modified or designed specifically for drag racing may also incorporate a transbrake as part of a manual valve body. Activated by electrical solenoid control, a transbrake simultaneously engages the first and reverse gears, locking the transmission and preventing the input shaft from turning. This allows the driver of the car to raise the engine RPM against the resistance of the torque converter, then launch the car by simply releasing the transbrake switch.

Comparison with Manual Transmission

Most cars sold in North America since the 1950s have been available with an automatic transmission based on the fact that the three major American car manufacturers had started using automatics. Conversely, in Europe a manual gearbox is standard, with 20% of drivers opting for an automatic transmission. In some Asian markets and in Australia, automatic transmissions have become very popular since the 1980s.

Vehicles equipped with automatic transmissions are not as complex to drive. Consequently, in some jurisdictions, drivers who have passed their driving test in a vehicle with an automatic transmission will not be licensed to drive a manual transmission vehicle. Conversely, a manual license will allow the driver to drive vehicles with either transmission. Countries in which such driving license restrictions are applied include some states in Australia, Austria, Belgium, Belize, China, Croatia, Denmark, Dominican Republic, Estonia, Finland, France, Germany, Hong Kong, Hungary, India, Indonesia, Ireland, Israel, Japan, Latvia, Lebanon, Lithuania, Macau, Mauritius, the Netherlands, New Zealand (restricted licence only), Norway, Philippines, Poland, Portugal, Qatar, Romania, Russia (as of April 2014), Saudi Arabia (as of March 2012), Singapore, Slovenia, South Africa, South Korea, Spain, Sweden, Switzerland, Taiwan, Trinidad and Tobago, United Arab Emirates and the United Kingdom.

A conventional manual transmission is frequently the base equipment in a car, with the option being an automated transmission such as a conventional automatic, semi-automatic, or CVT.

Effects on Vehicle Control

Cornering

Unexpected gear changes can affect the attitude of a vehicle in marginal conditions.

Maintaining Constant Speed

Torque converters and CVT transmissions make changes in vehicle speed less apparent by the engine noise, as they decouple the engine speed from vehicle speed.

Lockup torque converters that engage and disengage at certain speeds can make these speeds unstable — the transmission wastes less power above the speed at which the torque converter locks up, usually causing more power to the wheels for the same throttle input.

Controlling Wheelspin

Torque converters respond quickly to loss of traction (torque) by an increased speed of the driving wheels for the same engine speed. Thus, under most conditions, where the static friction is higher than the kinetic friction, the engine speed must be brought down to counteract wheelspin when it has occurred, requiring a stronger or quicker throttle reduction by the driver than with a manual transmission, making wheelspin harder to control. This is most apparent in driving conditions with much higher static friction than kinetic, such as packed hard snow (that turns to ice by friction work), or snow on top of ice.

Climbing Steep Slippery Slopes

In situations where a driver with a manual transmission can't afford a gearshift, in fear of losing too much speed to reach a hilltop, automatic transmissions are at a great advantage — whereas the manual driver depends on finding a gear that is not too low to enter the bottom of the hill at the necessary speed, but not too high to stall the engine at the top of the hill, sometimes an impossible task, this is a non-issue with automatic transmissions, not just because gearshifts are quick, but they typically maintain some power on the driving wheels during the gearshift.

Energy Efficiency

Earlier hydraulic automatic transmissions were almost always less energy efficient than manual transmissions due mainly to viscous and pumping losses, both in the torque converter and the hydraulic actuators. 21% is the loss on a 3 speed Chrysler Torqueflite compared to a modern GM 6L80 automatic. A relatively small amount of energy is required to pressurize the hydraulic control system, which uses fluid pressure to determine the correct shifting patterns and operate the various automatic clutch mechanisms. However, with technological developments some modern Continuously variable transmission are more fuel efficient than their manual counterparts and modern 8 speed automatics are within 5% as efficient as a manual gearbox.

Manual transmissions use a mechanical clutch to transmit torque, rather than a torque converter, thus avoiding the primary source of loss in an automatic transmission. Manual transmissions also avoid the power requirement of the hydraulic control system, by relying on the human muscle power of the vehicle operator to disengage the clutch and actuate the gear levers, and the mental power of the operator to make appropriate gear ratio selections. Thus the manual transmission requires very little engine power to function, with the main power consumption due to drag from the gear train being immersed in the lubricating oil of the gearbox.

The on-road acceleration of an automatic transmission can occasionally exceed that of an otherwise identical vehicle equipped with a manual transmission in turbocharged diesel applications. Turbo-boost is normally lost between gear changes in a manual whereas in an automatic the accelerator pedal can remain fully depressed. This however is still largely dependent upon the number and optimal spacing of gear ratios for each unit, and whether or not the elimination of spooldown/accelerator lift off represent a significant enough gain to counter the slightly higher power consumption of the automatic transmission itself.

Automatic Transmission Models

Some of the best known automatic transmission families include:

- General Motors — Dynaflow, Powerglide, Turboglide, "Turbo-Hydramatic" TH350, TH400 and 700R4, 4L60-E, 4L80-E, Holden Trimatic

- Ford: Cruise-O-Matic, C4, CD4E, C6, AOD/AODE, E4OD, ATX, AXOD/AX4S/AX4N

- Cummins 68 RFE (fitted to the Ram diesel segment)

- Chrysler: TorqueFlite 727 and 904, A500, A518, 45RFE, 545RFE

- BorgWarner (later Aisin AW)

- ZF Friedrichshafen automatic transmissions

- Mercedes-Benz transmissions

- Allison Transmission

- Voith Voith Turbo

- Aisin AW; Aisin AW is a Japanese automotive parts supplier, known for its automatic transmissions and navigation systems

- Honda

- Nissan/Jatco

- Volkswagen Group — 01M

- Drivetrain Systems International (DSI) — M93, M97 and M74 4-speeds, M78 and M79 6-speeds

- Hyundai Hyundai Powertech — 4F12, 4F16, 4F23 4-Speeds, 5F25, 5F16, 5F23 5-Speeds, 6F17, 6F26, 6F40 6-Speeds, 8R40, 8R50 8-Speeds, Mini Cooper — Automatic or manual transmission, all models

Automatic transmission families are usually based on Ravigneaux, Lepelletier, or Simpson planetary gearsets. Each uses some arrangement of one or two central sun gears, and a ring gear, with differing arrangements of planet gears that surround the sun and mesh with the ring. An exception to this is the Hondamatic line from Honda, which uses sliding gears on parallel axes like a manual transmission without any planetary gearsets. Although the Honda is quite different from all other automatics, it is also quite different from an automated manual transmission (AMT).

Many of the above AMTs exist in modified states, which were created by racing enthusiasts and their mechanics by systematically re-engineering the transmission to achieve higher levels of performance. These are known as "performance transmissions". Example of manufacturers of high performance transmissions are General Motors and Ford.

Manufacturing Engineering

Manufacturing engineering is a discipline of engineering dealing with various manufacturing sciences and practices including the research, design and development of systems, processes, machines, tools, and equipment. The manufacturing engineer's primary focus is to turn raw materials into a new or updated product in the most economic, efficient, and effective way possible.

Overview

This field also deals with the integration of different facilities and systems for producing quality products (with optimal expenditure) by applying the principles of physics and the results of manufacturing systems studies, such as the following:

• Craft or Guild	• Computer integrated manufacturing	• Agile manufacturing
• Putting-out system	• Computer-aided technologies in manufacturing	• Rapid manufacturing
• British factory system	• Just in time manufacturing	• Prefabrication
• American system of manufacturing	• Lean manufacturing	• Ownership
• Soviet collectivism in manufacturing	• Flexible manufacturing	• Fabrication
• Mass production	• Mass customization	• Publication

A set of six-axis robots used for welding.

Manufacturing engineers develop and create physical artifacts, production processes, and technology. It is a very broad area which includes the design and development of products. Manufacturing engineering is considered to be a subdiscipline of industrial engineering/systems engineering and has very strong overlaps with mechanical engineering. Manufacturing engineers' success or failure directly impacts the advancement of technology and the spread of innovation. This field of manufacturing engineering emerged from tool and die discipline in the early 20th century. It expanded greatly from the 1960s when industrialized countries introduced factories with:

1. Numerical control machine tools and automated systems of production.

2. Advanced statistical methods of quality control: These factories were pioneered by the American electrical engineer William Edwards Deming, who was initially ignored by his home country. The same methods of quality control later turned Japanese factories into world leaders in cost-effectiveness and production quality.

3. Industrial robots on the factory floor, introduced in the late 1970s: These computer-controlled welding arms and grippers could perform simple tasks such as attaching a car door quickly and flawlessly 24 hours a day. This cut costs and improved production speed.

History

The history of manufacturing engineering can be traced to factories in the mid 19th century USA and 18th century UK. Although large home production sites and workshops were established in China, ancient Rome and the Middle East, the Venice Arsenal provides one of the first examples of a factory in the modern sense of the word. Founded in 1104 in the Republic of Venice several hundred years before the Industrial Revolution, this factory mass-produced ships on assembly lines using manufactured parts. The Venice Arsenal apparently produced nearly one ship every day and, at its height, employed 16,000 people.

Many historians regard Matthew Boulton's Soho Manufactory (established in 1761 in Birmingham) as the first modern factory. Similar claims can be made for John Lombe's silk mill in Derby (1721), or Richard Arkwright's Cromford Mill (1771). The Cromford Mill was purpose-built to accommodate the equipment it held and to take the material through the various manufacturing processes.

Ford assembly line, 1913.

One historian, Murno Gladst, contends that the first factory was in Potosí. The Potosi factory took advantage of the abundant silver that was mined nearby and processed silver ingot slugs into coins.

British colonies in the 19th century built factories simply as buildings where a large number of workers gathered to perform hand labor, usually in textile production. This proved more efficient for the administration and distribution of materials to individual workers than earlier methods of manufacturing, such as cottage industries or the putting-out system.

Cotton mills used inventions such as the steam engine and the power loom to pioneer the industrial factories of the 19th century, where precision machine tools and replaceable parts allowed greater efficiency and less waste. This experience formed the basis for the later studies of manufacturing engineering. Between 1820 and 1850, non-mechanized factories supplanted traditional artisan shops as the predominant form of manufacturing institution.

Henry Ford further revolutionized the factory concept and thus manufacturing engineering in the early 20th century with the innovation of mass production. Highly specialized workers situated alongside a series of rolling ramps would build up a product such as (in Ford's case) an automobile. This concept dramatically decreased production costs for virtually all manufactured goods and brought about the age of consumerism.

Modern Developments

Modern manufacturing engineering studies include all intermediate processes required for the production and integration of a product's components.

Some industries, such as semiconductor and steel manufacturers use the term "fabrication" for these processes.

KUKA industrial robots being used at a bakery for food production

Automation is used in different processes of manufacturing such as machining and welding. Automated manufacturing refers to the application of automation to produce goods in a factory. The main advantages of automated manufacturing for the manufacturing process are realized with effective implementation of automation and include: higher consistency and quality, reduction of lead times, simplification of production, reduced handling, improved work flow, and improved worker morale.

Robotics is the application of mechatronics and automation to create robots, which are often used in manufacturing to perform tasks that are dangerous, unpleasant, or repetitive. These robots may be of any shape and size, but all are preprogrammed and interact physically with the world. To create a robot, an engineer typically employs kinematics (to determine the robot's range of motion) and mechanics (to determine the stresses within the robot). Robots are used extensively in manufacturing engineering.

Robots allow businesses to save money on labor, perform tasks that are either too dangerous or too precise for humans to perform economically, and to ensure better quality. Many companies employ assembly lines of robots, and some factories are so robotized that they can run by themselves. Outside the factory, robots have been employed in bomb disposal, space exploration, and many other fields. Robots are also sold for various residential applications.

Education

Certification Programs

Manufacturing engineers possess a bachelor's degree in engineering with a major in manufacturing engineering. The length of study for such a degree is usually four to five years followed by five more years of professional practice to qualify as a professional engineer. Working as a manufacturing engineering technologist involves a more applications-oriented qualification path.

Academic degrees for manufacturing engineers are usually the Bachelor of Engineering, [BE] or [BEng], and the Bachelor of Science, [BS] or [BSc]. For manufacturing technologists the required degrees are Bachelor of Technology [B.TECH] or Bachelor of Applied Science [BASc] in Manufacturing, depending upon the university. Master's degrees in engineering manufacturing include Master of Engineering [ME] or [MEng] in Manufacturing, Master of Science [M.Sc] in Manufacturing Management, Master of Science [M.Sc] in Industrial and Production Management, and Master of Science [M.Sc] as well as Master of Engineering [ME] in Design, which is a subdiscipline of manufacturing. Doctoral [PhD] or [DEng] level courses in manufacturing are also available depending on the university.

The undergraduate degree curriculum generally includes courses in physics, mathematics, computer science, project management, and specific topics in mechanical and manufacturing engineering. Initially such topics cover most, if not all, of the subdisciplines of manufacturing engineering. Students then choose to specialize in one or more subdisciplines towards the end of their degree work.

Syllabus

The foundational curriculum for a bachelor's degree in manufacturing engineering is very similar to that for mechanical engineering, and includes:

- Statics and dynamics

- Strength of materials and solid mechanics

- Instrumentation and measurement

- Applied thermodynamics, heat transfer, energy conversion, and HVAC

- Fluid mechanics and fluid dynamics

- Mechanism design (including kinematics and dynamics)

- Manufacturing technology or processes

- Hydraulics and pneumatics

- Mathematics - in particular, calculus, differential equations, statistics, and linear algebra.

- Engineering design and graphics

- Circuit Analysis

- Lean manufacturing

- Mechatronics and control theory

- Automation and reverse engineering

- Quality assurance and control

- Material science

- Drafting, CAD (including solid modeling), and CAM, etc.

A bachelor's degree in these two areas will typically differ only by a few specialized classes, although the mechanical engineering degree requires more mathematics expertise.

Manufacturing Engineering Certification

Certification and licensure:

In some countries, "professional engineer" is the term for registered or licensed engineers who are permitted to offer their professional services directly to the public. Professional Engineer, abbreviated (PE - USA) or (PEng - Canada), is the designation for licensure in North America. In order to qualify for this license, a candidate needs a bachelor's degree from an ABET recognized university in the USA, a passing score on a state examination, and four years of work experience usually gained via a structured internship. In the USA, more recent graduates have the option of dividing this licensure process into two segments. The Fundamentals of Engineering (FE) exam is often taken immediately after graduation and the Principles and Practice of Engineering exam is taken after four years of working in a chosen engineering field.

Society of Manufacturing Engineers (SME) certifications (USA):

The SME administers qualifications specifically for the manufacturing industry. These are not degree level qualifications and are not recognized at the professional engineering level. The following discussion deals with qualifications in the USA only. Qualified candidates for the Certified Manufacturing Technologist Certificate (CMfgT) must pass a three-hour, 130-question multiple-choice exam. The exam covers math, manufacturing processes, manufacturing management, automation, and related subjects. Additionally, a candidate must have at least four years of combined education and manufacturing-related work experience.

Certified Manufacturing Engineer (CMfgE) is an engineering qualification administered by the Society of Manufacturing Engineers, Dearborn, Michigan, USA. Candidates qualifying for a Certified Manufacturing Engineer credential must pass a four-hour, 180 question multiple-choice exam which covers more in-depth topics than does the CMfgT exam. CMfgE candidates must also have eight years of combined education and manufacturing-related work experience, with a minimum of four years of work experience.

Certified Engineering Manager (CEM). The Certified Engineering Manager Certificate is also designed for engineers with eight years of combined education and manufacturing experience. The test is four hours long and has 160 multiple-choice questions. The CEM certification exam covers business processes, teamwork, responsibility, and other management-related categories.

Modern Tools

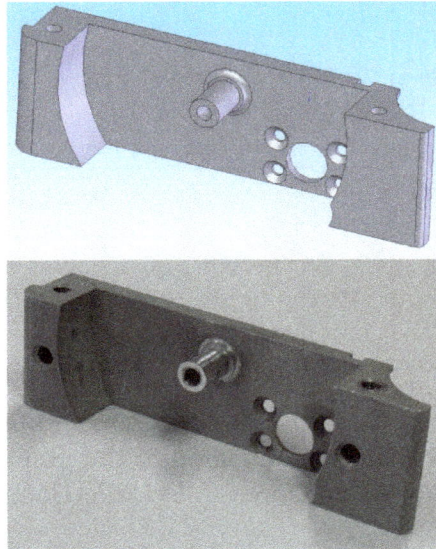

CAD model and CNC machined part

Many manufacturing companies, especially those in industrialized nations, have begun to incorporate computer-aided engineering (CAE) programs into their existing design and analysis processes, including 2D and 3D solid modeling computer-aided design (CAD). This method has many benefits, including easier and more exhaustive visualization of products, the ability to create virtual assemblies of parts, and ease of use in designing mating interfaces and tolerances.

Other CAE programs commonly used by product manufacturers include product life cycle management (PLM) tools and analysis tools used to perform complex simulations. Analysis tools may be used to predict product response to expected loads, including fatigue life and manufacturability. These tools include finite element analysis (FEA), computational fluid dynamics (CFD), and computer-aided manufacturing (CAM).

Using CAE programs, a mechanical design team can quickly and cheaply iterate the design process to develop a product that better meets cost, performance, and other constraints. No physical prototype need be created until the design nears completion, allowing hundreds or thousands of designs to be evaluated, instead of relatively few. In addition, CAE analysis programs can model complicated physical phenomena which cannot be solved by hand, such as viscoelasticity, complex contact between mating parts, or non-Newtonian flows.

Just as manufacturing engineering is linked with other disciplines, such as mechatronics, multidisciplinary design optimization (MDO) is also being used with other CAE programs to automate and improve the iterative design process. MDO tools wrap around existing CAE processes, allowing product evaluation to continue even after the analyst goes home for the day. They also utilize sophisticated optimization algorithms to more intelligently explore possible designs, often finding better, innovative solutions to difficult multidisciplinary design problems.

Subdisciplines

Mechanics

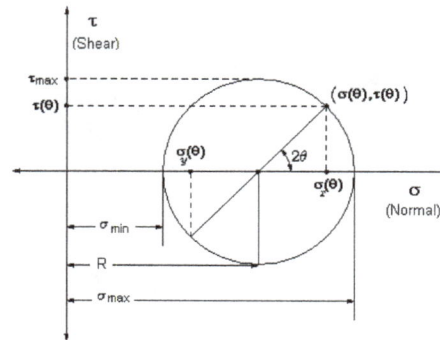

Mohr's circle, a common tool to study stresses in a mechanical element

Mechanics, in the most general sense, is the study of forces and their effects on matter. Typically, engineering mechanics is used to analyze and predict the acceleration and deformation (both elastic and plastic) of objects under known forces (also called loads) or stresses. Subdisciplines of mechanics include:

- Statics, the study of non-moving bodies under known loads

- Dynamics (or kinetics), the study of how forces affect moving bodies

- Mechanics of materials, the study of how different materials deform under various types of stress

- Fluid mechanics, the study of how fluids react to forces

- Continuum mechanics, a method of applying mechanics that assumes that objects are continuous (rather than discrete)

If the engineering project were to design a vehicle, statics might be employed to design the frame of the vehicle in order to evaluate where the stresses will be most intense. Dynamics might be used when designing the car's engine to evaluate the forces in the pistons and cams as the engine cycles. Mechanics of materials might be used to choose appropriate materials for the manufacture of the frame and engine. Fluid mechanics might be used to design a ventilation system for the vehicle or to design the intake system for the engine.

Kinematics

Kinematics is the study of the motion of bodies (objects) and systems (groups of objects), while ignoring the forces that cause the motion. The movement of a crane and the oscillations of a piston in an engine are both simple kinematic systems. The crane is a type of open kinematic chain, while the piston is part of a closed four-bar linkage. Engineers typically use kinematics in the design and analysis of mechanisms. Kinematics can be used to find the possible range of motion for a given mechanism, or, working in reverse, can be used to design a mechanism that has a desired range of motion.

Drafting

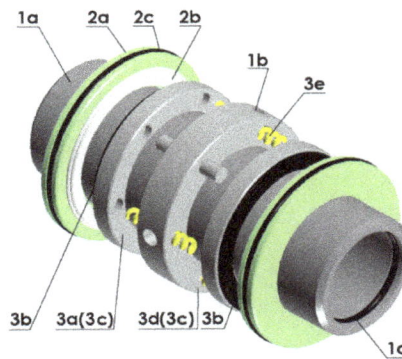

A CAD model of a mechanical double seal

Drafting or technical drawing is the means by which manufacturers create instructions for manufacturing parts. A technical drawing can be a computer model or hand-drawn schematic showing all the dimensions necessary to manufacture a part, as well as assembly notes, a list of required materials, and other pertinent information. A U.S engineer or skilled worker who creates technical drawings may be referred to as a drafter or draftsman. Drafting has historically been a two-dimensional process, but computer-aided design (CAD) programs now allow the designer to create in three dimensions.

Instructions for manufacturing a part must be fed to the necessary machinery, either manually, through programmed instructions, or through the use of a computer-aided manufacturing (CAM) or combined CAD/CAM program. Optionally, an engineer may also manually manufacture a part using the technical drawings, but this is becoming an increasing rarity with the advent of computer numerically controlled (CNC) manufacturing. Engineers primarily manufacture parts manually in the areas of applied spray coatings, finishes, and other processes that cannot economically or practically be done by a machine.

Drafting is used in nearly every subdiscipline of mechanical and manufacturing engineering, and by many other branches of engineering and architecture. Three-dimensional models created using CAD software are also commonly used in finite element analysis (FEA) and computational fluid dynamics (CFD).

Machine Tools and Metal Fabrication

Machine tools employ some sort of tool that does the cutting or shaping. All machine tools have some means of constraining the workpiece and provide a guided movement of the parts of the machine. Metal fabrication is the building of metal structures by cutting, bending, and assembling processes.

Computer Integrated Manufacturing

Computer-integrated manufacturing (CIM) is the manufacturing approach of using computers to control the entire production process. Computer-integrated manufacturing is used in automotive, aviation, space, and ship building industries.

Mechatronics

Training FMS with learning robot SCORBOT-ER 4u, workbench CNC mill and CNC lathe

Mechatronics is an engineering discipline that deals with the convergence of electrical, mechanical and manufacturing systems. Such combined systems are known as electromechanical systems and are widespread. Examples include automated manufacturing systems, heating, ventilation and air-conditioning systems, and various aircraft and automobile subsystems.

The term mechatronics is typically used to refer to macroscopic systems, but futurists have predicted the emergence of very small electromechanical devices. Already such small devices, known as Microelectromechanical systems (MEMS), are used in automobiles to initiate the deployment of airbags, in digital projectors to create sharper images, and in inkjet printers to create nozzles for high-definition printing. In the future it is hoped that such devices will be used in tiny implantable medical devices and to improve optical communication.

Textile Engineering

Textile engineering courses deal with the application of scientific and engineering principles to the design and control of all aspects of fiber, textile, and apparel processes, products, and machinery. These include natural and man-made materials, interaction of materials with machines, safety and health, energy conservation, and waste and pollution control. Additionally, students are given experience in plant design and layout, machine and wet process design and improvement, and designing and creating textile products. Throughout the textile engineering curriculum, students take classes from other engineering and disciplines including: mechanical, chemical, materials and industrial engineering.

Advanced Composite Materials

Advanced composite materials (engineering) (ACMs) are also known as advanced polymer matrix composites. These are generally characterized or determined by unusually high strength fibres with unusually high stiffness, or modulus of elasticity characteristics, compared to other materials, while bound together by weaker matrices. Advanced composite materials have broad, proven applications, in the aircraft, aerospace, and sports equipment sectors. Even more specifically ACMs are very attractive for aircraft and aerospace structural parts. Manufacturing ACMs is a multibillion-dollar industry worldwide. Composite products range from skateboards to components of the space shuttle. The industry can be generally divided into two basic segments, industrial composites and advanced composites.

Employment

Manufacturing engineering is just one facet of the engineering industry. Manufacturing engineers enjoy improving the production process from start to finish. They have the ability to keep the whole production process in mind as they focus on a particular portion of the process. Successful students in manufacturing engineering degree programs are inspired by the notion of starting with a natural resource, such as a block of wood, and ending with a usable, valuable product, such as a desk, produced efficiently and economically.

Manufacturing engineers are closely connected with engineering and industrial design efforts. Examples of major companies that employ manufacturing engineers in the United States include General Motors Corporation, Ford Motor Company, Chrysler, Boeing, Gates Corporation and Pfizer. Examples in Europe include Airbus, Daimler, BMW, Fiat, Navistar International, and Michelin Tyre.

Industries where manufacturing engineers are generally employed include:

- Aerospace industry
- Automotive industry
- Chemical industry
- Computer industry
- Food processing industry
- Garment industry
- Pharmaceutical industry
- Pulp and paper industry
- Toy industry

Frontiers of Research

Flexible Manufacturing Systems

A typical FMS system

A flexible manufacturing system (FMS) is a manufacturing system in which there is some amount of flexibility that allows the system to react to changes, whether predicted or unpredicted. This flexibility is generally considered to fall into two categories, both of which have numerous subcategories. The first category, machine flexibility, covers the system's ability to be changed to produce new product types and the ability to change the order of operations executed on a part. The second category, called routing flexibility, consists of the ability to use multiple machines to perform the same operation on a part, as well as the system's ability to absorb large-scale changes, such as in volume, capacity, or capability.

Most FMS systems comprise three main systems. The work machines, which are often automated CNC machines, are connected by a material handling system to optimize parts flow, and to a central control computer, which controls material movements and machine flow. The main advantages of an FMS is its high flexibility in managing manufacturing resources like time and effort in order to manufacture a new product. The best application of an FMS is found in the production of small sets of products from a mass production.

Computer Integrated Manufacturing

Computer-integrated manufacturing (CIM) in engineering is a method of manufacturing in which the entire production process is controlled by computer. Traditionally separated process methods are joined through a computer by CIM. This integration allows the processes to exchange information and to initiate actions. Through this integration, manufacturing can be faster and less error-prone, although the main advantage is the ability to create automated manufacturing processes. Typically CIM relies on closed-loop control processes based on real-time input from sensors. It is also known as flexible design and manufacturing.

Friction Stir Welding

Friction stir welding was discovered in 1991 by The Welding Institute (TWI). This innovative steady state (non-fusion) welding technique joins previously un-weldable materials, including several aluminum alloys. It may play an important role in the future construction of airplanes, potentially replacing rivets. Current uses of this technology to date include: welding the seams of the aluminum main space shuttle external tank, the Orion Crew Vehicle test article, Boeing Delta II and Delta IV Expendable Launch Vehicles and the SpaceX Falcon 1 rocket; armor plating for amphibious assault ships; and welding the wings and fuselage panels of the new Eclipse 500 aircraft from Eclipse Aviation, among an increasingly growing range of uses.

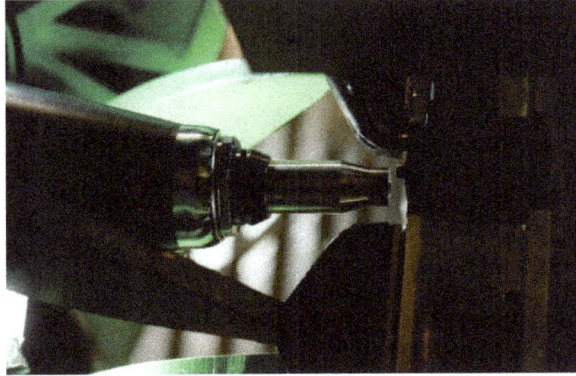

Close-up view of a friction stir weld tack tool

Other areas of research are Product Design, MEMS (Micro-Electro-Mechanical Systems), Lean Manufacturing, Intelligent Manufacturing Systems, Green Manufacturing, Precision Engineering, Smart Materials, etc.

g-force

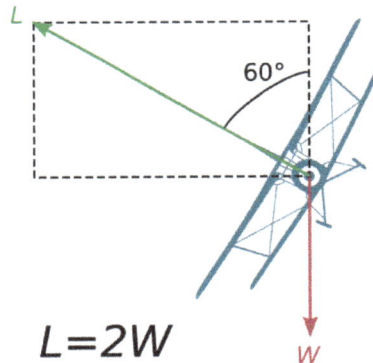

In straight and level flight, lift (L) equals weight (W). In a banked turn of 60°, lift equals double the weight (L=2W). The pilot experiences 2 g and a doubled weight. The steeper the bank, the greater the g-forces.

g-force (with g from gravitational) is a measurement of the type of acceleration that causes weight. Despite the name, it is incorrect to consider g-force a fundamental force, as "g-force" (lower case character) is a type of acceleration that can be measured with an accelerometer. Since g-force accelerations indirectly produce weight, any g-force can be described as a "weight per unit mass". When the g-force acceleration is produced by the surface of one object being pushed by the surface of another object, the reaction-force to this push produces an equal and opposite weight for every unit of an object's mass. The types of forces involved are transmitted through objects by interior mechanical stresses. The g-force acceleration (save for certain electromagnetic force influences) is the cause of an object's acceleration in relation to free-fall.

This top-fuel dragster can accelerate from zero to 160 kilometres per hour (99 mph) in 0.86 seconds. This is a horizontal acceleration of 5.3 g. Combined with the vertical g-force in the stationary case the Pythagorean theorem yields a g force of 5.4 g.

The g-force acceleration experienced by an object is due to the vector sum of all non-gravitational and non-electromagnetic forces acting on an object's freedom to move. In practice, as noted, these are surface-contact forces between objects. Such forces cause stresses and strains on objects, since they must be transmitted from an object surface. Because of these strains, large g-forces may be destructive.

Gravitation acting alone does not produce a g-force, even though g-forces are expressed in multiples of the acceleration of a standard gravity. Thus, the standard gravitational acceleration at the Earth's surface produces g-force only indirectly, as a result of resistance to it by mechanical forces. These mechanical forces actually produce the g-force acceleration on a mass. For example, the 1 g force on an object sitting on the Earth's surface is caused by mechanical force exerted in the upward direction by the ground, keeping the object from going into free-fall. The upward contact-force from the ground ensures that an object at rest on the Earth's surface is accelerating relative to the free-fall condition. (Free fall is the path that the object would follow when falling freely toward the Earth's center). Stress inside the object is ensured from the fact that the ground contact forces are transmitted only from the point of contact with the ground.

Objects allowed to free-fall in an inertial trajectory under the influence of gravitation only, feel no g-force acceleration, a condition known as zero-g (which means zero g-force). This is demonstrated by the "zero-g" conditions inside a freely falling elevator falling toward the Earth's center (in vacuum), or (to good approximation) conditions inside a spacecraft in Earth orbit. These are examples of coordinate acceleration (a change in velocity) without a sensation of weight. The experience of no g-force (zero-g), however it is produced, is synonymous with weightlessness.

In the absence of gravitational fields, or in directions at right angles to them, proper and coordinate accelerations are the same, and any coordinate acceleration must be produced by a corresponding g-force acceleration. An example here is a rocket in free space, in which simple changes in velocity are produced by the engines, and produce g-forces on the rocket and passengers.

Unit and Measurement

The unit of measure of acceleration in the International System of Units (SI) is m/s². However, to distinguish acceleration relative to free-fall from simple acceleration (rate of change of velocity), the unit g (or g) is often used. One g is the acceleration due to gravity at the Earth's surface and is the standard gravity (symbol: g_n), defined as 9.80665 metres per second squared, or equivalently 9.80665 newtons of force per kilogram of mass. Note that the unit definition does not vary with location—the g-force when standing on the moon is about 0.181 g.

The unit g is not one of the SI units, which uses "g" for gram. Also "g" should not be confused with "G", which is the standard symbol for the gravitational constant. This notation is commonly used in aviation, especially in acrobatic or combat military aviation, to describe the increased forces that must be overcome by pilots in order to remain conscious and not G-LOC (G-induced loss of consciousness). For example, it is often said an F-16 fighter jet is able to sustain up to 9 G's for a limited time.

Measurement of g-force is typically achieved using an accelerometer. In certain cases, g-forces may be measured using suitably calibrated scales. Specific force is another name that has been used for g-force.

Acceleration and Forces

Newton's third law: law of reciprocal actions

The term g-force is technically incorrect as it is a measure of acceleration, not force. While acceleration is a vector quantity, g-force accelerations ("g-forces" for short) are often expressed as a scalar, with positive g-forces pointing downward (indicating upward acceleration), and negative g-forces pointing upward. Thus, a g-force is a vector acceleration. It is an acceleration that must be produced by a mechanical force, and cannot be produced by simple gravitation. Objects acted upon only by gravitation, experience (or "feel") no g-force, and are weightless.

G-forces, when multiplied by a mass upon which they act, are associated with a certain type of mechanical force in the correct sense of the term force, and this force produces compressive stress and tensile stress. Such forces result in the operational sensation of weight, but the equation carries a sign change due to the definition of positive weight in the direction downward, so the direction of weight-force is opposite to the direction of g-force acceleration:

Weight = mass × −g-force

The reason for the minus sign is that the actual force (i.e., measured weight) on an object produced by a g-force is in the opposite direction to the sign of the g-force, since in physics, weight is not the force that produces the acceleration, but rather the equal-and-opposite reaction force to it. If the direction upward is taken as positive (the normal cartesian convention) then positive g-force (an acceleration vector that points upward) produces a force/weight on any mass, that acts downward (an example is positive-g acceleration of a rocket launch, producing downward weight). In the same way, a negative-g force is an acceleration vector downward (the negative direction on the y axis), and this acceleration downward produces a weight-force in a direction upward (thus pulling a pilot upward out of the seat, and forcing blood toward the head of a normally oriented pilot).

If a g-force (acceleration) is vertically upward and is applied by the ground (which is accelerating through space-time) or applied by the floor of an elevator to a standing person, most of the body experiences compressive stress which at any height, if multiplied by the area, is the related mechanical force, which is the product of the g-force and the supported mass (the mass above the level of support, including arms hanging down from above that level). At the same time, the arms themselves experience a tensile stress, which at any height, if multiplied by the area, is again the related mechanical force, which is the product of the g-force and the mass hanging below the point of mechanical support. The mechanical resistive force spreads from points of contact with the floor or supporting structure, and gradually decreases toward zero at the unsupported ends (the top in the case of support from below, such as a seat or the floor, the bottom for a hanging part of the body or object). With compressive force counted as negative tensile force, the rate of change of the tensile force in the direction of the g-force, per unit mass (the change between parts of the object such that the slice of the object between them has unit mass), is equal to the g-force plus the non-gravitational external forces on the slice, if any (counted positive in the direction opposite to the g-force).

For a given g-force the stresses are the same, regardless of whether this g-force is caused by mechanical resistance to gravity, or by a coordinate-acceleration (change in velocity) caused by a mechanical force, or by a combination of these. Hence, for people all mechanical forces feels exactly the same whether they cause coordinate acceleration or not. For objects likewise, the question of whether they can withstand the mechanical g-force without damage is the same for any type of g-force. For example, upward acceleration (e.g., increase of speed when going up or decrease of speed when going down) on Earth feels the same as being stationary on a celestial body with a higher surface gravity. Again, one should note that gravitation acting alone does not produce any g-force; g-force is only produced from mechanical pushes and pulls. For a free body (one that is free to move in space) such g-forces only arise as the "inertial" path that is the natural effect of gravitation, or the natural effect of the inertia of mass, is modified. Such modification may only arise from influences other than gravitation.

Examples of important situations involving g-forces include:

- The g-force acting on a stationary object resting on the Earth's surface is 1 g (upwards) and results from the resisting reaction of the Earth's surface bearing upwards equal to an acceleration of 1 g, and is equal and opposite to gravity. The number 1 is approximate, depending on location.

- The g-force acting on an object in any weightless environment such as free-fall in a vacuum is 0 g.

- The g-force acting on an object under acceleration can be much greater than 1 g, for example, the dragster pictured at top right can exert a horizontal g-force of 5.3 when accelerating.

- The g-force acting on an object under acceleration may be downwards, for example when cresting a sharp hill on a roller coaster.

- If there are no other external forces than gravity, the g-force in a rocket is the thrust per unit mass. Its magnitude is equal to the thrust-to-weight ratio times g, and to the consumption of delta-v per unit time.

- In the case of a shock, e.g., a collision, the g-force can be very large during a short time.

A classic example of negative g-force is in a fully inverted roller coaster which is accelerating (changing velocity) toward the ground. In this case, the roller coaster riders are accelerated toward the ground faster than gravity would accelerate them, and are thus pinned upside down in their seats. In this case, the mechanical force exerted by the seat causes the g-force by altering the path of the passenger downward in a way that differs from gravitational acceleration. The difference in downward motion, now faster than gravity would provide, is caused by the push of the seat, and it results in a g-force toward the ground.

All "coordinate accelerations" (or lack of them), are described by Newton's laws of motion as follows:

The Second Law of Motion, the law of acceleration states that: F = ma., meaning that a force F acting on a body is equal to the mass m of the body times its acceleration a.

The Third Law of Motion, the law of reciprocal actions states that: all forces occur in pairs, and these two forces are equal in magnitude and opposite in direction. Newton's third law of motion means that not only does gravity behave as a force acting downwards on, say, a rock held in your hand but also that the rock exerts a force on the Earth, equal in magnitude and opposite in direction.

This acrobatic airplane is pulling up in a +g maneuver; the pilot is experiencing several g's of inertial acceleration in addition to the force of gravity. The cumulative vertical axis forces acting upon his body make him momentarily 'weigh' many times more than normal.

In an airplane, the pilot's seat can be thought of as the hand holding the rock, the pilot as the rock. When flying straight and level at 1 g, the pilot is acted upon by the force of gravity. His weight (a downward force) is 725 newtons (163 lb$_f$). In accordance with Newton's third law, the plane and the seat underneath the pilot provides an equal and opposite force pushing upwards with a force of 725 N (163 lb$_f$). This mechanical force provides the 1.0 g-force upward proper acceleration on the pilot, even though this velocity in the upward direction does not change (this is similar to the situation of a person standing on the ground, where the ground provides this force and this g-force).

If the pilot were suddenly to pull back on the stick and make his plane accelerate upwards at 9.8 m/s^2, the total gforce on his body is 2 g, half of which comes from the seat pushing the pilot to resist gravity, and half from the seat pushing the pilot to cause his upward acceleration—a change in velocity which also is a proper acceleration because it also differs from a free fall trajectory. Considered in the frame of reference of the plane his body is now generating a force of 1,450 N (330 lb$_f$) downwards into his seat and the seat is simultaneously pushing upwards with an equal force of 1,450 N (330 lb$_f$).

Unopposed acceleration due to mechanical forces, and consequentially g-force, is experienced whenever anyone rides in a vehicle because it always causes a proper acceleration, and (in the absence of gravity) also always a coordinate acceleration (where velocity changes). Whenever the vehicle changes either direction or speed, the occupants feel lateral (side to side) or longitudinal (forward and backwards) forces produced by the mechanical push of their seats.

The expression "1 g = 9.80665 m/s^2" means that for every second that elapses, velocity changes 9.80665 meters per second ($\equiv$35.30394 km/h). This rate of change in velocity can also be denoted as 9.80665 (meter per second) per second, or 9.80665 m/s^2. For example: An acceleration of 1 g equates to a rate of change in velocity of approximately 35 kilometres per hour (22 mph) for each second that elapses. Therefore, if an automobile is capable of braking at 1 g and is traveling at 35 kilometres per hour (22 mph) it can brake to a standstill in one second and the driver will experience a deceleration of 1 g. The automobile traveling at three times this speed, 105 km/h (65 mph), can brake to a standstill in three seconds.

In the case of an increase in speed from 0 to v with constant acceleration within a distance of s this acceleration is v^2/(2s).

Preparing an object for g-tolerance (not getting damaged when subjected to a high g-force) is called g-hardening. This may apply to, e.g., instruments in a projectile shot by a gun.

Human Tolerance

Human tolerances depend on the magnitude of the g-force, the length of time it is applied, the direction it acts, the location of application, and the posture of the body.

The human body is flexible and deformable, particularly the softer tissues. A hard slap on the face may briefly impose hundreds of g locally but not produce any real damage; a constant 16 g for a minute, however, may be deadly. When vibration is experienced, relatively low peak g levels can be severely damaging if they are at the resonance frequency of organs and connective tissues.

To some degree, g-tolerance can be trainable, and there is also considerable variation in innate ability between individuals. In addition, some illnesses, particularly cardiovascular problems, reduce g-tolerance.

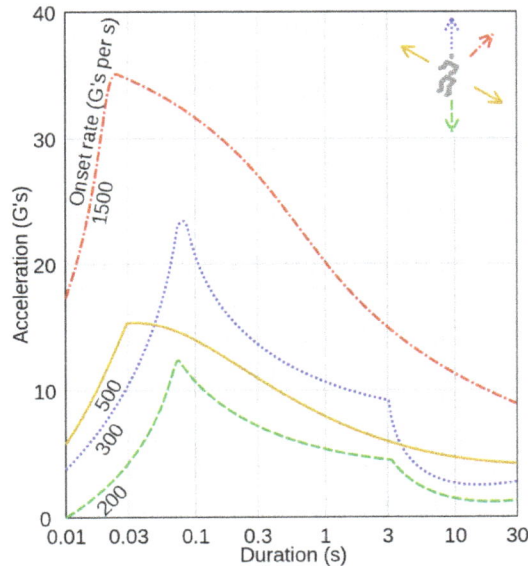

Semilog graph of the limits of tolerance of humans to linear acceleration

Vertical

Aircraft pilots (in particular) sustain g-forces along the axis aligned with the spine. This causes significant variation in blood pressure along the length of the subject's body, which limits the maximum g-forces that can be tolerated.

Positive, or "upward" g, drives blood downward to the feet of a seated or standing person (more naturally, the feet and body may be seen as being driven by the upward force of the floor and seat, upward around the blood). Resistance to positive g varies. A typical person can handle about 5 g_o (49 m/s^2) (meaning some people might pass out when riding a higher-g roller coaster, which in some cases exceeds this point) before losing consciousness, but through the combination of special g-suits and efforts to strain muscles—both of which act to force blood back into the brain—modern pilots can typically handle a sustained 9 g_o (88 m/s^2).

In aircraft particularly, vertical g-forces are often positive (force blood towards the feet and away from the head); this causes problems with the eyes and brain in particular. As positive vertical g-force is progressively increased (such as in a centrifuge) the following symptoms may be experienced:

- Grey-out, where the vision loses hue, easily reversible on levelling out.

- Tunnel vision, where peripheral vision is progressively lost.

- Blackout, a loss of vision while consciousness is maintained, caused by a lack of blood to the head.

- G-LOC, a g-force induced loss of consciousness.

- Death, if g-forces are not quickly reduced, death can occur.

Resistance to "negative" or "downward" g, which drives blood to the head, is much lower. This limit is typically in the -2 to -3 g_0 (-20 to -29 m/s^2) range. This condition is sometimes referred to as red out where vision is figuratively reddened due to the blood laden lower eyelid being pulled into the field of vision Negative g is generally unpleasant and can cause damage. Blood vessels in the eyes or brain may swell or burst under the increased blood pressure, resulting in degraded sight or even blindness.

Horizontal

The human body is better at surviving g-forces that are perpendicular to the spine. In general when the acceleration is forwards (subject essentially lying on their back, colloquially known as "eyeballs in") a much higher tolerance is shown than when the acceleration is backwards (lying on their front, "eyeballs out") since blood vessels in the retina appear more sensitive in the latter direction.

Early experiments showed that untrained humans were able to tolerate a range of accelerations depending on the time of exposure. This ranged from as much as 20 g for less than 10 seconds, to 10 g for 1 minute, and 6 g for 10 minutes for both eyeballs in and out. These forces were endured with cognitive facilities intact, as subjects were able to perform simple physical and communication tasks. The tests were determined to not cause long or short term harm although tolerance was quite subjective, with only the most motivated non-pilots capable of completing tests. The record for peak experimental horizontal g-force tolerance is held by acceleration pioneer John Stapp, in a series of rocket sled deceleration experiments culminating in a late 1954 test in which he was clocked in a little over a second from a land speed of Mach 0.9. He survived a peak "eyeballs-out" acceleration of 46.2 times the acceleration of gravity, and more than 25 g for 1.1 seconds, proving that the human body is capable of this. Stapp lived another 45 years to age 89 without any ill effects.

Short Duration Shock, Impact, and Jerk

Impact and mechanical shock are usually used to describe a high kinetic energy, short term excitation. A shock pulse is often measured by its peak acceleration in g-s and the pulse duration. Vibration is a periodic oscillation which can also be measured in g-s as well as frequency. The dynamics of these phenomena are what distinguish them from the g-forces caused by a relatively longer term accelerations.

After a free fall from a height the shock on an object during impact is g, where is the distance covered during the impact. For example, a stiff and compact object dropped from 1 m that impacts over a distance of 1 mm is subjected to a 1000 g deceleration.

Jerk is the rate of change of acceleration. In SI units, jerk is expressed as m/s^3; it can also be expressed in standard gravity per second (g/s; 1 g/s $\approx$ 9.81 m/s^3).

Other Biological Responses

Recent research carried out on extremophiles in Japan involved a variety of bacteria including E. coli and Paracoccus denitrificans being subject to conditions of extreme gravity. The bacteria were cultivated while being rotated in an ultracentrifuge at high speeds corresponding to 403,627 g. Paracoccus denitrificans was one of the bacteria which displayed not only survival but also robust cellular growth under these conditions of hyperacceleration which are usually only to be found in cosmic environments, such as on very massive stars or in the shock waves of supernovas. Analysis showed that the small size of prokaryotic cells is essential for successful growth under hypergravity. The research has implications on the feasibility of panspermia.

Typical Examples

Example	g-force*
The gyro rotors in Gravity Probe B and the free-floating proof masses in the TRIAD I navigation satellite	0 g
A ride in the Vomit Comet (parabolic flight)	≈ 0 g
Standing on the Moon at its equator	0.1654 g
Standing on the Earth at sea level–standard	1 g
Saturn V moon rocket just after launch	1.14 g
Bugatti Veyron from 0 to 100 km/h in 2.4 s	1.55 g[†]
Space Shuttle, maximum during launch and reentry	3 g
Gravitron amusement ride	2.5-3 g
High-g roller coasters	3.5–6.3 g
Top Fuel drag racing world record of 4.4 s over 1/4 mile	4.2 g
First World War Aircraft Sopwith Camel, Fokker D.VII, Fokker Dr.1, SPAD S.XIII, Nieuport 17 in a steep dive or back or front looping.	4.5–7 g
Formula One car, maximum under heavy braking	5.4 g
Formula One car, peak lateral in turns	5–6 g
Luge, maximum expected at the Whistler Sliding Centre	5.2 g
Standard, full aerobatics certified glider	+7/−5 g
Apollo 16 on reentry	7.19 g
Maximum permitted g-force turn in Red Bull Air Race plane	10g
Maximum for human on a rocket sled	46.2 g
Sprint missile	100 g
Brief human exposure survived in crash	> 100 g
Space gun with a barrel length of 1 km and a muzzle velocity of 6 km/s, as proposed by Quicklaunch (assuming constant acceleration)	1,800 g
Shock capability of mechanical wrist watches	> 5,000 g
V8 Formula One engine, maximum piston acceleration	8,600 g
Rating of electronics built into military artillery shells	15,500 g
Analytical ultracentrifuge spinning at 60,000 rpm, at the bottom of the analysis cell (7.2 cm)	300,000 g
Mean acceleration of a proton in the Large Hadron Collider	190,000,000 g
Gravitational acceleration at the surface of a typical neutron star	2.0×10^{11} g
Acceleration from a wakefield plasma accelerator	8.9×10^{20} g

* Including contribution from resistance to gravity.

† Directed 40 degrees from horizontal.

Measurement using an Accelerometer

The Superman: Escape from Krypton roller coaster at Six Flags Magic Mountain provides 6.5 seconds of ballistic weightlessness.

An accelerometer, in its simplest form, is a damped mass on the end of a spring, with some way of measuring how far the mass has moved on the spring in a particular direction, called an 'axis'.

Accelerometers are often calibrated to measure g-force along one or more axes. If a stationary, single-axis accelerometer is oriented so that its measuring axis is horizontal, its output will be 0 g, and it will continue to be 0 g if mounted in an automobile traveling at a constant velocity on a level road. When the driver presses on the brake or gas pedal, the accelerometer will register positive or negative acceleration.

If the accelerometer is rotated by 90° so that it is vertical, it will read +1 g upwards even though stationary. In that situation, the accelerometer is subject to two forces: the gravitational force and the ground reaction force of the surface it is resting on. Only the latter force can be measured by the accelerometer, due to mechanical interaction between the accelerometer and the ground. The reading is the acceleration the instrument would have if it were exclusively subject to that force.

A three-axis accelerometer will output zerog on all three axes if it is dropped or otherwise put into a ballistic trajectory (also known as an inertial trajectory), so that it experiences "free fall," as do astronauts in orbit (astronauts experience small tidal accelerations called microgravity, which are neglected for the sake of discussion here). Some amusement park rides can provide several seconds at near-zero g. Riding NASA's "Vomit Comet" provides near-zero g for about 25 seconds at a time.

Fuel Efficiency

Fuel efficiency is a form of thermal efficiency, meaning the efficiency of a process that converts chemical potential energy contained in a carrier fuel into kinetic energy or work. Overall fuel efficiency may vary per device, which in turn may vary per application fuel efficiency, especially fossil fuel power plants or industries dealing with combustion, such as ammonia production during the Haber process.

In the context of transport, fuel economy is the energy efficiency of a particular vehicle, given as a ratio of distance traveled per unit of fuel consumed. It is dependent on engine efficiency, transmission design, and tire design. Fuel economy is expressed in miles per gallon (mpg) in the USA and usually also in the UK (imperial gallon); there is sometimes confusion as the imperial gallon is 20% larger than the US gallon so that mpg values are not directly comparable. In countries using the metric system fuel economy is stated as "fuel consumption" in liters per 100 kilometers (L/100 km). Litres per mil are used in Norway and Sweden.

Fuel consumption is a more accurate measure of a vehicle's performance because it is a linear relationship while fuel economy leads to distortions in efficiency improvements.

Weight-specific efficiency (efficiency per unit weight) may be stated for freight, and passenger-specific efficiency (vehicle efficiency per passenger).

Vehicle Design

Fuel efficiency is dependent on many parameters of a vehicle, including its engine parameters, aerodynamic drag, weight, and rolling resistance. There have been advances in all areas of vehicle design in recent decades.

Hybrid vehicles use two or more power sources for propulsion. In many designs, a small combustion engine is combined with electric motors. Kinetic energy which would otherwise be lost to heat during braking is recaptured as electrical power to improve fuel efficiency. Engines automatically shut off when vehicles come to a stop and start again when the accelerator is pressed preventing wasted energy from idling.

Fleet Efficiency

Fleet efficiency describes the average efficiency of a population of vehicles. Technological advances in efficiency may be offset by a change in buying habits with a propensity to heavier vehicles, which are less efficient, all else being equal.

Energy Efficiency Terminology

Energy efficiency is similar to fuel efficiency but the input is usually in units of energy such as British thermal units (BTU), megajoules (MJ), gigajoules (GJ), kilocalories (kcal), or kilowatt-hours (kW·h). The inverse of "energy efficiency" is "energy intensity", or the amount of input energy required for a unit of output such as MJ/passenger-km (of passenger transport), BTU/ton-mile or kJ/t-km (of freight transport), GJ/t (for production of steel and other materials), BTU/(kW·h) (for electricity generation), or litres/100 km (of vehicle travel). Litres per 100 km is also a measure of "energy intensity" where the input is measured by the amount of fuel and the output is measured by the distance travelled. For example: Fuel economy in automobiles.

Given a heat value of a fuel, it would be trivial to convert from fuel units (such as litres of gasoline) to energy units (such as MJ) and conversely. But there are two problems with comparisons made using energy units:

- There are two different heat values for any hydrogen-containing fuel which can differ by several percent.

- When comparing transportation energy costs, it must be remembered that a kilowatt hour of electric energy may require an amount of fuel with heating value of 2 or 3 kilowatt hours to produce it.

Energy Content of Fuel

The specific energy content of a fuel is the heat energy obtained when a certain quantity is burned (such as a gallon, litre, kilogram). It is sometimes called the heat of combustion. There exists two different values of specific heat energy for the same batch of fuel. One is the high (or gross) heat of combustion and the other is the low (or net) heat of combustion. The high value is obtained when, after the combustion, the water in the exhaust is in liquid form. For the low value, the exhaust has all the water in vapor form (steam). Since water vapor gives up heat energy when it changes from vapor to liquid, the liquid water value is larger since it includes the latent heat of vaporization of water. The difference between the high and low values is significant, about 8 or 9%. This accounts for most of the apparent discrepancy in the heat value of gasoline. In the U.S. (and the table) the high heat values have traditionally been used, but in many other countries, the low heat values are commonly used.

Fuel type	MJ/L	MJ/kg	BTU/imp gal	BTU/US gal	Research octane number (RON)
Regular gasoline/petrol	34.8	~47	150,100	125,000	Min. 91
Premium gasoline/petrol		~46			Min. 95
Autogas (LPG) (60% propane and 40% butane)	25.5–28.7	~51			108–110
Ethanol	23.5	31.1	101,600	84,600	129
Methanol	17.9	19.9	77,600	64,600	123
Gasohol (10% ethanol and 90% gasoline)	33.7	~45	145,200	121,000	93/94
E85 (85% ethanol and 15% gasoline)	25.2	~33	108,878	90,660	100–105
Diesel	38.6	~48	166,600	138,700	N/A
Biodiesel	35.1	39.9	151,600	126,200	N/A
Vegetable oil (using 9.00 kcal/g)	34.3	37.7	147,894	123,143	
Aviation gasoline	33.5	46.8	144,400	120,200	80-145
Jet fuel, naphtha	35.5	46.6	153,100	127,500	N/A to turbine engines
Jet fuel, kerosene	37.6	~47	162,100	135,000	N/A to turbine engines
Liquefied natural gas	25.3	~55	109,000	90,800	
Liquid hydrogen	9.3	~130	40,467	33,696	

Neither the gross heat of combustion nor the net heat of combustion gives the theoretical amount of mechanical energy (work) that can be obtained from the reaction. (This is given by the change in Gibbs free energy, and is around 45.7 MJ/kg for gasoline.) The actual amount of mechanical work obtained from fuel (the inverse of the specific fuel consumption) depends on the engine. A figure of 17.6 MJ/kg is possible with a gasoline engine, and 19.1 MJ/kg for a diesel engine.

Fuel Efficiency of Motor Vehicles

The fuel efficiency of motor vehicles can be expressed in more ways:

- Fuel consumption is the amount of fuel used per unit distance; for example, litres per 100 kilometres (L/100 km). In this case, the lower the value, the more economic a vehicle is (the less fuel it needs to travel a certain distance); this is the measure generally used across Europe (except the UK, Denmark and The Netherlands), New Zealand, Australia and Canada. Also in Uruguay, Paraguay, Guatemala, Colombia, China, and Madagascar., as also in post-Soviet space.

- Fuel economy is the distance travelled per unit volume of fuel used; for example, kilometres per litre (km/L) or miles per gallon (MPG), where 1 MPG (imperial) ≈ 0.354006 km/L. In this case, the higher the value, the more economic a vehicle is (the more distance it can travel with a certain volume of fuel). This measure is popular in the USA and the UK (mpg), but in Europe, India, Japan, South Korea and Latin America the metric unit km/L is used instead.

Converting from mpg or to L/100 km (or vice versa) involves the use of the reciprocal function, which is not distributive. Therefore, the average of two fuel economy numbers gives different values if those units are used, because one of the functions is reciprocal, thus not linear. If two people calculate the fuel economy average of two groups of cars with different units, the group with better fuel economy may be one or the other. However, from the point of energy used as a shared method of measure, the result shall be the same in both the cases.

The formula for converting to miles per US gallon (exactly 3.785411784 L) from L/100 km is , where is value of L/100 km. For miles per Imperial gallon (exactly 4.54609 L) the formula is .

In parts of Europe, the two standard measuring cycles for "litre/100 km" value are "urban" traffic with speeds up to 50 km/h from a cold start, and then "extra urban" travel at various speeds up to 120 km/h which follows the urban test. A combined figure is also quoted showing the total fuel consumed in divided by the total distance traveled in both tests. A reasonably modern European supermini and many mid-size cars, including station wagons, may manage motorway travel at 5 L/100 km (47 mpg US/56 mpg imp) or 6.5 L/100 km in city traffic (36 mpg US/43 mpg imp), with carbon dioxide emissions of around 140 g/km.

An average North American mid-size car travels 21 mpg (US) (11 L/100 km) city, 27 mpg (US) (9 L/100 km) highway; a full-size SUV usually travels 13 mpg (US) (18 L/100 km) city and 16 mpg (US) (15 L/100 km) highway. Pickup trucks vary considerably; whereas a 4 cylinder-engined light pickup can achieve 28 mpg (8 L/100 km), a V8 full-size pickup with extended cabin only travels 13 mpg (US) (18 L/100 km) city and 15 mpg (US) (15 L/100 km) highway.

The average fuel economy is higher in Europe due to the higher cost of fuel. In the UK, a gallon of gas without tax would cost US$1.97, but with taxes cost US$6.06 in 2005. The average cost in the United States was US$2.61. Consumers prefer "muscle cars" but choose more fuel efficient ones when gas prices increase.

European-built cars are generally more fuel-efficient than US vehicles. While Europe has many higher efficiency diesel cars, European gasoline vehicles are on average also more efficient than gasoline-powered vehicles in the USA. Most European vehicles cited in the CSI study run on diesel engines, which tend to achieve greater fuel efficiency than gas engines. Selling those cars in the United States is difficult because of emission standards, notes Walter McManus, a fuel economy expert at the University of Michigan Transportation Research Institute. "For the most part, European diesels don't meet U.S. emission standards", McManus said in 2007. Another reason why many European models are not marketed in the United States is that labor unions object to having the big 3 import any new foreign built models regardless of fuel economy while laying off workers at home.

An example of European cars' capabilities of fuel economy is the microcar Smart Fortwo cdi, which can achieve up to 3.4 l/100 km (69.2 mpg US) using a turbocharged three-cylinder 41 bhp (30 kW) Diesel engine. The Fortwo is produced by Daimler AG and is currently only sold by one company in the United States. Furthermore, the current (and to date already 10-year-old) world record in fuel economy of production cars is held by the Volkswagen Group, with special production models (labeled "3L") of the Volkswagen Lupo and the Audi A2, consuming as little as 3 L/100 km (94 mpg$_{-imp}$; 78 mpg$_{-US}$).

Diesel engines generally achieve greater fuel efficiency than petrol (gasoline) engines. Passenger car diesel engines have energy efficiency of up to 41% but more typically 30%, and petrol engines of up to 37.3%, but more typically 20%. That is one of the reasons why diesels have better fuel efficiency than equivalent petrol cars. A common margin is 25% more miles per gallon for an efficient turbodiesel.

For example, the current model Skoda Octavia, using Volkswagen engines, has a combined European fuel efficiency of 41.3 mpg for the 105 bhp (78 kW) petrol engine and 52.3 mpg for the 105 bhp (78 kW) — and heavier — diesel engine. The higher compression ratio is helpful in raising the energy efficiency, but diesel fuel also contains approximately 10% more energy per unit volume than gasoline which contributes to the reduced fuel consumption for a given power output.

In 2002, the United States had 85,174,776 trucks, and averaged 13.5 miles per US gallon (17.4 L/100 km; 16.2 mpg$_{-imp}$). Large trucks, over 33,000 pounds (15,000 kg), averaged 5.7 miles per US gallon (41 L/100 km; 6.8 mpg$_{-imp}$).

Truck fuel economy					
GVWR lbs	Number	Percentage	Average miles per truck	fuel economy	Percentage of fuel use
6,000 lbs and less	51,941,389	61.00%	11,882	17.6	42.70%
6,001 – 10,000 lbs	28,041,234	32.90%	12,684	14.3	30.50%
Light truck subtotal	79,982,623	93.90%	12,163	16.2	73.20%
10,001 – 14,000 lbs	691,342	0.80%	14,094	10.5	1.10%
14,001 – 16,000 lbs	290,980	0.30%	15,441	8.5	0.50%

16,001 – 19,500 lbs	166,472	0.20%	11,645	7.9	0.30%
19,501 – 26,000 lbs	1,709,574	2.00%	12,671	7	3.20%
Medium truck subtotal	2,858,368	3.40%	13,237	8	5.20%
26,001 – 33,000 lbs	179,790	0.20%	30,708	6.4	0.90%
33,001 lbs and up	2,153,996	2.50%	45,739	5.7	20.70%
Heavy truck subtotal	2,333,786	2.70%	44,581	5.8	21.60%
Total	85,174,776	100.00%	13,088	13.5	100.00%

The average economy of automobiles in the United States in 2002 was 22.0 miles per US gallon (10.7 L/100 km; 26.4 mpg$_{-imp}$). By 2010 this had increased to 23.0 miles per US gallon (10.2 L/100 km; 27.6 mpg$_{-imp}$). Average fuel economy in the United States gradually declined until 1973, when it reached a low of 13.4 miles per US gallon (17.6 L/100 km; 16.1 mpg$_{-imp}$) and gradually has increased since, as a result of higher fuel cost. A study indicates that a 10% increase in gas prices will eventually produce a 2.04% increase in fuel economy. One method by car makers to increase fuel efficiency is lightweighting in which lighter-weight materials are substituted in for improved engine performance and handling.

Fuel Efficiency in Microgravity

How fuel combusts affects how much energy is produced. The National Aeronautics and Space Administration (NASA) has investigated fuel consumption in microgravity.

The common distribution of a flame under normal gravity conditions depends on convection, because soot tends to rise to the top of a flame, such as in a candle, making the flame yellow. In microgravity or zero gravity, such as an environment in outer space, convection no longer occurs, and the flame becomes spherical, with a tendency to become more blue and more efficient. There are several possible explanations for this difference, of which the most likely one given is the hypothesis that the temperature is evenly distributed enough that soot is not formed and complete combustion occurs., National Aeronautics and Space Administration, April 2005. Experiments by NASA in microgravity reveal that diffusion flames in microgravity allow more soot to be completely oxidised after they are produced than diffusion flames on Earth, because of a series of mechanisms that behaved differently in microgravity when compared to normal gravity conditions.LSP-1 experiment results, National Aeronautics and Space Administration, April 2005. Premixed flames in microgravity burn at a much slower rate and more efficiently than even a candle on Earth, and last much longer.

Transportation

Vehicle Efficiency and Transportation Pollution

Fuel efficiency directly affects emissions causing pollution by affecting the amount of fuel used. However, it also depends on the fuel source used to drive the vehicle concerned. Cars for example, can run on a number of fuel types other than gasoline, such as natural gas, LPG or biofuel or electricity which creates various quantities of atmospheric pollution.

A kilogram of carbon, whether contained in petrol, diesel, kerosene, or any other hydrocarbon fuel in a vehicle, leads to approximately 3.6 kg of CO_2 emissions. Due to the carbon content of gasoline, its combustion emits 2.3 kg/l (19.4 lb/US gal) of CO_2; since diesel fuel is more energy dense per unit volume, diesel emits 2.6 kg/l (22.2 lb/US gal). This figure is only the CO_2 emissions of the final fuel product and does not include additional CO_2 emissions created during the drilling, pumping, transportation and refining steps required to produce the fuel. Additional measures to reduce overall emission includes improvements to the efficiency of air conditioners, lights and tires.

Driving Technique

Many drivers have the potential to improve their fuel efficiency significantly. These five basic fuel-efficient driving techniques can be effective. Simple things such as keeping tires properly inflated, having a vehicle well-maintained and avoiding idling can dramatically improve fuel efficiency.

There is a growing community of enthusiasts known as hypermilers who develop and practice driving techniques to increase fuel efficiency and reduce consumption. Hypermilers have broken records of fuel efficiency, for example, achieving 109 miles per gallon in a Prius. In non-hybrid vehicles these techniques are also beneficial, with fuel efficiencies of up to 59 MPG in a Honda Accord or 30 MPG in an Acura MDX.

Advanced Technology Improvements to Improve Fuel Efficiency

The most efficient machines for converting energy to rotary motion are electric motors, as used in electric vehicles. However, electricity is not a primary energy source so the efficiency of the electricity production has also to be taken into account. Currently railway trains can be powered using electricity, delivered through an additional running rail, overhead catenary system or by on-board generators used in diesel-electric locomotives as common on the US and UK rail networks. Pollution produced from centralised generation of electricity is emitted at a distant power station, rather than "on site". Pollution can be reduced by using more railway electrification and low carbon power for electricity. Some railways, such as the French SNCF and Swiss federal railways derive most, if not 100% of their power, from hydroelectric or nuclear power stations, therefore atmospheric pollution from their rail networks is very low. This was reflected in a study by AEA Technology between a Eurostar train and airline journeys between London and Paris, which showed the trains on average emitting 10 times less CO_2, per passenger, than planes, helped in part by French nuclear generation.

In the future, hydrogen cars may be commercially available. Toyota is test marketing hydrogen fuel cell powered vehicles in southern California where a series of hydrogen fueling stations has been established. Powered either through chemical reactions in a fuel cell that create electricity to drive very efficient electrical motors or by directly burning hydrogen in a combustion engine (near identically to a natural gas vehicle, and similarly compatible with both natural gas and gasoline); these vehicles promise to have near-zero pollution from the tailpipe (exhaust pipe). Potentially the atmospheric pollution could be minimal, provided the hydrogen is made by electrolysis using electricity from non-polluting sources such as solar, wind or hydroelectricity or nuclear. Commercial hydrogen production uses fossil fuels and produces more carbon dioxide than hydrogen.

Because there are pollutants involved in the manufacture and destruction of a car and the production, transmission and storage of electricity and hydrogen, the use of the label "zero pollution" should be understood as applying only to the car's conversion of stored energy into transportation.

In 2004, a consortium of major auto-makers — BMW, General Motors, Honda, Toyota and Volkswagen/Audi — came up with "Top Tier Detergent Gasoline Standard" to gasoline brands in the US and Canada that meet their minimum standards for detergent content and do not contain metallic additives. Top Tier gasoline contains higher levels of detergent additives in order to prevent the build-up of deposits (typically, on fuel injector and intake valve) known to reduce fuel economy and engine performance.

Safety Engineering

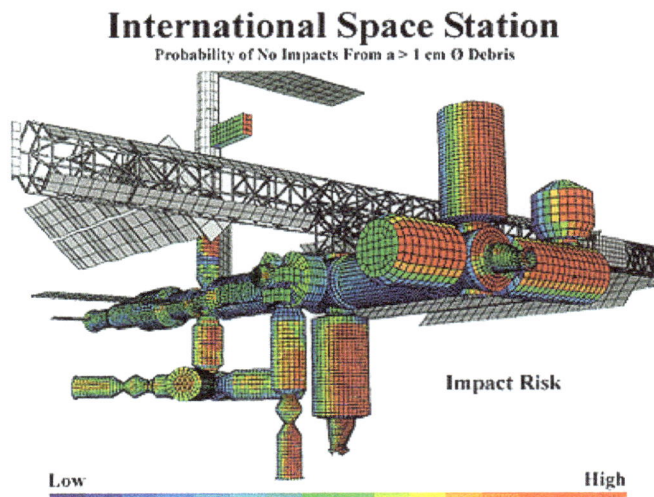

NASA's illustration showing high impact risk areas for the International Space Station

Safety engineering is an engineering discipline which assures that engineered systems provide acceptable levels of safety. It is strongly related to industrial engineering/systems engineering, and the subset system safety engineering. Safety engineering assures that a life-critical system behaves as needed, even when components fail.

Analysis Techniques

Analysis techniques can be split into two categories: qualitative and quantitative methods. Both approaches share the goal of finding causal dependencies between a hazard on system level and failures of individual components. Qualitative approaches focus on the question "What must go wrong, such that a system hazard may occur?", while quantitative methods aim at providing estimations about probabilities, rates and/or severity of consequences.

The complexity of the technical systems such as Improvements of Design and Materials, Planned Inspections, Fool-proof design, and Backup Redundancy decreases risk and increases the cost. The risk can be decreased to ALARA (as low as reasonably achievable) or ALAPA (as low as practically achievable) levels.

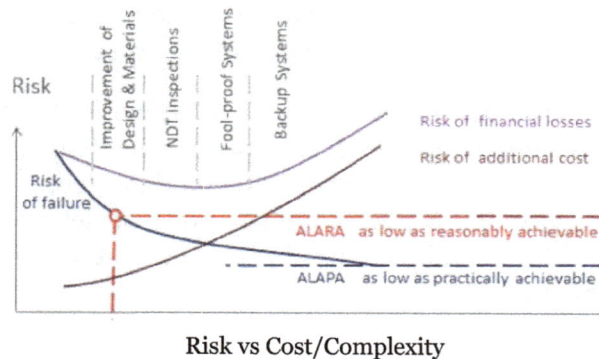

Risk vs Cost/Complexity

Traditionally, safety analysis techniques rely solely on skill and expertise of the safety engineer. In the last decade model-based approaches have become prominent. In contrast to traditional methods, model-based techniques try to derive relationships between causes and consequences from some sort of model of the system.

Traditional Methods for Safety Analysis

The two most common fault modeling techniques are called failure mode and effects analysis and fault tree analysis. These techniques are just ways of finding problems and of making plans to cope with failures, as in probabilistic risk assessment. One of the earliest complete studies using this technique on a commercial nuclear plant was the WASH-1400 study, also known as the Reactor Safety Study or the Rasmussen Report.

Failure Modes and Effects Analysis

Failure Mode and Effects Analysis (FMEA) is a bottom-up, inductive analytical method which may be performed at either the functional or piece-part level. For functional FMEA, failure modes are identified for each function in a system or equipment item, usually with the help of a functional block diagram. For piece-part FMEA, failure modes are identified for each piece-part component (such as a valve, connector, resistor, or diode). The effects of the failure mode are described, and assigned a probability based on the failure rate and failure mode ratio of the function or component. This quantiazation is difficult for software ---a bug exists or not, and the failure models used for hardware components do not apply. Temperature and age and manufacturing variability affect a resistor; they do not affect software.

Failure modes with identical effects can be combined and summarized in a Failure Mode Effects Summary. When combined with criticality analysis, FMEA is known as Failure Mode, Effects, and Criticality Analysis or FMECA, pronounced "fuh-MEE-kuh".

Fault Tree Analysis

Fault tree analysis (FTA) is a top-down, deductive analytical method. In FTA, initiating primary events such as component failures, human errors, and external events are traced through Boolean logic gates to an undesired top event such as an aircraft crash or nuclear reactor core melt. The intent is to identify ways to make top events less probable, and verify that safety goals have been achieved.

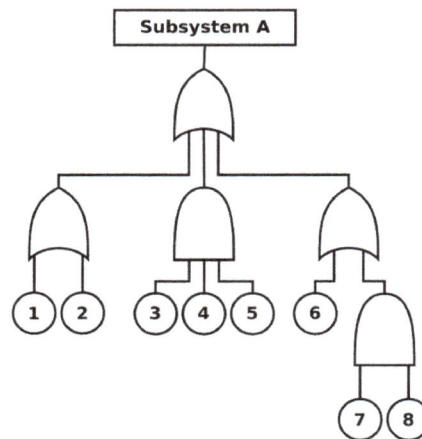

A fault tree diagram

Fault trees are a logical inverse of success trees, and may be obtained by applying de Morgan's theorem to success trees (which are directly related to reliability block diagrams).

FTA may be qualitative or quantitative. When failure and event probabilities are unknown, qualitative fault trees may be analyzed for minimal cut sets. For example, if any minimal cut set contains a single base event, then the top event may be caused by a single failure. Quantitative FTA is used to compute top event probability, and usually requires computer software such as CAFTA from the Electric Power Research Institute or SAPHIRE from the Idaho National Laboratory.

Some industries use both fault trees and event trees. An event tree starts from an undesired initiator (loss of critical supply, component failure etc.) and follows possible further system events through to a series of final consequences. As each new event is considered, a new node on the tree is added with a split of probabilities of taking either branch. The probabilities of a range of "top events" arising from the initial event can then be seen.

Safety Certification

Usually a failure in safety-certified systems is acceptable if, on average, less than one life per 10^9 hours of continuous operation is lost to failure.{as per FAA document AC 25.1309-1A} Most Western nuclear reactors, medical equipment, and commercial aircraft are certified to this level. The cost versus loss of lives has been considered appropriate at this level (by FAA for aircraft systems under Federal Aviation Regulations).

Preventing Failure

Once a failure mode is identified, it can usually be mitigated by adding extra or redundant equipment to the system. For example, nuclear reactors contain dangerous radiation, and nuclear reactions can cause so much heat that no substance might contain them. Therefore, reactors have emergency core cooling systems to keep the temperature down, shielding to contain the radiation, and engineered barriers (usually several, nested, surmounted by a containment building) to prevent accidental leakage. Safety-critical systems are commonly required to permit no single event or component failure to result in a catastrophic failure mode.

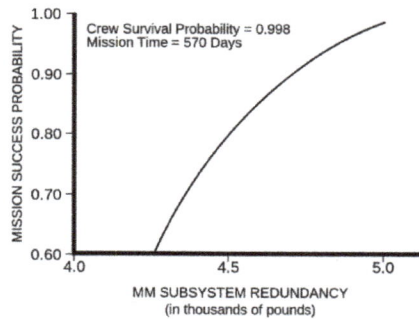

A NASA graph shows the relationship between the survival of a crew of astronauts and the amount of redundant equipment in their spacecraft (the "MM", Mission Module).

Most biological organisms have a certain amount of redundancy: multiple organs, multiple limbs, etc.

For any given failure, a fail-over or redundancy can almost always be designed and incorporated into a system.

There are two categories of techniques to reduce the probability of failure: Fault avoidance techniques increase the reliability of individual items (increased design margin, de-rating, etc.). Fault tolerance techniques increase the reliability of the system as a whole (redundancies, barriers, etc.).

Safety and Reliability

Safety engineering and reliability engineering have much in common, but safety is not reliability. If a medical device fails, it should fail safely; other alternatives will be available to the surgeon. If the engine on a single-engine aircraft fails, there is no backup. Electrical power grids are designed for both safety and reliability; telephone systems are designed for reliability, which becomes a safety issue when emergency (e.g. US "911") calls are placed.

Probabilistic risk assessment has created a close relationship between safety and reliability. Component reliability, generally defined in terms of component failure rate, and external event probability are both used in quantitative safety assessment methods such as FTA. Related probabilistic methods are used to determine system Mean Time Between Failure (MTBF), system availability, or probability of mission success or failure. Reliability analysis has a broader scope than safety analysis, in that non-critical failures are considered. On the other hand, higher failure rates are considered acceptable for non-critical systems.

Safety generally cannot be achieved through component reliability alone. Catastrophic failure probabilities of 10^{-9} per hour correspond to the failure rates of very simple components such as resistors or capacitors. A complex system containing hundreds or thousands of components might be able to achieve a MTBF of 10,000 to 100,000 hours, meaning it would fail at 10^{-4} or 10^{-5} per hour. If a system failure is catastrophic, usually the only practical way to achieve 10^{-9} per hour failure rate is through redundancy.

When adding equipment is impractical (usually because of expense), then the least expensive form of design is often "inherently fail-safe". That is, change the system design so its failure modes are not catastrophic. Inherent fail-safes are common in medical equipment, traffic and railway signals, communications equipment, and safety equipment.

The typical approach is to arrange the system so that ordinary single failures cause the mechanism to shut down in a safe way (for nuclear power plants, this is termed a passively safe design, although more than ordinary failures are covered). Alternately, if the system contains a hazard source such as a battery or rotor, then it may be possible to remove the hazard from the system so that its failure modes cannot be catastrophic. The U.S. Department of Defense Standard Practice for System Safety (MIL–STD–882) places the highest priority on elimination of hazards through design selection.

One of the most common fail-safe systems is the overflow tube in baths and kitchen sinks. If the valve sticks open, rather than causing an overflow and damage, the tank spills into an overflow. Another common example is that in an elevator the cable supporting the car keeps spring-loaded brakes open. If the cable breaks, the brakes grab rails, and the elevator cabin does not fall.

Some systems can never be made fail safe, as continuous availability is needed. For example, loss of engine thrust in flight is dangerous. Redundancy, fault tolerance, or recovery procedures are used for these situations (e.g. multiple independent controlled and fuel fed engines). This also makes the system less sensitive for the reliability prediction errors or quality induced uncertainty for the separate items. On the other hand, failure detection & correction and avoidance of common cause failures becomes here increasingly important to ensure system level reliability.

Engine Power

Engine power or horsepower is the maximum power that an engine can put out. It can be expressed in kilowatts or horsepower. The power output depends on the size and design of the engine, but also on the speed at which it is running and the load or torque. Maximum power is achieved at relatively high speeds and at high load.

Examples

Various Engines

The following tables give some examples of engine power for a wide variety of engines and vehicles.

Heat Engine/Heat pump type	Peak Power Output		Power-to-weight ratio		Example Use
Wärtsilä RTA96-C 14-cylinder two-stroke Turbo Diesel engine	80,080 kW	108,920 hp	0.03 kW/kg	0.02 hp/lb	Emma Mærsk container ship
Suzuki 538 cc V2 4-stroke gas (petrol) outboard Otto engine	19 kW	25 hp	0.27 kW/kg	0.16 hp/lb	Runabout boats
DOE/NASA/0032-28 Mod 2 502 cc gas (petrol) Stirling engine	62.3 kW	83.5 hp	0.30 kW/kg	0.18 hp/lb	Chevrolet Celebrity[·] 1985
GM 6.6 L Duramax LMM (LYE option) V8 Turbo Diesel engine	246 kW	330 hp	0.65 kW/kg	0.40 hp/lb	Chevrolet Kodiak[·], GMC Topkick[·]

Junkers Jumo 205A opposed-piston two-stroke Diesel engine	647 kW	867 hp	1.1 kW/kg	0.66 hp/lb	Ju 86C-1 airliner, B&V Ha 139 floatplane
GE LM2500+ marine turboshaft Brayton gas turbine	30,200 kW	40,500 hp	1.31 kW/kg	0.80 hp/lb	GTS Millennium cruiseship, QM2 ocean liner
Mazda 13B-MSP Renesis 1.3 L Wankel engine	184 kW	247 hp	1.5 kW/kg	0.92 hp/lb	Mazda RX-8[·]
PW R-4360 71.5 L 28-cylinder supercharged Radial engine	3,210 kW	4,300 hp	1.83 kW/kg	1.11 hp/lb	B-50 Superfortress, Convair B-36 / C-97 Stratofreighter, C-119 Flying Boxcar / Hughes H-4 Hercules "Spruce Goose"
Wright R-3350 54.57 L 18-c s/c Turbo-compound Radial engine	2,535 kW	3,400 hp	2.09 kW/kg	1.27 hp/lb	B-29 Superfortress, Douglas DC-7 / C-97 S/f prototype, Kaiser-Frazer C-119F
O.S. Engines 49-PI Type II 4.97 cc UAV Wankel engine	0.934 kW	1.252 hp	2.8 kW/kg	1.7 hp/lb	Model aircraft, Radio-controlled aircraft
GE LM6000 marine turboshaft Brayton gas turbine	44,700 kW	59,900 hp	5.67 kW/kg	3.38 hp/lb	Peaking power plant
GE CF6-80C2 Brayton high-bypass turbofan jet engine					Boeing 747[·], 767, Airbus A300
BMW V10 3L P84/5 2005 gas (petrol) Otto engine	690 kW	925 hp	7.5 kW/kg	4.6 hp/lb	Williams FW27 car[·], Formula One auto racing
GE90-115B Brayton turbofan jet engine	83,164 kW	111,526 hp	10.0 kW/kg	6.10 hp/lb	Boeing 777
PWR RS-24 (SSME) Block II H_2 Brayton turbopump	63,384 kW	85,000 hp	138 kW/kg	84 hp/lb	Space Shuttle (STS-110 and later) [·]
PWR RS-24 (SSME) Block I H_2 Brayton turbopump	53,690 kW	72,000 hp	153 kW/kg	93 hp/lb	Space Shuttle

Notable Low Ratio, (Listed as Weight to Power)

Vehicle	Power	Vehicle Weight	Weight to Power ratio
Benz Patent Motorwagen 954 cc 1886	560 W / 0.75 bhp	265 kg / 584 lb	2.1 W/kg / 779 lb/hp
Stephenson's Rocket 0-2-2 steam locomotive with tender 1829	15 kW / 20 bhp	4,320 kg / 9524 lb	3.5 W/kg / 476 lb/hp
CBQ Zephyr streamliner diesel locomotive with railcars 1934	492 kW / 660 bhp	94 t / 208,000 lb	5.21 W/kg / 315 lb/hp
Alberto Contador's Verbier climb 2009 Tour de France on Specialized bike	420 W / 0.56 bhp	62 kg / 137 lb	6.7 W/kg / 245 lb/hp
Force Motors Minidor Diesel 499 cc auto rickshaw	6.6 kW / 8.8 bhp	700 kg / 1543 lb	9 W/kg / 175 lb/hp

Vehicle	Power	Weight	Ratio
PRR Q2 4-4-6-4 steam locomotive with tender 1944	5,956 kW / 7,987 bhp	475.9 t / 1,049,100 lb	12.5 W/kg / 131 lb/hp
Mercedes-Benz Citaro O530BZ H_2 fuel cell bus 2002	205 kW / 275 bhp	14,500 kg / 32,000 lb	14.1 W/kg / 116 lb/hp
TGV BR Class 373 high-speed Eurostar Trainset 1993	12,240 kW / 16,414 bhp	816 t / 1,798,972 lb	15 W/kg / 110 lb/hp
General Dynamics M1 Abrams Main battle tank 1980	1,119 kW / 1500 bhp	55.7 t / 122,800 lb	20.1 W/kg / 81.9 lb/hp
BR Class 43 high-speed diesel electric locomotive 1975	1,678 kW / 2,250 bhp	70.25 t / 154,875 lb	23.9 W/kg / 69 lb/hp
GE AC6000CW diesel electric locomotive 1996	4,660 kW / 6,250 bhp	192 t / 423,000 lb	24.3 W/kg / 68 lb/hp
BR Class 55 Napier Deltic diesel electric locomotive 1961	2,460 kW / 3,300 bhp	101 t / 222,667 lb	24.4 W/kg / 68 lb/hp
International CXT 2004	164 kW / 220 bhp	6,577 kg / 14500 lb	25 W/kg / 66 lb/hp
Ford Model T 2.9 L flex-fuel 1908	15 kW / 20 bhp	540 kg / 1,200 lb	28 W/kg / 60 lb/hp
TH!NK City 2008	30 kW / 40 bhp	1038 kg / 2,288 lb	28.9 W/kg / 56.9 lb/hp
Messerschmitt KR200 Kabinenroller 191 cc 1955	6 kW / 8.2 bhp	230 kg / 506 lb	30 W/kg / 50 lb/hp
Wright Flyer 1903	9 kW / 12 bhp	274 kg / 605 lb	33 W/kg / 50 lb/hp
Tata Nano 624 cc 2008	26 kW / 35 bhp	635 kg / 1,400 lb	41.0 W/kg / 40 lb/hp
Bombardier JetTrain high-speed gas turbine-electric locomotive 2000	3,750 kW / 5,029 bhp	90,750 kg / 200,000 lb	41.2 W/kg / 39.8 lb/hp
Suzuki MightyBoy 543 cc 1988	23 kW / 31 bhp	550 kg / 1,213 lb	42 W/kg / 39 lb/hp
Mitsubishi i MiEV 2009	47 kW / 63 bhp	1,080 kg / 2,381 lb	43.5 W/kg / 37.8 lb/hp
Holden FJ 2,160 cc 1953	44.7 kW / 60 bhp	1,021 kg / 2,250 lb	43.8 W/kg / 37.5 lb/hp
Chevrolet Kodiak/GMC Topkick LYE 6.6 L 2005	246 kW / 330 bhp	5126 kg / 11,300 lb	48 W/kg / 34.2 lb/hp
DOE/NASA/0032-28 Chevrolet Celebrity 502 cc ASE Mod II 1985	62.3 kW / 83.5 bhp	1,297 kg / 2,860 lb	48.0 W/kg / 34.3 lb/hp
Suzuki Alto 796 cc 2000	35 kW / 46 bhp	720 kg / 1,587 lb	49 W/kg / 35 lb/hp
Land Rover Defender 2.4 L 1990	90 kW / 121 bhp	1,837 kg / 4,050 lb	49 W/kg / 33 lb/hp

Common Power, (Listed as Weight to Power)

Vehicle	Power	Vehicle Weight	Weight-to-Power ratio
Toyota Prius 1.8 L 2010 (petrol only)	73 kW / 98 bhp	1,380 kg / 3,042 lb	53 W/kg / 31 lb/hp

Bajaj Platina Naked 100 cc 2006	6 kW / 8 bhp	113 kg / 249 lb	53 W/kg / 31 lb/hp
Subaru R2 type S 2003	47 kW / 63 bhp	830 kg / 1,830 lb	57 W/kg / 29 lb/hp
Ford Fiesta ECOnetic 1.6 L TDCi 5dr 2009	66 kW / 89 bhp	1,155 kg / 2,546 lb	57 W/kg / 29 lb/hp
Volvo C30 1.6D DRIVe S/S 3dr Hatch 2010	80 kW / 108 bhp	1,347 kg / 2,970 lb	59.4 W/kg / 27.5 lb/hp
Ford Focus ECOnetic 1.6 L TDCi 5dr Hatch 2009	81 kW / 108 bhp	1,357 kg / 2,992 lb	59.7 W/kg / 27 lb/hp
Ford Focus 1.8 L Zetec S TDCi 5dr Hatch 2009	84 kW / 113 bhp	1,370 kg / 3,020 lb	61 W/kg / 27 lb/hp
Honda FCX Clarity 4 kg Hydrogen 2008	100 kW / 134 bhp	1,600 kg / 3,528 lb	63 W/kg / 26 lb/hp
Hummer H1 6.6 L V8 2006	224 kW / 300 bhp	3,559 kg / 7,847 lb	63 W/kg / 26 lb/hp
Audi A2 1.4 L TDI 90 type S 2003	66 kW / 89 bhp	1,030 kg / 2,270 lb	64 W/kg / 25 lb/hp
Opel/Vauxhall/Holden/Chevrolet Astra 1.7 L CTDi 125 2010	92 kW / 123 bhp	1,393 kg / 3,071 lb	66 W/kg / 24.9 lb/hp
Mini (new) Cooper 1.6D 2007	81 kW / 108 bhp	1,185 kg / 2,612 lb	68 W/kg / 24 lb/hp
Toyota Prius 1.8 L 2010 (electric boost)	100 kW / 134 bhp	1,380 kg / 3,042 lb	72 W/kg / 23 lb/hp
Ford Focus 2.0 L Zetec S TDCi 5dr Hatch 2009	100 kW / 134 bhp	1,370 kg / 3,020 lb	73 W/kg / 23 lb/hp
General Motors EV1 electric car Gen II 1998	102.2 kW / 137 bhp	1,400 kg / 3,086 lb	73 W/kg / 23 lb/hp
Toyota Venza I4 2.7 L FWD 2009	136 kW / 182 bhp	1,706 kg / 3,760 lb	80 W/kg / 20.7 lb/hp
Ford Focus 2.0 L Zetec S 5dr Hatch 2009	107 kW / 143 bhp	1,327 kg / 2,926 lb	81 W/kg / 20 lb/hp
Fiat Grande Punto 1.6 L Multijet 120 2005	88 kW / 118 bhp	1,075 kg / 2,370 lb	82 W/kg / 20 lb/hp
Mini (classic) 1275GT 1969	57 kW / 76 bhp	686 kg / 1,512 lb	83 W/kg / 20 lb/hp
Opel/Vauxhall/Holden/Chevrolet Astra 2.0 L CTDi 160 2010	118 kW / 158 bhp	1,393 kg / 3,071 lb	85 W/kg / 19.4 lb/hp
Ford Focus 2.0 auto 2007	104.4 kW / 140 bhp	1,198 kg / 2,641 lb	87.1 W/kg / 19 lb/hp
Subaru Legacy/Liberty 2.0R 2005	121 kW / 162 bhp	1,370 kg / 3,020 lb	88 W/kg / 19 lb/hp
Subaru Outback 2.5i 2008	130.5 kW / 175 bhp	1,430 kg / 3,153 lb	91 W/kg / 18 lb/hp
Smart Fortwo 1.0 L Brabus 2009	72 kW / 97 bhp	780 kg / 1,720 lb	92 W/kg / 18 lb/hp
Toyota Venza V6 3.5 L AWD 2009	200 kW / 268 bhp	1,835 kg / 4,045 lb	109 W/kg / 15 lb/hp

Toyota Venza I4 2.7 L FWD 2009 with Lotus mass reduction	136 kW / 182 bhp	1,210 kg / 2,667 lb	112.2 W/kg / 14.7 lb/hp
Toyota Hilux V6 DOHC 4 L 4×2 Single Cab Pickup ute 2009	175 kW / 235 bhp	1,555 kg / 3,428 lb	112.5 W/kg / 14.6 lb/hp
Toyota Venza V6 3.5 L FWD 2009	200 kW / 268 bhp	1,755 kg / 3,870 lb	114 W/kg / 14.4 lb/hp

Performance Luxury, Roadsters and Mild Sports, (Listed as Weight to Power)

Increased engine performance is a consideration, but also other features associated with luxury vehicles. Longitudinal engines are common. Bodies vary from hot hatches, sedans (saloons), coupés, convertibles and roadsters. Mid-range dual-sport and cruiser motorcycles tend to have similar power-to-weight ratios.

Vehicle	Power	Vehicle Weight	Weight-to-power ratio
Honda Accord sedan V6 2011	202 kW / 271 bhp	1630 kg / 3593 lb	124 W/kg / 13.26 lb/hp
Mini (new) Cooper 1.6T S JCW 2008	155 kW / 208 bhp	1205 kg / 2657 lb	129 W/kg / 13 lb/hp
Mazda RX-8 1.3 L Wankel 2003	173 kW / 232 bhp	1309 kg / 2888 lb	141 W/kg / 12 lb/hp
Holden Statesman/Caprice / Buick Park Avenue / Daewoo Veritas 6 L V8 2007	270 kW / 362 bhp	1891 kg / 4170 lb	143 W/kg / 12 lb/hp
Kawasaki KLR650 Gasoline DualSport 650 cc	26 kW / 35 bhp	182 kg / 401 lb	143 W/kg / 11 lb/hp
NATO HTC M1030M1 Diesel/Jet fuel DualSport 670 cc	26 kW / 35 bhp	182 kg / 401 lb	143 W/kg / 11 lb/hp
Harley-Davidson FLSTF Softail Fat Boy Cruiser 1,584 cc 2009	47 kW / 63 bhp	324 kg / 714 lb	145 W/kg / 11.3 lb/hp
BMW 7 Series 760Li 6 L V12 2006	327 kW / 439 bhp	2250 kg / 4960 lb	145 W/kg / 11 lb/hp
Subaru Impreza WRX STi 2.0 L 2008	227 kW / 304 bhp	1530 kg / 3373 lb	148 W/kg / 11 lb/hp
Honda S2000 roadster 1999	183.88 kW / 240 bhp	1250 kg / 2723 lb	150 W/kg / 11 lb/hp
GMH HSV Clubsport / GMV VXR8 / GMC CSV CR8 / Pontiac G8 6 L V8 2006	317 kW / 425 bhp	1831 kg / 4037 lb	173 W/kg / 9.5 lb/hp
Tesla Roadster 2011	215 kW / 288 bhp	1235 kg / 2723 lb	174 W/kg / 9.5 lb/hp

Sports Vehicles and Aircraft, (Listed as Weight to Power)

Power-to-weight ratio is an important vehicle characteristic that affects the acceleration and handling - and therefore the driving enjoyment - of any sports vehicle. Aircraft also depend on high power-to-weight ratio to achieve sufficient lift.

Vehicle	Power	Vehicle Weight	Weight-to-power ratio
Lotus Elise SC 2008	163 kW / 218 bhp	910 kg / 2006 lb	179 W/kg / 9 lb/hp
Ferrari Testarossa 1984	291 kW / 390 bhp	1506 kg / 3320 lb	193 W/kg / 9 lb/hp
Artega GT	220 kW / 300 bhp	1100 kg / 2425 lb	200 W/kg / 8 lb/hp
Lotus Exige GT3 2006	202.1 kW / 271 bhp	980 kg / 2160 lb	206 W/kg / 8 lb/hp
Chevrolet Corvette C6	321 kW / 430 bhp	1441 kg / 3177 lb	223 W/kg / 7 lb/hp
Suzuki V-Strom 650 V-twin DualSport 650 cc	50 kW / 67 bhp	194 kg / 427 lb	258 W/kg / 6.4 lb/hp
Chevrolet Corvette C6 Z06	376 kW / 505 bhp	1421 kg / 3133 lb	265 W/kg / 6.2 lb/hp
Porsche 911 GT2 2007	390 kW / 523 bhp	1440 kg / 3200 lb	271 W/kg / 6.1 lb/hp
Lamborghini Murciélago LP 670-4 SV 2009	493 kW / 661 bhp	1550 kg / 3417 lb	318 W/kg / 5.1 lb/hp
McLaren F1 GT 1997	467.6 kW / 627 bhp	1220 kg / 2690 lb	403 W/kg / 4.3 lb/hp
Bombardier Dash 8 Q400 turboprop airliner	7,562 kW / 10,142 bhp	17,185 kg / 37,888 lb	440 W/kg / 3.7 lb/hp
Supermarine Spitfire Fighter aircraft 1936	1,096 kW / 1,470 bhp	2,309 kg / 5,090 lb	475 W/kg / 3.46 lb/hp
Messerschmitt Bf 109 Fighter aircraft 1935	1,085 kW / 1,455 bhp	2,247 kg / 4,954 lb	483 W/kg / 3.40 lb/hp
Thunderbolt Land speed record car	3504 kW / 4700 bhp	7 t / 15432 lb	500 W/kg / 3.28 lb/hp
Ferrari FXX 2005	597 kW / 801 bhp	1155 kg / 2546 lb	517 W/kg / 3.2 lb/hp
Polaris Industries Assault Snowmobile 2009	115 kW / 154 bhp	221 kg / 487 lb	523 W/kg / 3.16 lb/hp
Ultima GTR 720 2006	536.9 kW / 720 bhp	920 kg / 2183 lb	583 W/kg / 3 lb/hp
Honda CBR1000RR 2009	133 kW / 178 bhp	199 kg / 439 lb	668 W/kg / 2.5 lb/hp
Ariel Atom 500 V8 2011	372 kW / 500 bhp	550 kg / 1212 lb	676.3 W/kg / 2.45 lb/hp
BMW S1000RR 2009	144 kW / 193 bhp	207.7 kg / 458 lb	693.3 W/kg / 2.37 lb/hp
Peugeot 208 T16 Pikes Peak 2013	652 kW / 875 bhp	875 kg / 1930 lb	745 W/kg / 2.2 lb/hp
KillaCycle Drag racing electric motorcycle	260 kW / 350 bhp	281 kg / 619 lb	925 W/kg / 1.77 lb/hp
MTT Turbine Superbike 2008	213.3 kW / 286 bhp	227 kg / 500 lb	940 W/kg / 1.75 lb/hp
Vyrus 987 C3 4V V supercharged motorcycle 2010	157.3 kW / 211 bhp	158 kg / 348.3 lb	996 W/kg / 1.65 lb/hp
BMW Williams FW27 Formula One 2005	690 kW / 925 bhp	600 kg / 1323 lb	1150 W/kg / 1.43 lb/hp
Honda RC211V MotoGP 2004-6	176.73 kW / 237 bhp	148 kg / 326 lb	1194 W/kg / 1.37 lb/hp

Boeing 747-300[dead link] at Mach 0.84 cruise, 35,000 ft altitude	245 MW / 328,656 bhp	178.1 t / 392,800 lb	1376 W/kg / 1.20 lb/hp
John Force Racing Funny Car NHRA Drag Racing 2008	5,963.60 kW / 8,000 bhp	1043 kg / 2,300 lb	5717 W/kg / 0.30 lb/hp
Koenigsegg One:1 2015	1000 kW / 1341 bhp	1310 kg / 2888 lb	763.35 W/kg / 2.15 lb/hp

Revolutions Per Minute

Revolutions per minute (abbreviated rpm, RPM, rev/min, r/min) is a measure of the frequency of rotation, specifically the number of rotations around a fixed axis in one minute. It is used as a measure of rotational speed of a mechanical component. In the French language, tr/mn (tours par minute) is the common abbreviation. The German language uses the abbreviation U/min or u/min (Umdrehungen pro Minute).

International System of Units

According to the International System of Units (SI), rpm is not a unit. This is because the word revolution is a semantic annotation rather than a unit. The annotation is instead done as a subscript of the formula sign if needed. Because of the measured physical quantity, the formula sign has to be f for (rotational) frequency and ω or Ω for angular velocity. The corresponding basic SI derived unit is s^{-1} or Hz. When measuring angular speed, the unit radians per second is used.

$$1\,\text{rad/s} \leftrightarrow \frac{1}{2\pi}\text{Hz}$$

$$\leftrightarrow \frac{60}{2\pi}\text{rpm}$$

$$1\,\text{rpm} \leftrightarrow \frac{1}{60}\text{Hz}$$

$$\leftrightarrow \frac{2\pi}{60}\text{rad/s}$$

$$1\,\text{Hz} \leftrightarrow 2\pi\,\text{rad/s}$$

$$\leftrightarrow 60\,\text{rpm}$$

Here the sign $\leftrightarrow$ (correspondent) is used instead of = (equal). Formally, hertz (Hz) and radian per second (rad/s) are two different names for the same SI unit, s^{-1}. However, they are used for two different but proportional ISQ quantities: frequency and angular frequency (angular speed, magnitude of angular velocity). The conversion between a frequency f (measured in hertz) and an angular velocity ω (measured in radians per second) are:

$$\omega = 2\pi f, \quad f = \frac{\omega}{2\pi}.$$

Thus a disc rotating at 60 rpm is said to be rotating at either 2π rad/s or 1 Hz, where the former measures the angular velocity and the latter reflects the number of revolutions per second.

If the non-SI unit rpm is considered a unit of frequency, then $1\,\text{rpm} = \dfrac{1}{60}\,\text{Hz}$. If it instead is considered a unit of angular velocity and the word "revolution" is considered to mean 2π radians, then $1\,\text{rpm} = \dfrac{2\pi}{60}\,\text{rad/s}$.

Examples

- On many kinds of disc recording media, the rotational speed of the medium under the read head is a standard given in rpm. Gramophone (phonograph) records, for example, typically rotate steadily at 16 $\frac{2}{3}$, 33 $\frac{1}{3}$, 45 or 78 rpm ($\frac{5}{18}$, $\frac{5}{9}$, $\frac{3}{4}$, or 1.3 Hz respectively).

- Modern ultrasonic dental drills can rotate at up to 800,000 rpm (13.3 kHz).

- The "second" hand of a conventional analogue clock rotates at 1 rpm.

- Audio CD players read their discs at a precise, constant rate (4.3218 Mbit/s of raw physical data for 1.4112 Mbit/s (176.4 kB/s) of usable audio data) and thus must vary the disc's rotational speed from 8 Hz (480 rpm) when reading at the innermost edge, to 3.5 Hz (210 rpm) at the outer edge.

- DVD players also usually read discs at a constant linear rate. The disc's rotational speed varies from 25.5 Hz (1530 rpm) when reading at the innermost edge, to 10.5 Hz (630 rpm) at the outer edge.

- A washing machine's drum may rotate at 500 to 2000 rpm (8–33 Hz) during the spin cycles.

- A power generation turbine (with a 2 pole alternator) rotates at 3000 rpm (50 Hz) or 3600 rpm (60 Hz), depending on country.

- Modern automobile engines are typically operated around 2000–3000 rpm (33–50 Hz) when cruising, with a minimum (idle) speed around 750–900 rpm (12.5–15 Hz), and an upper limit anywhere from 4500 to 10,000 rpm (75–166 Hz) for a road car or nearly 20,000 rpm for racing engines such as those in Formula 1 cars (currently limited to 15,000 rpm). The exhaust note of V8 F1 cars have a much higher pitch than an I4 engine, because each of the cylinders of a four-stroke engine fires once for every two revolutions of the crankshaft. Thus an eight-cylinder engine turning 300 times per second will have an exhaust note of 1200 Hz.

- A piston aircraft engine typically rotates at a rate between 2000 and 3000 rpm (30–50 Hz).

- Computer hard drives typically rotate at 5400 or 7200 rpm (90 or 120 Hz), the most common speeds for the ATA or SATA-based drives in consumer models. High-performance drives (used in fileservers and enthusiast-gaming PCs) rotate at 10,000 or 15,000 rpm (160 or 250 Hz), usually with higher-level SATA, SCSI or Fibre Channel interfaces and smaller platters to allow these higher speeds, the reduction in storage capacity and ultimate outer-edge speed paying off in much quicker access time and average transfer speed thanks to the high spin rate. Until recently, lower-end and power-efficient laptop drives could be found with 4200 or even 3600 rpm spindle speeds (70 and 60 Hz), but these have fallen out of favour due to their lower performance, improvements in energy efficiency in faster models and the takeup of solid-state drives for use in slimline and ultraportable laptops. Similar to CD and DVD media, the amount of data that can be stored or read for each turn of the disc is greater at the outer edge than near the spindle; however, hard drives keep a constant rotational speed so the effective data rate is faster at the edge (conventionally, the "start" of the disc, opposite to CD/DVD).

- Floppy disc drives typically ran at a constant 300 or occasionally 360 rpm (a relatively slow 5 or 6 Hz) with a constant per-revolution data density, which was simple and inexpensive to implement, though inefficient. Some designs such as those used with older Apple computers (Lisa, early Macintosh, later II's) were more complex and used variable rotational speeds and per-track storage density (at a constant read/record rate) to store more data per disc; for example, between 394 rpm (with 12 sectors per track) and 590 rpm (8 sectors) with the Mac's 800 KB double-density drive at a constant 39.4 KB/s (max) – vs. 300 rpm, 720 KB and 23 KB/s (max) for double-density drives in other machines.

- A Zippe-type centrifuge for enriching uranium spins at 90,000 rpm (1,500 Hz) or faster.

- Gas turbine engines rotate at tens of thousands of rpm. JetCat model aircraft turbines are capable of over 100,000 rpm (1,700 Hz) with the fastest reaching 165,000 rpm (2,750 Hz).

- A Flywheel energy storage system works at 60,000–200,000 rpm (1–3 kHz) range using a passively magnetic levitated flywheel in vacuum. The choice of the flywheel material is not the most dense, but the one that pulverises the most safely, at surface speeds about 7 times the speed of sound.

- A typical 80 mm, 30 CFM computer fan will spin at 2,600–3,000 rpm on 12 V DC power.

- A millisecond pulsar can have near 50,000 rpm.

- A turbocharger can reach 290,000 rpm (4.8 kHz), while 80,000–200,000 rpm (1–3 kHz) is common.

- Molecular microbiology - molecular engines. The rotation rates of bacterial flagella have been measured to be 10,200 rpm (170 Hz) for Salmonella typhimurium, 16,200 rpm (270 Hz) for Escherichia coli, and up to 102,000 rpm (1,700 Hz) for polar flagellum of Vibrio alginolyticus, allowing the latter organism to move in simulated natural conditions at a maximum speed of 540 mm per hour.

References

- Rosen (Ed.), Erwin M. (1975). The Peterson automotive troubleshooting & repair manual. Grosset & Dunlap, Inc. ISBN 978-0-448-11946-5.

- George Bibel. Beyond the Black Box: the Forensics of Airplane Crashes. Johns Hopkins University Press, 2008. ISBN 0-8018-8631-7.

- Snyder, John Beltz (2014-07-10). "2015 Nissan Leaf gets B mode standard, new MorningSky Blue color". AutoBlog. Retrieved 2016-03-21.

- Martin, Douglas (16 November 1999). "John Paul Stapp, 89, Is Dead; 'The Fastest Man on Earth'". The New York Times. Retrieved 29 October 2016.

- Bornschlegl, Susanne (2012). Ready for SIL 4: Modular Computers for Safety-Critical Mobile Applications (pdf). MEN Mikro Elektronik. Retrieved 2015-09-21.

- Ofria, Charles. "A short course on automatic transmissions". CarParts.com. JC Whitney. Archived from the original on 6 October 2014. Retrieved 6 October 2014.

- "NASA Physiological Acceleration Systems". Web.archive.org. 2008-05-20. Archived from the original on 2008-05-20. Retrieved 2012-12-25.

Propulsion: An Integrated Study

Propulsion is a way of generating force leading to movement. The features related to propulsion are ground propulsion, maglev, marine propulsion, spacecraft propulsion and jet propulsion. The major components of propulsion are discussed in this chapter.

Propulsion

Armadillo Aerospace's quad rocket vehicle showing visible banding (shock diamonds) in the exhaust plume from its propulsion system

Propulsion is a means of creating force leading to movement. The term is derived from two Latin words: pro, meaning before or forward; and pellere, meaning to drive. A propulsion system consists of a source of mechanical power, and a propulsor (means of converting this power into propulsive force).

A technological system uses an engine or motor as the power source, and wheels and axles, propellers, or a propulsive nozzle to generate the force. Components such as clutches or gearboxes may be needed to connect the motor to axles, wheels, or propellors.

Biological propulsion systems use an animal's muscles as the power source, and limbs such as wings, fins or legs as the propulsors.

Vehicular Propulsion

Air Propulsion

An aircraft propulsion system generally consists of an aircraft engine and some means to generate thrust, such as a propeller or a propulsive nozzle.

A turboprop-engined Tupolev Tu-95

An aircraft propulsion system must achieve two things. First, the thrust from the propulsion system must balance the drag of the airplane when the airplane is cruising. And second, the thrust from the propulsion system must exceed the drag of the airplane for the airplane to accelerate. In fact, the greater the difference between the thrust and the drag, called the excess thrust, the faster the airplane will accelerate.

Some aircraft, like airliners and cargo planes, spend most of their life in a cruise condition. For these airplanes, excess thrust is not as important as high engine efficiency and low fuel usage. Since thrust depends on both the amount of gas moved and the velocity, we can generate high thrust by accelerating a large mass of gas by a small amount, or by accelerating a small mass of gas by a large amount. Because of the aerodynamic efficiency of propellers and fans, it is more fuel efficient to accelerate a large mass by a small amount, which is why high-bypass turbofans and turboprops are commonly used on cargo planes and airliners.

Some aircraft, like fighter planes or experimental high speed aircraft, require very high excess thrust to accelerate quickly and to overcome the high drag associated with high speeds. For these airplanes, engine efficiency is not as important as very high thrust. Modern military aircraft typically employ afterburners on a low bypass turbofan core. Future hypersonic aircraft will employ some type of ramjet or rocket propulsion.

Ground

Wheels are commonly used in ground propulsion

Ground propulsion is any mechanism for propelling solid bodies along the ground, usually for the purposes of transportation. The propulsion system often consists of a combination of an engine or motor, a gearbox and wheel and axles in standard applications.

Maglev

Transrapid 09 at the Emsland test facility in Germany

Maglev (derived from magnetic levitation) is a system of transportation that uses magnetic levitation to suspend, guide and propel vehicles with magnets rather than using mechanical methods, such as wheels, axles and bearings. With maglev a vehicle is levitated a short distance away from a guide way using magnets to create both lift and thrust. Maglev vehicles are claimed to move more smoothly and quietly and to require less maintenance than wheeled mass transit systems. It is claimed that non-reliance on friction also means that acceleration and deceleration can far surpass that of existing forms of transport. The power needed for levitation is not a particularly large percentage of the overall energy consumption; most of the power used is needed to overcome air resistance (drag), as with any other high-speed form of transport.

Marine

A view of a ship's engine room

Marine propulsion is the mechanism or system used to generate thrust to move a ship or boat across water. While paddles and sails are still used on some smaller boats, most modern ships are propelled by mechanical systems consisting a motor or engine turning a propeller, or less frequently, in jet drives, an impeller. Marine engineering is the discipline concerned with the design of marine propulsion systems.

Steam engines were the first mechanical engines used in marine propulsion, but have mostly been replaced by two-stroke or four-stroke diesel engines, outboard motors, and gas turbine engines on faster ships.Nuclear reactors producing steam are used to propel warships and icebreakers, and there have been attempts to utilize them to power commercial vessels.Electric motors have been used on submarines and electric boats and have been proposed for energy-efficient propulsion. Recent development in liquified natural gas (LNG) fueled engines are gaining recognition for their low emissions and cost advantages.

Space

A remote camera captures a close-up view of a Space Shuttle main engine during a test firing at the John C. Stennis Space Center in Hancock County, Mississippi

Spacecraft propulsion is any method used to accelerate spacecraft and artificial satellites. There are many different methods. Each method has drawbacks and advantages, and spacecraft propulsion is an active area of research. However, most spacecraft today are propelled by forcing a gas from the back/rear of the vehicle at very high speed through a supersonic de Laval nozzle. This sort of engine is called a rocket engine.

All current spacecraft use chemical rockets (bipropellant or solid-fuel) for launch, though some (such as the Pegasus rocket and SpaceShipOne) have used air-breathing engines on their first stage. Most satellites have simple reliable chemical thrusters (often monopropellant rockets) or resistojet rockets for orbital station-keeping and some use momentum wheels for attitude control. Soviet bloc satellites have used electric propulsion for decades, and newer Western geo-orbiting spacecraft are starting to use them for north-south stationkeeping and orbit raising. Interplanetary vehicles mostly use chemical rockets as well, although a few have used ion thrusters and Hall effect thrusters (two different types of electric propulsion) to great success.

Cable

A cable car is any of a variety of transportation systems relying on cables to pull vehicles along or lower them at a steady rate. The terminology also refers to the vehicles on these systems. The cable car vehicles are motor-less and engine-less and they are pulled by a cable that is rotated by a motor off-board.

Animal

A bee in flight.

Animal locomotion, which is the act of self-propulsion by an animal, has many manifestations, including running, swimming, jumping and flying. Animals move for a variety of reasons, such as to find food, a mate, or a suitable microhabitat, and to escape predators. For many animals the ability to move is essential to survival and, as a result, selective pressures have shaped the locomotion methods and mechanisms employed by moving organisms. For example, migratory animals that travel vast distances (such as the Arctic tern) typically have a locomotion mechanism that costs very little energy per unit distance, whereas non-migratory animals that must frequently move quickly to escape predators (such as frogs) are likely to have costly but very fast locomotion. The study of animal locomotion is typically considered to be a sub-field of biomechanics.

Locomotion requires energy to overcome friction, drag, inertia, and gravity, though in many circumstances some of these factors are negligible. In terrestrial environments gravity must be overcome, though the drag of air is much less of an issue. In aqueous environments however, friction (or drag) becomes the major challenge, with gravity being less of a concern. Although animals with natural buoyancy need not expend much energy maintaining vertical position, some will naturally sink and must expend energy to remain afloat. Drag may also present a problem in flight, and the aerodynamically efficient body shapes of birds highlight this point. Flight presents a different problem from movement in water however, as there is no way for a living organism to have lower density than air. Limbless organisms moving on land must often contend with surface friction, but do not usually need to expend significant energy to counteract gravity.

Newton's third law of motion is widely used in the study of animal locomotion: if at rest, to move forwards an animal must push something backwards. Terrestrial animals must push the solid ground, swimming and flying animals must push against a fluid or gas (either water or air). The effect of forces during locomotion on the design of the skeletal system is also important, as is the interaction between locomotion and muscle physiology, in determining how the structures and effectors of locomotion enable or limit animal movement.

Ground Propulsion

Ground propulsion is any mechanism for propelling solid bodies along the ground, usually for the purposes of transportation. The propulsion system often consists of a combination of an engine or motor, a gearbox and wheel and axles (or caterpillar tracks) in standard applications.

The primary and most natural type of propulsion is the use of muscle power. Vehicles drawn by an animal have nearly disappeared nowadays. There are some communities left, where one can make a living solely relying on the techniques employing muscle energy in a very efficient form like bicycles, wheelchairs etc.. But civilization employs other types of methods to transport bodies with higher velocities, in the last two centuries up into the 1970s mostly steam engines. Now the main focus is on

- internal-combustion engines
- electric motors (which includes linear motors being part of the track)

or combinations of those. Turbines are rarely used because of the small part load efficiency, although land speed record cars do use them.

Also other types with external combustion like the Stirling engine or without combustion like the fuel cell are only used in planes.

It is different in transmission that has a manual or automatic.

Maglev

SCMaglev test track in the Yamanashi Prefecture, Japan

Maglev (derived from magnetic levitation) is a transport method that uses magnetic levitation to move vehicles without making contact with the ground. With maglev, a vehicle travels along a guideway using magnets to create both lift and propulsion, thereby reducing friction by a great extent and allowing very high speeds. In itself, maglev technology includes no moving parts.

Transrapid 09 at the Emsland test facility in Germany

Maglev trains move more smoothly and more quietly than wheeled mass transit systems. The power needed for levitation is typically not a large percentage of its overall energy consumption; most goes to overcome drag, as with other high-speed transport. Maglev trains hold the speed record for trains.

Compared to conventional trains, differences in construction affect the economics of maglev trains, making them much more efficient. For high-speed trains with wheels, wear and tear from friction from wheels on rails accelerates equipment wear and prevents high speeds. Conversely, maglev systems have been much more expensive to construct, offsetting lower maintenance costs.

Despite decades of research and development, only three commercial maglev transport systems are in operation, while one more is under construction. In April 2004, Shanghai's Transrapid system began commercial operations. In March 2005, Japan began operation of its relatively low-speed HSST "Linimo" line in time for the 2005 World Expo. In its first three months, the Linimo line carried over 10 million passengers. South Korea became the world's fourth country to succeed in commercializing maglev technology with the Incheon Airport Maglev beginning commercial operation on February 3, 2016.

Development

In the late 1940s, the British electrical engineer Eric Laithwaite, a professor at Imperial College London, developed the first full-size working model of the linear induction motor. He became professor of heavy electrical engineering at Imperial College in 1964, where he continued his successful development of the linear motor. Since linear motors do not require physical contact between the vehicle and guideway, they became a common fixture on advanced transportation systems in the 1960s and 70s. Laithwaite joined one such project, the tracked hovercraft, although the project was cancelled in 1973.

The linear motor was naturally suited to use with maglev systems as well. In the early 1970s, Laithwaite discovered a new arrangement of magnets, the magnetic river, that allowed a single linear motor to produce both lift and forward thrust, allowing a maglev system to be built with a single set of magnets. Working at the British Rail Research Division in Derby, along with teams at several civil engineering firms, the "transverse-flux" system was developed into a working system.

The first commercial maglev people mover was simply called "MAGLEV" and officially opened in 1984 near Birmingham, England. It operated on an elevated 600-metre (2,000 ft) section of monorail track between Birmingham Airport and Birmingham International railway station, running at speeds up to 42 km/h (26 mph). The system was closed in 1995 due to reliability problems.

History

First Maglev Patent

High-speed transportation patents were granted to various inventors throughout the world. Early United States patents for a linear motor propelled train were awarded to German inventor Alfred Zehden. The inventor was awarded U.S. Patent 782,312 (14 February 1905) and U.S. Patent RE12,700 (21 August 1907). In 1907, another early electromagnetic transportation system was developed by F. S. Smith. A series of German patents for magnetic levitation trains propelled by linear motors were awarded to Hermann Kemper between 1937 and 1941. An early maglev train was described in U.S. Patent 3,158,765, "Magnetic system of transportation", by G. R. Polgreen (25 August 1959). The first use of "maglev" in a United States patent was in "Magnetic levitation guidance system" by Canadian Patents and Development Limited.

New York, United States, 1913

Emile Bachelet, of Mount Vernon, N. Y., demonstrated a prototype of a magnetic levitating railway car.

New York, United States, 1968

In 1968, while delayed in traffic on the Throgs Neck Bridge, James Powell, a researcher at Brookhaven National Laboratory (BNL), thought of using magnetically levitated transportation. Powell and BNL colleague Gordon Danby worked out a MagLev concept using static magnets mounted on a moving vehicle to induce electrodynamic lifting and stabilizing forces in specially shaped loops, such as figure of 8 coils on a guideway.

Hamburg, Germany, 1979

Transrapid 05 was the first maglev train with longstator propulsion licensed for passenger transportation. In 1979, a 908 m track was opened in Hamburg for the first International Transportation Exhibition (IVA 79). Interest was sufficient that operations were extended three months after the exhibition finished, having carried more than 50,000 passengers. It was reassembled in Kassel in 1980.

Birmingham, United Kingdom, 1984–95

The world's first commercial maglev system was a low-speed maglev shuttle that ran between the airport terminal of Birmingham International Airport and the nearby Birmingham International railway station between 1984 and 1995. Its track length was 600 metres (2,000 ft), and trains levitated at an altitude of 15 millimetres (0.59 in), levitated by electromagnets, and propelled with linear induction motors. It operated for nearly eleven years, but obsolescence problems with the electronic systems made it progressively unreliable as years passed. One of the original cars is now on display at Railworld in Peterborough, together with the RTV31 hover train vehicle. Another is on display at the National Railway Museum in York.

The Birmingham International Maglev shuttle

Several favourable conditions existed when the link was built:

- The British Rail Research vehicle was 3 tonnes and extension to the 8 tonne vehicle was easy.

- Electrical power was available.

- The airport and rail buildings were suitable for terminal platforms.

- Only one crossing over a public road was required and no steep gradients were involved.

- Land was owned by the railway or airport.

- Local industries and councils were supportive.

- Some government finance was provided and because of sharing work, the cost per organization was low.

After the system closed in 1995, the original guideway lay dormant until 2003, when a replacement cable-hauled, the AirRail Link Cable Liner people mover was opened.

Emsland, Germany, 1984–2012

Transrapid at the Emsland test facility

Transrapid, a German maglev company, had a test track in Emsland with a total length of 31.5 kilometres (19.6 mi). The single-track line ran between Dörpen and Lathen with turning loops at each end. The trains regularly ran at up to 420 kilometres per hour (260 mph). Paying passengers were carried as part of the testing process. The construction of the test facility began in 1980 and finished in 1984. In 2006, the Lathen maglev train accident occurred killing 23 people, found to have been caused by human error in implementing safety checks. From 2006 no passengers were carried. At the end of 2011 the operation licence expired and was not renewed, and in early 2012 demolition permission was given for its facilities, including the track and factory.

Japan, 1969–Present

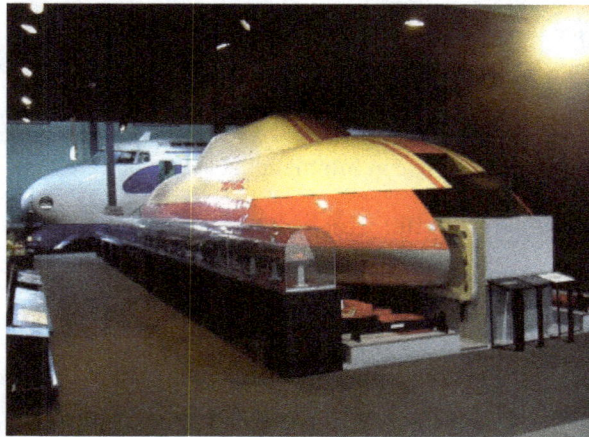

JNR ML500 at a test track in Miyazaki, Japan, on 21 December 1979 travelled at 517 km/h (321 mph), authorized by Guinness World Records.

Japan operates two independently developed maglev trains. One is HSST (and its descendant, the Linimo line) by Japan Airlines and the other, which is more well-known, is SCMaglev by the Central Japan Railway Company.

The development of the latter started in 1969. Miyazaki test track regularly hit 517 km/h (321 mph) by 1979. After an accident that destroyed the train, a new design was selected. In Okazaki, Japan (1987), the SCMaglev took a test ride at the Okazaki exhibition. Tests through the 1980s continued in Miyazaki before transferring to a far larger test track, 20 km (12 mi) long, in Yamanashi in 1997.

Development of HSST started in 1974, based on technologies introduced from Germany. In Tsukuba, Japan (1985), the HSST-03 (Linimo) became popular in spite of its 30 km/h (19 mph) at the Tsukuba World Exposition. In Saitama, Japan (1988), the HSST-04-1 was revealed at the Saitama exhibition performed in Kumagaya. Its fastest recorded speed was 300 km/h (190 mph).

Vancouver, Canada and Hamburg, Germany, 1986–88

In Vancouver, Canada, the HSST-03 by HSST Development Corporation (Japan Airlines and Sumitomo Corporation) was exhibited at Expo 86 and ran on a 400-metre (0.25 mi) test track that provided guests with a ride in a single car along a short section of track at the fairgrounds. It was removed after the fair and debut at the Aoi Expo in 1987 and now on static display at Okazaki Minami Park.

HSST-03 at Okazaki Minami Park

In Hamburg, Germany, the TR-07 was exhibited at the international traffic exhibition (IVA88) in 1988.

Berlin, Germany, 1989–91

In West Berlin, the M-Bahn was built in the late 1980s. It was a driverless maglev system with a 1.6 km (0.99 mi) track connecting three stations. Testing with passenger traffic started in August 1989, and regular operation started in July 1991. Although the line largely followed a new elevated alignment, it terminated at Gleisdreieck U-Bahn station, where it took over an unused platform for a line that formerly ran to East Berlin. After the fall of the Berlin Wall, plans were set in motion to reconnect this line (today's U2). Deconstruction of the M-Bahn line began only two months after regular service began. It was called the Pundai project and was completed in February 1992.

South Korea, 1993–present

Korea's Incheon Airport Maglev, the world's fourth commercially operating maglev.

In 1993, Korea completed the development of its own maglev train, shown off at the Taejŏn Expo '93, which was developed further into a full-fledged maglev capable of travelling up to 110 km/h in 2006. This final model was incorporated in the Incheon Airport Maglev which opened on February

3, 2016, making Korea the world's fourth country to operate its own self-developed maglev after the United Kingdom's Birmingham International Airport, Germany's Berlin M-Bahn, and Japan's Linimo. It links Incheon International Airport to the Yongyu Station and Leisure Complex while crossing Yeongjong island. It offers a transfer to the Seoul Metropolitan Subway at AREX's Incheon International Airport Station and is offered free of charge to anyone to ride, operating between 9am and 6pm every 15 minutes. Operating hours are to be raised in the future.

The maglev system was co-developed by the Korea Institute of Machinery and Materials (KIMM) and Hyundai Rotem. It is 6.1 kilometres (3.8 mi) long, with six stations and a 110 km/h (68 mph) operating speed.

Two more stages are planned of 9.7 km and 37.4 km. Once completed it will become a circular line.

Hyundai Rotem is exporting its Maglev technology to Russia's Leningrad MagLev System, the first overseas customer who will be getting the first urban commuter Maglev system in Europe.

Technology

In the public imagination, "maglev" often evokes the concept of an elevated monorail track with a linear motor. Maglev systems may be monorail or dual rail and not all monorail trains are maglevs. Some railway transport systems incorporate linear motors but use electromagnetism only for propulsion, without levitating the vehicle. Such trains have wheels and are not maglevs. Maglev tracks, monorail or not, can also be constructed at grade (i.e. not elevated). Conversely, non-maglev tracks, monorail or not, can be elevated too. Some maglev trains do incorporate wheels and function like linear motor-propelled wheeled vehicles at slower speeds but "take off" and levitate at higher speeds.

MLX01 Maglev train Superconducting magnet bogie

The two notable types of maglev technology are:

- Electromagnetic suspension (EMS), electronically controlled electromagnets in the train attract it to a magnetically conductive (usually steel) track.

- Electrodynamic suspension (EDS) uses superconducting electromagnets or strong permanent magnets that create a magnetic field, which induces currents in nearby metallic conductors when there is relative movement, which pushes and pulls the train towards the designed levitation position on the guide way.

Another technology, which was designed, proven mathematically, peer-reviewed, and patented, but is, as of May 2015, unbuilt, is magnetodynamic suspension (MDS). It uses the attractive magnetic force of a permanent magnet array near a steel track to lift the train and hold it in place. Other technologies such as repulsive permanent magnets and superconducting magnets have seen some research.

Electromagnetic Suspension

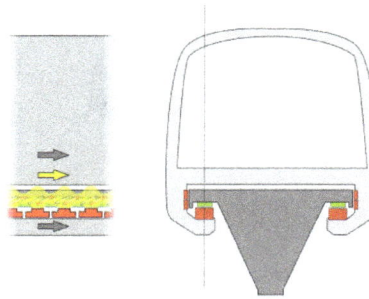

Electromagnetic suspension (EMS) is used to levitate the Transrapid on the track, so that the train can be faster than wheeled mass transit systems

In electromagnetic suspension (EMS) systems, the train levitates above a steel rail while electromagnets, attached to the train, are oriented toward the rail from below. The system is typically arranged on a series of C-shaped arms, with the upper portion of the arm attached to the vehicle, and the lower inside edge containing the magnets. The rail is situated inside the C, between the upper and lower edges.

Magnetic attraction varies inversely with the cube of distance, so minor changes in distance between the magnets and the rail produce greatly varying forces. These changes in force are dynamically unstable – a slight divergence from the optimum position tends to grow, requiring sophisticated feedback systems to maintain a constant distance from the track, (approximately 15 millimetres (0.59 in)).

The major advantage to suspended maglev systems is that they work at all speeds, unlike electrodynamic systems, which only work at a minimum speed of about 30 km/h (19 mph). This eliminates the need for a separate low-speed suspension system, and can simplify track layout. On the downside, the dynamic instability demands fine track tolerances, which can offset this advantage. Eric Laithwaite was concerned that to meet required tolerances, the gap between magnets and rail would have to be increased to the point where the magnets would be unreasonably large. In practice, this problem was addressed through improved feedback systems, which support the required tolerances.

Electrodynamic Suspension (Eds)

In electrodynamic suspension (EDS), both the guideway and the train exert a magnetic field, and the train is levitated by the repulsive and attractive force between these magnetic fields. In some configurations, the train can be levitated only by repulsive force. In the early stages of maglev development at the Miyazaki test track, a purely repulsive system was used instead of the later repulsive and attractive EDS system. The magnetic field is produced either by superconducting magnets (as in JR–Maglev) or by an array of permanent magnets (as in Inductrack). The repulsive and attractive force in the track is created by an induced magnetic field in wires or other conducting strips in the track. A major advantage of EDS maglev systems is that they are dynamically stable – changes in distance between the track and the magnets creates strong forces to return the system to its original position. In addition, the attractive force varies in the opposite manner, providing the same adjustment effects. No active feedback control is needed.

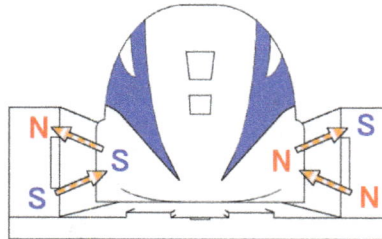

The Japanese SCMaglev's EDS suspension is powered by the magnetic fields induced either side of the vehicle by the passage of the vehicle's superconducting magnets.

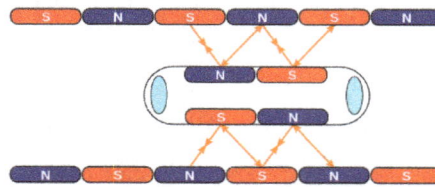

EDS Maglev propulsion via propulsion coils

However, at slow speeds, the current induced in these coils and the resultant magnetic flux is not large enough to levitate the train. For this reason, the train must have wheels or some other form of landing gear to support the train until it reaches take-off speed. Since a train may stop at any location, due to equipment problems for instance, the entire track must be able to support both low- and high-speed operation.

Another downside is that the EDS system naturally creates a field in the track in front and to the rear of the lift magnets, which acts against the magnets and creates magnetic drag. This is generally only a concern at low speeds (This is one of the reasons why JR abandoned a purely repulsive system and adopted the sidewall levitation system.) At higher speeds other modes of drag dominate.

The drag force can be used to the electrodynamic system's advantage, however, as it creates a varying force in the rails that can be used as a reactionary system to drive the train, without the need for a separate reaction plate, as in most linear motor systems. Laithwaite led development of such "traverse-flux" systems at his Imperial College laboratory. Alternatively, propulsion coils on the guideway are used to exert a force on the magnets in the train and make the train move forward. The propulsion coils that exert a force on the train are effectively a linear motor: an alternating current through the coils generates a continuously varying magnetic field that moves forward along the track. The frequency of the alternating current is synchronized to match the speed of the train. The offset between the field exerted by magnets on the train and the applied field creates a force moving the train forward.

Tracks

The term "maglev" refers not only to the vehicles, but to the railway system as well, specifically designed for magnetic levitation and propulsion. All operational implementations of maglev technology make minimal use of wheeled train technology and are not compatible with conventional rail tracks. Because they cannot share existing infrastructure, maglev systems must be designed as standalone systems. The SPM maglev system is inter-operable with steel rail tracks and would permit maglev vehicles and conventional trains to operate on the same tracks. MAN in Germany also designed a maglev system that worked with conventional rails, but it was never fully developed.

Evaluation

Each implementation of the magnetic levitation principle for train-type travel involves advantages and disadvantages.

Technology	Pros	Cons
E M S (Electromagnetic suspension)	Magnetic fields inside and outside the vehicle are less than EDS; proven, commercially available technology; high speeds (500 km/h (310 mph)); no wheels or secondary propulsion system needed.	The separation between the vehicle and the guideway must be constantly monitored and corrected due to the unstable nature of electromagnetic attraction; to the system's inherent instability and the required constant corrections by outside systems may induce vibration.
E D S (Electrodynamic suspension)	Onboard magnets and large margin between rail and train enable highest recorded speeds (603 km/h (375 mph)) and heavy load capacity; demonstrated successful operations using high-temperature superconductors in its onboard magnets, cooled with inexpensive liquid nitrogen.	Strong magnetic fields on the train would make the train unsafe for passengers with pacemakers or magnetic data storage media such as hard drives and credit cards, necessitating the use of magnetic shielding; limitations on guideway inductivity limit maximum speed; vehicle must be wheeled for travel at low speeds.
Inductrack System (Permanent Magnet Passive Suspension)	Failsafe Suspension—no power required to activate magnets; Magnetic field is localized below the car; can generate enough force at low speeds (around 5 km/h (3.1 mph)) for levitation; given power failure cars stop safely; Halbach arrays of permanent magnets may prove more cost-effective than electromagnets.	Requires either wheels or track segments that move for when the vehicle is stopped. Under development (as of 2008); No commercial version or full scale prototype.

Neither Inductrack nor the Superconducting EDS are able to levitate vehicles at a standstill, although Inductrack provides levitation at much lower speed; wheels are required for these systems. EMS systems are wheel-free.

The German Transrapid, Japanese HSST (Linimo), and Korean Rotem EMS maglevs levitate at a standstill, with electricity extracted from guideway using power rails for the latter two, and wirelessly for Transrapid. If guideway power is lost on the move, the Transrapid is still able to generate levitation down to 10 km/h (6.2 mph) speed, using the power from onboard batteries. This is not the case with the HSST and Rotem systems.

Propulsion

EMS systems such as HSST/Linimo can provide both levitation and propulsion using an onboard linear motor. But EDS systems and some EMS systems such as Transrapid levitate but do not propel. Such systems need some other technology for propulsion. A linear motor (propulsion coils) mounted in the track is one solution. Over long distances coil costs could be prohibitive.

Stability

Earnshaw's theorem shows that no combination of static magnets can be in a stable equilibrium. Therefore a dynamic (time varying) magnetic field is required to achieve stabilization. EMS systems rely on active electronic stabilization that constantly measures the bearing distance and adjusts the electromagnet current accordingly. EDS systems rely on changing magnetic fields to create currents, which can give passive stability.

Because maglev vehicles essentially fly, stabilisation of pitch, roll and yaw is required. In addition to rotation, surge (forward and backward motions), sway (sideways motion) or heave (up and down motions) can be problematic.

Superconducting magnets on a train above a track made out of a permanent magnet lock the train into its lateral position. It can move linearly along the track, but not off the track. This is due to the Meissner effect and flux pinning.

Guidance System

Some systems use Null Current systems (also sometimes called Null Flux systems). These use a coil that is wound so that it enters two opposing, alternating fields, so that the average flux in the loop is zero. When the vehicle is in the straight ahead position, no current flows, but any moves off-line create flux that generates a field that naturally pushes/pulls it back into line.

Evacuated Tubes

Some systems (notably the Swissmetro system) propose the use of vactrains—maglev train technology used in evacuated (airless) tubes, which removes air drag. This has the potential to increase speed and efficiency greatly, as most of the energy for conventional maglev trains is lost to aerodynamic drag.

One potential risk for passengers of trains operating in evacuated tubes is that they could be exposed to the risk of cabin depressurization unless tunnel safety monitoring systems can repressurize the tube in the event of a train malfunction or accident though since trains are likely to operate at or near the Earth's surface, emergency restoration of ambient pressure should be straightforward. The RAND Corporation has depicted a vacuum tube train that could, in theory, cross the Atlantic or the USA in ~21 minutes.

Energy use

Energy for maglev trains is used to accelerate the train. Energy may be regained when the train slows down via regenerative braking. It also levitates and stabilises the train's movement. Most of the energy is needed to overcome "air drag". Some energy is used for air conditioning, heating, lighting and other miscellany.

At low speeds the percentage of power used for levitation can be significant, consuming up to 15% more power than a subway or light rail service. For short distances the energy used for acceleration might be considerable.

The power used to overcome air drag increases with the cube of the velocity and hence dominates at high speed. The energy needed per unit distance increases by the square of the velocity and the time decreases linearly. For example, 2.5 times more power is needed to travel at 400 km/h than 300 km/h.

Comparison with Conventional Trains

Maglev transport is non-contact and electric powered. It relies less or not at all on the wheels, bearings and axles common to wheeled rail systems.

- Speed: Maglev allows higher top speeds than conventional rail, but experimental wheel-based high-speed trains have demonstrated similar speeds.

- Maintenance: Maglev trains currently in operation have demonstrated the need for minimal guideway maintenance. Vehicle maintenance is also minimal (based on hours of operation, rather than on speed or distance traveled). Traditional rail is subject to mechanical wear and tear that increases exponentially with speed, also increasing maintenance.

- Weather: Maglev trains are little affected by snow, ice, severe cold, rain or high winds. However, they have not operated in the wide range of conditions that traditional friction-based rail systems have operated. Maglev vehicles accelerate and decelerate faster than mechanical systems regardless of the slickness of the guideway or the slope of the grade because they are non-contact systems.

- Track: Maglev trains are not compatible with conventional track, and therefore require custom infrastructure for their entire route. By contrast conventional high-speed trains such as the TGV are able to run, albeit at reduced speeds, on existing rail infrastructure, thus reducing expenditure where new infrastructure would be particularly expensive (such as the final approaches to city terminals), or on extensions where traffic does not justify new infrastructure. John Harding, former chief maglev scientist at the Federal Railroad Administration, claimed that separate maglev infrastructure more than pays for itself with higher levels of all-weather operational availability and nominal maintenance costs. These claims have yet to be proven in an intense operational setting and does not consider the increased maglev construction costs.

- Efficiency: Conventional rail is probably more efficient at lower speeds. But due to the lack of physical contact between the track and the vehicle, maglev trains experience no rolling resistance, leaving only air resistance and electromagnetic drag, potentially improving power efficiency. Some systems however such as the Central Japan Railway Company SC-Maglev use rubber tires at low speeds, reducing efficiency gains.

- Weight: The electromagnets in many EMS and EDS designs require between 1 and 2 kilowatts per ton. The use of superconductor magnets can reduce the electromagnets' energy consumption. A 50-ton Transrapid maglev vehicle can lift an additional 20 tons, for a total of 70 tons, which consumes 70-140 kW. Most energy use for the TRI is for propulsion and overcoming air resistance at speeds over 100 mph.

- Weight loading: High speed rail requires more support and construction for its concentrated wheel loading. Maglev cars are lighter and distribute weight more evenly.

- Noise: Because the major source of noise of a maglev train comes from displaced air rather than from wheels touching rails, maglev trains produce less noise than a conventional train at equivalent speeds. However, the psychoacoustic profile of the maglev may reduce this benefit: a study concluded that maglev noise should be rated like road traffic, while conventional trains experience a 5–10 dB "bonus", as they are found less annoying at the same loudness level.

- Braking: Braking and overhead wire wear have caused problems for the Fastech 360 rail Shinkansen. Maglev would eliminate these issues.

- Magnet reliability: Superconducting magnets are generally used to generate the powerful magnetic fields to levitate and propel the trains. These magnets must be kept below their critical temperatures (this ranges form 4.2 K to 77 K, depending on the material) . New alloys and manufacturing techniques in superconductors and cooling systems have helped addressed this issue.

- Control systems: No signalling systems are needed for high-speed rail, because such systems are computer controlled. Human operators cannot react fast enough to manage high-speed trains. High speed systems require dedicated rights of way and are usually elevated. Two maglev system microwave towers are in constant contact with trains. There is no need for train whistles or horns, either.

- Terrain: Maglevs are able to ascend higher grades, offering more routing flexibility and reduced tunneling.

Comparison with Aircraft

Differences between airplane and maglev travel:

- Efficiency: For maglev systems the lift-to-drag ratio can exceed that of aircraft (for example Inductrack can approach 200:1 at high speed, far higher than any aircraft). This can make maglev more efficient per kilometer. However, at high cruising speeds, aerodynamic drag is much larger than lift-induced drag. Jets take advantage of low air density at high altitudes to significantly reduce air drag. Hence despite their lift-to-drag ratio disadvantage, they can travel more efficiently at high speeds than maglev trains that operate at sea level.

- Routing: While aircraft can theoretically take any route between points, commercial air routes are rigidly defined. Maglevs offer competitive journey times over distances of 800 kilometres (500 miles) or less. Additionally, maglevs can easily serve intermediate destinations.

- Availability: Maglevs are little affected by weather.

- Safety: Maglevs offer a significant safety margin since maglevs do not crash into other maglevs or leave their guideways.

- Travel time: Maglevs do not face the extended security protocols faced by air travelers nor is time consumed for taxiing, or for queuing for take-off and landing.

Economics

The Shanghai maglev demonstration line cost US$1.2 billion to build. This total includes capital costs such as right-of-way clearing, extensive pile driving, on-site guideway manufacturing, in-situ pier construction at 25 metre intervals, a maintenance facility and vehicle yard, several switches, two stations, operations and control systems, power feed system, cables and inverters, and operational training. Ridership is not a primary focus of this demonstration line, since the Longyang Road station is on the eastern outskirts of Shanghai. Once the line is extended to South Shanghai Train station and Hongqiao Airport station, which may not happen because of economic reasons, ridership was expected to cover operation and maintenance costs and generate significant net revenue.

The South Shanghai extension was expected to cost approximately US$18 million per kilometre. In 2006 the German government invested $125 million in guideway cost reduction development that produced an all-concrete modular design that is faster to build and is 30% less costly. Other new construction techniques were also developed that put maglev at or below price parity with new high-speed rail construction.

The United States Federal Railroad Administration, in a 2005 report to Congress, estimated cost per mile of between $50m and $100m. The Maryland Transit Administration (MTA) Environmental Impact Statement estimated a pricetag at US$4.9 billion for construction, and $53 million a year for operations of its project.

The proposed Chuo Shinkansen maglev in Japan was estimated to cost approximately US$82 billion to build, with a route requiring long tunnels. A Tokaido maglev route replacing the current Shinkansen would cost 1/10 the cost, as no new tunnel would be needed, but noise pollution issues made this infeasible.

The only low-speed maglev (100 km/h or 62 mph) currently operational, the Japanese Linimo HSST, cost approximately US$100 million/km to build. Besides offering improved operation and maintenance costs over other transit systems, these low-speed maglevs provide ultra-high levels of operational reliability and introduce little noise and generate zero air pollution into dense urban settings.

As more maglev systems are deployed, experts expected construction costs to drop by employing new construction methods and from economies of scale.

Records

The highest recorded maglev speed is 603 km/h (375 mph), achieved in Japan by JR Central's L0 superconducting Maglev on 21 April 2015, 28 km/h (17 mph) faster than the conventional TGV wheel-rail speed record. However, the operational and performance differences between these two very different technologies is far greater. The TGV record was achieved accelerating down a 72.4 km (45.0 mi) slight decline, requiring 13 minutes. It then took another 77.25 km (48.00 mi) for the TGV to stop, requiring a total distance of 149.65 km (92.99 mi) for the test. The MLX01 record, however, was achieved on the 18.4 km (11.4 mi) Yamanashi test track – 1/8 the distance. No maglev or wheel-rail commercial operation has actually been attempted at speeds over 500 km/h.

History of Maglev Speed Records

Year	Country	Train	Speed	Notes
1971	West Germany	Prinzipfahrzeug	90 km/h (56 mph)	
1971	West Germany	TR-02 (TSST)	164 km/h (102 mph)	
1972	Japan	ML100	60 km/h (37 mph)	Manned
1973	West Germany	TR04	250 km/h (160 mph)	Manned
1974	West Germany	EET-01	230 km/h (140 mph)	Unmanned
1975	West Germany	Komet	401 km/h (249 mph)	by steam rocket propulsion, unmanned
1978	Japan	HSST-01	308 km/h (191 mph)	by supporting rockets propulsion, made in Nissan, unmanned
1978	Japan	HSST-02	110 km/h (68 mph)	Manned
1979-12-12	Japan	ML-500R	504 km/h (313 mph)	(unmanned) It succeeds in operation over 500 km/h for the first time in the world.
1979-12-21	Japan	ML-500R	517 km/h (321 mph)	(unmanned)
1987	West Germany	TR-06	406 km/h (252 mph)	(manned)
1987	Japan	MLU001	401 km/h (249 mph)	(manned)
1988	West Germany	TR-06	413 km/h (257 mph)	(manned)
1989	West Germany	TR-07	436 km/h (271 mph)	(manned)
1993	Germany	TR-07	450 km/h (280 mph)	(manned)
1994	Japan	MLU002N	431 km/h (268 mph)	(unmanned)

1997	● Japan	MLX01	531 km/h (330 mph)	(manned)
1997	● Japan	MLX01	550 km/h (340 mph)	(unmanned)
1999	● Japan	MLX01	552 km/h (343 mph)	(manned/five-car formation). Guinness authorization.
2003	● Japan	MLX01	581 km/h (361 mph)	(manned/three formation). Guinness authorization.
2015	● Japan	Lo	590 km/h (370 mph)	(manned/seven-car formation)
2015	● Japan	Lo	603 km/h (375 mph)	(manned/seven-car formation)

Systems

San Diego, USA

General Atomics has a 120-metre test facility in San Diego, that is used to test Union Pacific's 8 km (5.0 mi) freight shuttle in Los Angeles. The technology is "passive" (or "permanent"), using permanent magnets in a halbach array for lift and requiring no electromagnets for either levitation or propulsion. General Atomics received US$90 million in research funding from the federal government. They are also considering their technology for high-speed passenger services.

SCMaglev, Japan

Japan has a demonstration line in Yamanashi prefecture where test train SCMaglev Lo Series Shinkansen reached 603 km/h (375 mph), faster than any wheeled trains.

These trains use superconducting magnets, which allow for a larger gap, and repulsive/attractive-type electrodynamic suspension (EDS). In comparison Transrapid uses conventional electromagnets and attractive-type electromagnetic suspension (EMS).

On 15 November 2014, The Central Japan Railway Company ran eight days of testing for the experimental maglev Shinkansen train on its test track in Yamanashi Prefecture. One hundred passengers covered a 42.8 km (27-mile) route between the cities of Uenohara and Fuefuki, reaching speeds of up to 500 km/h (311 mph).

FTA's UMTD Program

In the US, the Federal Transit Administration (FTA) Urban Maglev Technology Demonstration program funded the design of several low-speed urban maglev demonstration projects. It assessed HSST for the Maryland Department of Transportation and maglev technology for the Colorado Department of Transportation. The FTA also funded work by General Atomics at California University of Pennsylvania to evaluate the MagneMotion M3 and of the Maglev2000 of Florida superconducting EDS system. Other US urban maglev demonstration projects of note are the LEVX in Washington State and the Massachusetts-based Magplane.

Southwest Jiaotong University, China

On 31 December 2000, the first crewed high-temperature superconducting maglev was tested successfully at Southwest Jiaotong University, Chengdu, China. This system is based on the principle that bulk high-temperature superconductors can be levitated stably above or below a permanent magnet. The load was over 530 kg (1,170 lb) and the levitation gap over 20 mm (0.79 in). The system uses liquid nitrogen to cool the superconductor.

Operational Systems

Shanghai Maglev

A maglev train coming out of the Pudong International Airport

The Shanghai Maglev Train, also known as the Transrapid, is the fastest commercial train currently in operation and has a top speed of 430 km/h (270 mph). The line was designed to connect Shanghai Pudong International Airport and the outskirts of central Pudong, Shanghai. It covers a distance of 30.5 kilometres (19.0 mi) in 8 minutes. The Shanghai system was labeled a white elephant by rivals.

In January 2001, the Chinese signed an agreement with Transrapid to build an EMS high-speed maglev line to link Pudong International Airport with Longyang Road Metro station on the eastern edge of Shanghai. This Shanghai Maglev Train demonstration line, or Initial Operating Segment (IOS), has been in commercial operations since April 2004 and now operates 115 daily trips (up from 110 in 2010) that traverse the 30 km (19 mi) between the two stations in 7 minutes, achieving a top speed of 431 km/h (268 mph) and averaging 266 km/h (165 mph). On a 12 November 2003 system commissioning test run, it achieved 501 km/h (311 mph), its designed top cruising speed. The Shanghai maglev is faster than Birmingham technology and comes with on-time – to the second – reliability greater than 99.97%.

Plans to extend the line to Shanghai South Railway Station and Hongqiao Airport on the western edge of Shanghai are on hold. After the Shanghai–Hangzhou Passenger Railway became operational in late 2010, the maglev extension became somewhat redundant and may be canceled.

Linimo (Tobu Kyuryo Line, Japan)

The commercial automated "Urban Maglev" system commenced operation in March 2005 in Aichi, Japan. The Tobu-kyuryo Line, otherwise known as the Linimo line, covers 9 km (5.6 mi). It has a minimum operating radius of 75 m (246 ft) and a maximum gradient of 6%. The linear-motor magnetically levitated train has a top speed of 100 km/h (62 mph). More than 10 million passengers used this "urban maglev" line in its first three months of operation. At 100 km/h (62 mph), it is sufficiently fast for frequent stops, has little or no noise impact on surrounding communities, can navigate short radius rights of way, and operates during inclement weather. The trains were designed by the Chubu HSST Development Corporation, which also operates a test track in Nagoya.

Linimo train approaching Banpaku Kinen Koen, towards Fujigaoka Station in March 2005

Incheon Airport Maglev

Maglev train departing Incheon International Airport Station.

The Incheon Airport Maglev beginning commercial operation on February 3, 2016. It was developed and built domestically. Compared to Linimo, it has a more futuristic design thanks to it being lighter with construction costs cut to half. It connects Incheon International Airport with Yongyu, cutting journey time.

The first maglev test trials using electromagnetic suspension opened to public was HML-03, made by Hyundai Heavy Industries for the Daejeon Expo in 1993, after five years of research and manufacturing two prototypes, HML-01 and HML-02. Government research on urban maglev using electromagnetic suspension began in 1994. The first operating urban maglev was UTM-02 in Daejeon beginning on 21 April 2008 after 14 years of development and one prototype; UTM-01. The train runs on a 1 km (0.62 mi) track between Expo Park and National Science Museum. Meanwhile UTM-02 conducted the world's first ever maglev simulation. However UTM-02 is still the second prototype of a final model. The final UTM model of Rotem's urban maglev, UTM-03, was scheduled to debut at the end of 2014 in Incheon's Yeongjong island where Incheon International Airport is located.

Changsha Maglev

The Hunan provincial government launched the construction of a maglev line between Changsha Huanghua International Airport and Changsha South Railway Station. Construction started in May 2014, completed by the end of 2015, trial running on 26 December 2015, and finally start trial operations on 6 May 2016.

Maglevs Under Construction

AMT test track – Powder Springs, Georgia

A second prototype system in Powder Springs, Georgia, USA, was built by American Maglev Technology, Inc. The test track is 610 m (2,000 feet) long with a 168.6 m (550 feet) curve. Vehicles are operated up to 60 km/h (37 mph), below the proposed operational maximum of 97 km/h (60 mph). A June 2013 review of the technology called for an extensive testing program to be carried out to ensure the system complies with various regulatory requirements including the American Society of Civil Engineers (ASCE) People Mover Standard. The review noted that the test track is too short to assess the vehicles' dynamics at the maximum proposed speeds.

Beijing S1 line

The Beijing municipal government is building China's first low-speed maglev line, the Line S1, BCR, using technology developed by Defense Technology University. It is a 10.2 km (6.3 mi) long S1-West commuter rail line, which, together with seven other conventional lines, began construction on 28 February 2011. The top speed will be 105 km/h (65 mph). This project was scheduled to be completed in 2015.

Tokyo – Nagoya – Osaka

The Chūō Shinkansen route (bold yellow and red line) and existing Tōkaidō Shinkansen route (thin blue line)

Construction of Chuo Shinkansen began in 2014. It was expected to begin operations by 2027.

The plan for the Chuo Shinkansen bullet train system was finalized based on the Law for Construction of Countrywide Shinkansen. The Linear Chuo Shinkansen Project aimed to operate the Superconductive Magnetically Levitated Train to connect Tokyo and Osaka by way of Nagoya, the capital city of Aichi, in approximately one hour at a speed of 500 km/h (310 mph). The full track between Tokyo and Osaka was to be completed in 2045.

Lo Series train type undergoing testing by the Central Japan Railway Company (JR Central) for eventual use on the Chūō Shinkansen line set a world speed record of 603 km/h (375 mph) on 21 April 2015. The trains are planned to run at a maximum speed of 505 km/h (314 mph), offering journey times of 40 minutes between Tokyo (Shinagawa Station) and Nagoya, and 1 hour 7 minutes between Tokyo and Osaka.

SkyTran – Tel Aviv (Israel)

Skytran announced it would build an elevated network of sky cars in Tel Aviv, Israel. The technology was developed by NASA with the support of Israel Aerospace Industries. The system was meant to be suspended from an elevated track. The vehicles would travel at 70 km/h (43 mph) although the commercial rollout was expected to offer much faster vehicles. A trial of the system was to be built with a test track on the campus of Israel Aerospace Industries. Once successful, a full commercial version of SkyTran was expected to be rolled out first in Tel Aviv. The trial was scheduled to be up and running by the end of 2015. The company stated that speeds of up to 240 km/h (150 mph) are achievable.

Proposed Maglev Systems

Many maglev systems have been proposed in North America, Asia and Europe. Many are in the early planning stages or were explicitly rejected.

Australia

Sydney-Illawarra

A maglev route was proposed between Sydney and Wollongong. The proposal came to prominence in the mid-1990s. The Sydney–Wollongong commuter corridor is the largest in Australia, with upwards of 20,000 people commuting each day. Current trains use the Illawarra line, between the cliff face of the Illawarra escarpment and the Pacific Ocean, with travel times about two hours. The proposal would cut travel times to 20 minutes.

Melbourne

In late 2008, a proposal was put forward to the Government of Victoria to build a privately funded and operated maglev line to service the Greater Melbourne metropolitan area in response to the Eddington Transport Report that did not investigate above-ground transport options. The maglev would service a population of over 4 million and the proposal was costed at A$8 billion.

However despite road congestion and Australia's highest roadspace per capita, the government dismissed the proposal in favour of road expansion including an A$8.5 billion road tunnel, $6 billion extension of the Eastlink to the Western Ring Road and a $700 million Frankston Bypass.

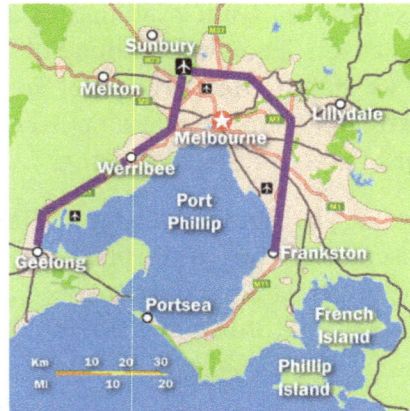

The proposed Melbourne maglev connecting the city of Geelong through Metropolitan Melbourne's outer suburban growth corridors, Tullamarine and Avalon domestic in and international terminals in under 20 min and on to Frankston, Victoria, in under 30 min.

Italy

A first proposal was formalized in April 2008, in Brescia, by journalist Andrew Spannaus who recommended a high speed connection between Malpensa airport to the cities of Milan, Bergamo and Brescia.

In March 2011 Nicola Oliva proposed a maglev connection between Pisa airport and the cities of Prato and Florence (Santa Maria Novella train station and Florence Airport). The travelling time would be reduced from the typical hour and a quarter to around twenty minutes. The second part of the line would be a connection to Livorno, to integrate maritime, aerial and terrestrial transport systems.

United Kingdom

London – Glasgow: A line was proposed in the United Kingdom from London to Glasgow with several route options through the Midlands, Northwest and Northeast of England. It was reported to be under favourable consideration by the government. The approach was rejected in the Government White Paper Delivering a Sustainable Railway published on 24 July 2007. Another high-speed link was planned between Glasgow and Edinburgh but the technology remained unsettled.

United States

Union Pacific freight conveyor: Plans are under way by American rail road operator Union Pacific to build a 7.9 km (4.9 mi) container shuttle between the ports of Los Angeles and Long Beach, with UP's intermodal container transfer facility. The system would be based on "passive" technology, especially well suited to freight transfer as no power is needed on board. The vehicle is a chassis that glides to its destination. The system is being designed by General Atomics.

California-Nevada Interstate Maglev: High-speed maglev lines between major cities of southern California and Las Vegas are under study via the California-Nevada Interstate Maglev Project. This plan was originally proposed as part of an I-5 or I-15 expansion plan, but the federal government ruled that it must be separated from interstate public work projects.

After the decision, private groups from Nevada proposed a line running from Las Vegas to Los Angeles with stops in Primm, Nevada; Baker, California; and other points throughout San Bernardino County into Los Angeles. Politicians expressed concern that a high-speed rail line out of state would carry spending out of state along with travelers.

Baltimore – Washington D.C. Maglev: A 64 km (40 mi) project has been proposed linking Camden Yards in Baltimore and Baltimore-Washington International (BWI) Airport to Union Station in Washington, D.C.

The Pennsylvania Project: The Pennsylvania High-Speed Maglev Project corridor extends from the Pittsburgh International Airport to Greensburg, with intermediate stops in Downtown Pittsburgh and Monroeville. This initial project was claimed to serve approximately 2.4 million people in the Pittsburgh metropolitan area. The Baltimore proposal competed with the Pittsburgh proposal for a US$90 million federal grant.

San Diego-Imperial County airport: In 2006 San Diego commissioned a study for a maglev line to a proposed airport located in Imperial County. SANDAG claimed that the concept would be an "airports [sic] without terminals", allowing passengers to check in at a terminal in San Diego ("satellite terminals"), take the train to the airport and directly board the airplane. In addition, the train would have the potential to carry freight. Further studies were requested although no funding was agreed.

Orlando International Airport to Orange County Convention Center: In December 2012 the Florida Department of Transportation gave conditional approval to a proposal by American Maglev to build a privately run 14.9-mile (24.0 km), 5-station line from Orlando International Airport to Orange County Convention Center. The Department requested a technical assessment and said there would be a request for proposals issued to reveal any competing plans. The route requires the use of a public right of way. If the first phase succeeded American Maglev would propose two further phases (of 4.9 and 19.4 miles (7.9 and 31.2 km)) to carry the line to Walt Disney World.

Puerto Rico

San Juan – Caguas: A 16.7-mile (26.8 km) maglev project was proposed linking Tren Urbano's Cupey Station in San Juan with two proposed stations in the city of Caguas, south of San Juan. The maglev line would run along Highway PR-52, connecting both cities. According to American Maglev project cost would be approximately US$380 million.

Germany

On 25 September 2007, Bavaria announced a high-speed maglev-rail service from Munich to its airport. The Bavarian government signed contracts with Deutsche Bahn and Transrapid with Siemens and ThyssenKrupp for the €1.85 billion project.

On 27 March 2008, the German Transport minister announced the project had been cancelled due to rising costs associated with constructing the track. A new estimate put the project between €3.2–3.4 billion.

Switzerland

SwissRapide: The SwissRapide AG together with the SwissRapide Consortium was planning and developing the first maglev monorail system for intercity traffic between the country's major cities. SwissRapide was to be financed by private investors. In the long-term, the SwissRapide Express was to connect the major cities north of the Alps between Geneva and St. Gallen, including Lucerne and Basel. The first projects were Bern – Zurich, Lausanne – Geneva as well as Zurich – Winterthur. The first line (Lausanne – Geneva or Zurich – Winterthur) could go into service as early as 2020.

Swissmetro: An earlier project, Swissmetro AG envisioned a partially evacuated underground maglev (a vactrain). As with SwissRapide, Swissmetro envisioned connecting the major cities in Switzerland with one another. In 2011, Swissmetro AG was dissolved and the IPRs from the organisation were passed onto the EPFL in Lausanne.

China

Shanghai – Hangzhou

China planned to extend the existing Shanghai Maglev Train, initially by some 35 kilometres to Shanghai Hongqiao Airport and then 200 kilometres to the city of Hangzhou (Shanghai-Hangzhou Maglev Train). If built, this would be the first inter-city maglev rail line in commercial service.

The project was controversial and repeatedly delayed. In May 2007 the project was suspended by officials, reportedly due to public concerns about radiation from the system. In January and February 2008 hundreds of residents demonstrated in downtown Shanghai that the line route came too close to their homes, citing concerns about sickness due to exposure to the strong magnetic field, noise, pollution and devaluation of property near to the lines. Final approval to build the line was granted on 18 August 2008. Originally scheduled to be ready by Expo 2010, plans called for completion by 2014. The Shanghai municipal government considered multiple options, including undergrounding the line to allay public fears. This same report stated that the final decision had to be approved by the National Development and Reform Commission.

In 2007 the Shanghai municipal government was considering build a factory in Nanhui district to produce low-speed maglev trains for urban use.

Shanghai – Beijing

A proposed line would have connected Shanghai to Beijing, over a distance of 1,300 kilometres (800 mi), at an estimated cost of £15.5bn. No projects had been revealed as of 2014.

India

Mumbai – Delhi
A project was presented to Indian railway minister (Mamata Banerjee) by an American company to connect Mumbai and Delhi. Then Prime Minister Manmohan Singh said that if the line project was successful the Indian government would build lines between other cities and also between Mumbai Central and Chhatrapati Shivaji International Airport.
Mumbai – Nagpur

The State of Maharashtra approved a feasibility study for a maglev train between Mumbai and Nagpur, some 1,000 km (620 mi) apart.
Chennai – Bangalore – Mysore
A detailed report was to be prepared and submitted by December 2012 for a line to connect Chennai to Mysore via Bangalore at a cost $26 million per kilometre, reaching speeds of 350 km/h.

Malaysia

A Consortium led by UEM Group Bhd and ARA Group, proposed Maglev technology to link Malaysian cities to Singapore. The idea was first mooted by YTL Group. Its technology partner then was said to be Siemens. High costs sank the proposal. The concept of a high-speed rail link from Kuala Lumpur to Singapore resurfaced. It was cited as a proposed "high impact" project in the Economic Transformation Programme (ETP) that was unveiled in 2010.

Iran

In May 2009, Iran and a German company signed an agreement to use maglev to link Tehran and Mashhad. The agreement was signed at the Mashhad International Fair site between Iranian Ministry of Roads and Transportation and the German company. The 900 km (560 mi) line allegedly could reduce travel time between Tehran and Mashhad to about 2.5 hours. Munich-based Schlegel Consulting Engineers said they had signed the contract with the Iranian ministry of transport and the governor of Mashad. "We have been mandated to lead a German consortium in this project," a spokesman said. "We are in a preparatory phase." The project could be worth between 10 billion and 12 billion euros, the Schlegel spokesman said.

Taiwan

Low speed maglev (urban maglev) is proposed for YangMingShan MRT Line for Taipei, a circular line connecting Taipei City to New Taipei City, and almost all other Taipei transport routes, but especially the access starved northern suburbs of Tien Mou and YangMingShan. From these suburbs to the city, transit times would be reduced by 70% or more compared to peak hours, and between Tien Mou and YangMingShan, from approx 20 minutes, to 3 minutes. Key to the line is YangMingShan Station, at 'Taipei level' in the mountain, 200M below YangMingShan (YangMing Mountain) Village, with 40 second high speed elevators to the Village.

Linimo or a similar system would be preferred, as being the core of Taipei's public transport system, it should run 24 hours a day. Also, in certain areas it would run within metres of apartments, so the near silent operation, and minimal maintenance requirements of maglev would be major advantages.

An extension of the line could run to Chiang Kai Shek Airport, and possibly on down the island, passing through major population centres, which the High Speed Rail must avoid. The minimal vibration of maglev would also be suitable to provide access Hsinchu Science Park, where sensitive silicon foundries are located. In the other direction, connection to the Tansui Line and to High Speed ferries at Tansui would provide overnight travel to Shanghai and Nagasaki, and to Busan or

Mokpo in South Korea, thus interconnecting the public transport systems of four countries, with great savings in fossil fuel consumption compared to flight.

YangMingShan MRT Line won the 'Engineering Excellence' Award, at the 2013 World Metro Summit in Shanghai.

Hong Kong

The Express Rail Link, previously known as the Regional Express, which will connect Kowloon with the territory's border with China, explored different technologies and designs in its planning stage, between Maglev and conventional highspeed railway, and if the latter was chosen, between a dedicated new route and sharing the tracks with the existing West Rail. Finally conventional highspeed with dedicated new route was chosen. It is expected to be operational in 2017.

Incidents

Two incidents involved fires. A Japanese test train in Miyazaki, MLU002, was completely consumed in a fire in 1991.

On 11 August 2006, a fire broke out on the commercial Shanghai Transrapid shortly after arriving at the Longyang terminal. People were evacuated without incident before the vehicle was moved about 1 kilometre to keep smoke from filling the station. NAMTI officials toured the SMT maintenance facility in November 2010 and learned that the cause of the fire was "thermal runaway" in a battery tray. As a result, SMT secured a new battery vendor, installed new temperature sensors and insulators and redesigned the trays.

On 22 September 2006, a Transrapid train collided with a maintenance vehicle on a test/publicity run in Lathen (Lower Saxony / north-western Germany). Twenty-three people were killed and ten were injured; these were the first maglev crash fatalities. The accident was caused by human error. Charges were brought against three Transrapid employees after a year-long investigation.

Marine Propulsion

A view of a ship's engine room

Marine propulsion is the mechanism or system used to generate thrust to move a ship or boat across water. While paddles and sails are still used on some smaller boats, most modern ships are propelled by mechanical systems consisting of an electric motor or engine turning a propeller, or less frequently, in pump-jets, an impeller. Marine engineering is the discipline concerned with the engineering design process of marine propulsion systems.

Marine steam engines were the first mechanical engines used in marine propulsion, however they have mostly been replaced by two-stroke or four-stroke diesel engines, outboard motors, and gas turbine engines on faster ships. Nuclear reactors producing steam are used to propel warships and icebreakers. Nuclear reactors to power commercial vessels has not been adopted by the marine industry. Electric motors using electric battery storage have been used for propulsion on submarines and electric boats and have been proposed for energy-efficient propulsion. Development in liquefied natural gas (LNG) fueled engines are gaining recognition for their low emissions and cost advantages. Stirling engines, which are more efficient, quieter, smoother running producing less harmful emissions than diesel engines, propel a number of small submarines. The Stirling engine has yet to be upscaled for larger surface ships.

Power Sources

Pre-mechanisation

A wind propelled fishing boat in Mozambique

Until the application of the coal-fired steam engine to ships in the early 19th century, oars or the wind were used to assist watercraft propulsion. Merchant ships predominantly used sail, but during periods when naval warfare depended on ships closing to ram or to fight hand-to-hand, galley were preferred for their manoeuvrability and speed. The Greek navies that fought in the Peloponnesian War used triremes, as did the Romans at the Battle of Actium. The development of naval gunnery from the 16th century onward meant that manoeuvrability took second place to broadside weight; this led to the dominance of the sail-powered warship over the following three centuries.

In modern times, human propulsion is found mainly on small boats or as auxiliary propulsion on sailboats. Human propulsion includes the push pole, rowing, and pedals.

Propulsion by sail generally consists of a sail hoisted on an erect mast, supported by stays, and controlled by lines made of rope. Sails were the dominant form of commercial propulsion until the late nineteenth century, and continued to be used well into the twentieth century on routes where wind was assured and coal was not available, such as in the South American nitrate trade. Sails are now generally used for recreation and racing, although innovative applications of kites/royals, turbosails, rotorsails, wingsails, windmills and SkySails's own kite buoy-system have been used on larger modern vessels for fuel savings.

Reciprocating Steam Engines

SS Ukkopekka uses a triple expansion steam engine

The development of piston-engined steamships was a complex process. Early steamships were fueled by wood, later ones by coal or fuel oil. Early ships used stern or side paddle wheels, while later ones used screw propellers.

The first commercial success accrued to Robert Fulton's North River Steamboat (often called Clermont) in US in 1807, followed in Europe by the 45-foot Comet of 1812. Steam propulsion progressed considerably over the rest of the 19th century. Notable developments include the steam surface condenser, which eliminated the use of sea water in the ship's boilers. This, along with improvements in boiler technology, permitted higher steam pressures, and thus the use of higher efficiency multiple expansion (compound) engines. As the means of transmitting the engine's power, paddle wheels gave way to more efficient screw propellers.

Steam Turbines

Steam turbines were fueled by coal or, later, fuel oil or nuclear power. The marine steam turbine developed by Sir Charles Algernon Parsons raised the power-to-weight ratio. He achieved publicity by demonstrating it unofficially in the 100-foot Turbinia at the Spithead Naval Review in 1897. This facilitated a generation of high-speed liners in the first half of the 20th century, and rendered the reciprocating steam engine obsolete; first in warships, and later in merchant vessels.

In the early 20th century, heavy fuel oil came into more general use and began to replace coal as the fuel of choice in steamships. Its great advantages were convenience, reduced manpower by removal of the need for trimmers and stokers, and reduced space needed for fuel bunkers.

In the second half of the 20th century, rising fuel costs almost led to the demise of the steam turbine. Most new ships since around 1960 have been built with diesel engines. The last major passenger ship built with steam turbines was the Fairsky, launched in 1984. Similarly, many steam ships were re-engined to improve fuel efficiency. One high-profile example was the 1968 built Queen Elizabeth 2 which had her steam turbines replaced with a diesel-electric propulsion plant in 1986.

Most new-build ships with steam turbines are specialist vessels such as nuclear-powered vessels, and certain merchant vessels (notably Liquefied Natural Gas (LNG) and coal carriers) where the cargo can be used as bunker fuel.

LNG Carriers

New LNG carriers (a high growth area of shipping) continue to be built with steam turbines. The natural gas is stored in a liquid state in cryogenic vessels aboard these ships, and a small amount of 'boil off' gas is needed to maintain the pressure and temperature inside the vessels within operating limits. The 'boil off' gas provides the fuel for the ship's boilers, which provide steam for the turbines, the simplest way to deal with the gas. Technology to operate internal combustion engines (modified marine two-stroke diesel engines) on this gas has improved, however, such engines are starting to appear in LNG carriers; with their greater thermal efficiency, less gas is burnt. Developments have also been made in the process of re-liquifying 'boil off' gas, letting it be returned to the cryogenic tanks. The financial returns on LNG are potentially greater than the cost of the marine-grade fuel oil burnt in conventional diesel engines, so the re-liquefaction process is starting to be used on diesel engine propelled LNG carriers. Another factor driving the change from turbines to diesel engines for LNG carriers is the shortage of steam turbine qualified seagoing engineers. With the lack of turbine powered ships in other shipping sectors, and the rapid rise in size of the worldwide LNG fleet, not enough have been trained to meet the demand. It may be that the days are numbered for marine steam turbine propulsion systems, even though all but sixteen of the orders for new LNG carriers at the end of 2004 were for steam turbine propelled ships.

The NS Savannah was the first nuclear-powered cargo-passenger ship

Nuclear-Powered Steam Turbines

In these vessels, the nuclear reactor heats water to create steam to drive the turbines. Due to low prices of diesel oil, nuclear propulsion is rare except in some Navy and specialist vessels such as icebreakers. In large aircraft carriers, the space formerly used for ship's bunkerage could be used instead to bunker aviation fuel. In submarines, the ability to run submerged at high speed and in relative quiet for long periods holds obvious advantages. A few cruisers have also employed nuclear power; as of 2006, the only ones remaining in service are the Russian Kirov class. An example of a non-military ship with nuclear marine propulsion is the Arktika class icebreaker with 75,000 shaft horsepower (55,930 kW). Commercial experiments such as the NS Savannah have so far proved uneconomical compared with conventional propulsion.

In recent times, there is some renewed interest in commercial nuclear shipping. Nuclear-powered cargo ships could lower costs associated with carbon dioxide emissions and travel at higher cruise speeds than conventional diesel powered vessels.

Reciprocating Diesel Engines

A modern diesel engine aboard a cargo ship

Most modern ships use a reciprocating diesel engine as their prime mover, due to their operating simplicity, robustness and fuel economy compared to most other prime mover mechanisms. The rotating crankshaft can be directly coupled to the propeller with slow speed engines, via a reduction gearbox for medium and high speed engines, or via an alternator and electric motor in diesel-electric vessels. The rotation of the crankshaft is connected to the camshaft or a hydraulic pump on an intelligent diesel.

The reciprocating marine diesel engine first came into use in 1903 when the diesel electric rivertanker Vandal was put into service by Branobel. Diesel engines soon offered greater efficiency than the steam turbine, but for many years had an inferior power-to-space ratio. The advent of turbocharging however hastened their adoption, by permitting greater power densities.

Diesel engines today are broadly classified according to

- Their operating cycle: two-stroke engine or four-stroke engine

- Their construction: crosshead, trunk, or opposed piston

- Their speed

    - Slow speed: any engine with a maximum operating speed up to 300 revolutions per minute (rpm), although most large two-stroke slow speed diesel engines operate below 120 rpm. Some very long stroke engines have a maximum speed of around 80 rpm. The largest, most powerful engines in the world are slow speed, two stroke, crosshead diesels.

    - Medium speed: any engine with a maximum operating speed in the range 300-900 rpm. Many modern four-stroke medium speed diesel engines have a maximum operating speed of around 500 rpm.

    - High speed: any engine with a maximum operating speed above 900 rpm.

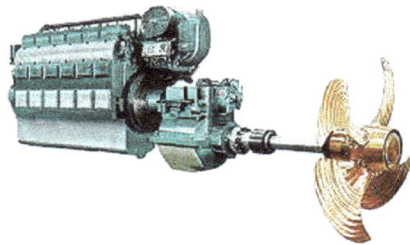

4-Stroke Marine Diesel Engine System

Most modern larger merchant ships use either slow speed, two stroke, crosshead engines, or medium speed, four stroke, trunk engines. Some smaller vessels may use high speed diesel engines.

The size of the different types of engines is an important factor in selecting what will be installed in a new ship. Slow speed two-stroke engines are much taller, but the footprint required is smaller than that needed for equivalently rated four-stroke medium speed diesel engines. As space above the waterline is at a premium in passenger ships and ferries (especially ones with a car deck), these ships tend to use multiple medium speed engines resulting in a longer, lower engine room than that needed for two-stroke diesel engines. Multiple engine installations also give redundancy in the event of mechanical failure of one or more engines, and the potential for greater efficiency over a wider range of operating conditions.

As modern ships' propellers are at their most efficient at the operating speed of most slow speed diesel engines, ships with these engines do not generally need gearboxes. Usually such propulsion systems consist of either one or two propeller shafts each with its own direct drive engine. Ships propelled by medium or high speed diesel engines may have one or two (sometimes more) propellers, commonly with one or more engines driving each propeller shaft through a gearbox. Where more than one engine is geared to a single shaft, each engine will most likely drive through a clutch, allowing engines not being used to be disconnected from the gearbox while others keep running. This arrangement lets maintenance be carried out while under way, even far from port.

LNG Engines

Shipping companies are required to comply with the International Maritime Organization (IMO) and the International Convention for the Prevention of Pollution from Ships emissions rules. Dual fuel engines are fueled by either marine grade diesel, heavy fuel oil, or liquefied natural gas (LNG). A Marine LNG Engine has multiple fuel options, allowing vessels to transit without relying on one type of fuel. Studies show that LNG is the most efficient of fuels, although limited access to LNG fueling stations limits the production of such engines. Vessels providing services in the LNG industry have been retrofitted with dual-fuel engines, and have been proved to be extremely effective. Benefits of dual-fuel engines include fuel and operational flexibility, high efficiency, low emissions, and operational cost advantages. Liquefied natural gas engines offer the marine transportation industry with an environmentally friendly alternative to provide power to vessels. In 2010, STX Finland and Viking Line signed an agreement to begin construction on what would be the largest environmentally friendly cruise ferry. Construction of NB 1376 will be completed in 2013. According to Viking Line, vessel NB 1376 will primarily be fueled by liquefied natural gas. Vessel NB 1376 nitrogen oxide emissions will be almost zero, and sulphur oxide emissions will be at least 80% below the International Maritime Organization's (IMO) standards. Company profits from tax cuts and operational cost advantages has led to the gradual growth of LNG fuel use in engines.

Gas Turbines

Combined marine propulsion
CODOG CODAG CODLAG CODAD COSAG COGOG COGAG COGAS CONAS IEP or IFEP

\Many warships built since the 1960s have used gas turbines for propulsion, as have a few passenger ships, like the jetfoil. Gas turbines are commonly used in combination with other types of engine. Most recently, the RMS Queen Mary 2 has had gas turbines installed in addition to diesel engines. Because of their poor thermal efficiency at low power (cruising) output, it is common for ships using them to have diesel engines for cruising, with gas turbines reserved for when higher speeds are needed. However, in the case of passenger ships the main reason for installing gas turbines has been to allow a reduction of emissions in sensitive environmental areas or while in port. Some warships, and a few modern cruise ships have also used steam turbines to improve the efficiency of their gas turbines in a combined cycle, where waste heat from a gas turbine exhaust is utilized to boil water and create steam for driving a steam turbine. In such combined cycles, thermal efficiency can be the same or slightly greater than that of diesel engines alone; however, the grade of fuel needed for these gas turbines is far more costly than that needed for the diesel engines, so the running costs are still higher.

Stirling Engines

Since the late 1980s, Swedish shipbuilder Kockums has built a number of successful Stirling engine powered submarines. The submarines store compressed oxygen to allow more efficient and cleaner external fuel combustion when submerged, providing heat for the Stirling engine's operation. The engines are currently used on submarines of the Gotland and Södermanland classes. and the Japanese Sōryū-class submarine. These are the first submarines to feature Stirling air-independent propulsion (AIP), which extends the underwater endurance from a few days to several weeks.

The heat sink of a Stirling engine is typically the ambient air temperature. In the case of medium to high power Stirling engines, a radiator is generally required to transfer the heat from the engine to the ambient air. Stirling marine engines have the advantage of using the ambient temperature water. Placing the cooling radiator section in seawater rather than ambient air allows for the radiator to be smaller. The engine's cooling water may be used directly or indirectly for heating and cooling purposes of the ship. The Stirling engine has potential for surface-ship propulsion, as the engine's larger physical size is less of a concern.

Screws

Marine propellers are also known as "screws". There are many variations of marine screw systems, including twin, contra-rotating, controllable-pitch, and nozzle-style screws. While smaller vessels tend to have a single screw, even very large ships such as tankers, container ships and bulk carriers may have single screws for reasons of fuel efficiency. Other vessels may have twin, triple or quadruple screws. Power is transmitted from the engine to the screw by way of a propeller shaft, which may or may not be connected to a gearbox.

Paddle Wheels

Left: original paddle wheel from a paddle steamer.
Right: detail of a paddle steamer.

The paddle wheel is a large wheel, generally built of a steel framework, upon the outer edge of which are fitted numerous paddle blades (called floats or buckets). The bottom quarter or so of the wheel travels underwater. Rotation of the paddle wheel produces thrust, forward or backward as required. More advanced paddle wheel designs have featured feathering methods that keep each paddle blade oriented closer to vertical while it is in the water; this increases efficiency. The upper part of a paddle wheel is normally enclosed in a paddlebox to minimise splashing.

Paddle wheels have been superseded by screws, which are a much more efficient form of propulsion. Nevertheless, paddle wheels have two advantages over screws, making them suitable for vessels in shallow rivers and constrained waters: first, they are less likely to be clogged by obstacles and debris; and secondly, when contra-rotating, they allow the vessel to spin around its own vertical axis. Some vessels had a single screw in addition to two paddle wheels, to gain the advantages of both types of propulsion.

Sailing

The purpose of sails is to use wind energy to propel the vessel, sled, board, vehicle or rotor.

Water Caterpillar

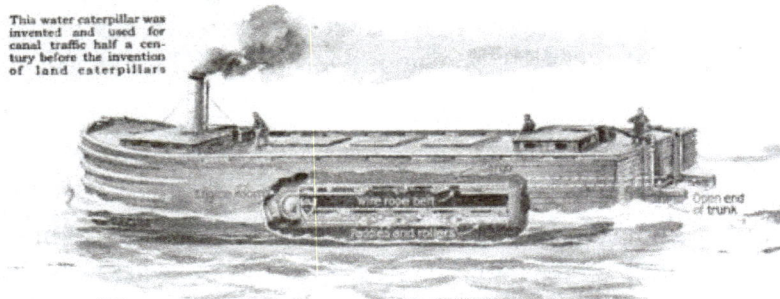

The water caterpillar boat propulsion system (Popular Science Monthly, December 1918)

An early uncommon means of boat propulsion was the water caterpillar. This moved a series of paddles on chains along the bottom of the boat to propel it over the water and preceded the development of tracked vehicles. The first water caterpillar was developed by Desblancs in 1782 and propelled by a steam engine. In the United States the first water caterpillar was patented in 1839 by William Leavenworth of New York.

Buoyancy

Underwater gliders convert buoyancy to thrust, using wings, or more recently hull shape (SeaExplorer Glider). Buoyancy is made alternatively negative and positive, generating tooth-saw profiles.

Spacecraft Propulsion

A remote camera captures a close-up view of a Space Shuttle Main Engine during a test firing at the John C. Stennis Space Center in Hancock County, Mississippi.

Spacecraft propulsion is any method used to accelerate spacecraft and artificial satellites. There are many different methods. Each method has drawbacks and advantages, and spacecraft propulsion is an active area of research. However, most spacecraft today are propelled by forcing a gas from the back/rear of the vehicle at very high speed through a supersonic de Laval nozzle. This sort of engine is called a rocket engine.

All current spacecraft use chemical rockets (bipropellant or solid-fuel) for launch, though some (such as the Pegasus rocket and SpaceShipOne) have used air-breathing engines on their first stage. Most satellites have simple reliable chemical thrusters (often monopropellant rockets) or resistojet rockets for orbital station-keeping and some use momentum wheels for attitude control. Soviet bloc satellites have used electric propulsion for decades, and newer Western geo-orbiting spacecraft are starting to use them for north-south stationkeeping and orbit raising. Interplanetary vehicles mostly use chemical rockets as well, although a few have used ion thrusters and Hall effect thrusters (two different types of electric propulsion) to great success.

Requirements

Artificial satellites must be launched into orbit and once there they must be placed in their nominal orbit. Once in the desired orbit, they often need some form of attitude control so that they are correctly pointed with respect to Earth, the Sun, and possibly some astronomical object of interest. They are also subject to drag from the thin atmosphere, so that to stay in orbit for a long period of time some form of propulsion is occasionally necessary to make small corrections (orbital stationkeeping). Many satellites need to be moved from one orbit to another from time to time, and this also requires propulsion. A satellite's useful life is usually over once it has exhausted its ability to adjust its orbit.

Spacecraft designed to travel further also need propulsion methods. They need to be launched out of the Earth's atmosphere just as satellites do. Once there, they need to leave orbit and move around.

For interplanetary travel, a spacecraft must use its engines to leave Earth orbit. Once it has done so, it must somehow make its way to its destination. Current interplanetary spacecraft do this with a series of short-term trajectory adjustments. In between these adjustments, the spacecraft simply falls freely along its trajectory. The most fuel-efficient means to move from one circular orbit to another is with a Hohmann transfer orbit: the spacecraft begins in a roughly circular orbit around the Sun. A short period of thrust in the direction of motion accelerates or decelerates the spacecraft into an elliptical orbit around the Sun which is tangential to its previous orbit and also to the orbit of its destination. The spacecraft falls freely along this elliptical orbit until it reaches its destination, where another short period of thrust accelerates or decelerates it to match the orbit of its destination. Special methods such as aerobraking or aerocapture are sometimes used for this final orbital adjustment.

Some spacecraft propulsion methods such as solar sails provide very low but inexhaustible thrust; an interplanetary vehicle using one of these methods would follow a rather different trajectory, either constantly thrusting against its direction of motion in order to decrease its distance from the Sun or constantly thrusting along its direction of motion to increase its distance from the Sun. The concept has been successfully tested by the Japanese IKAROS solar sail spacecraft.

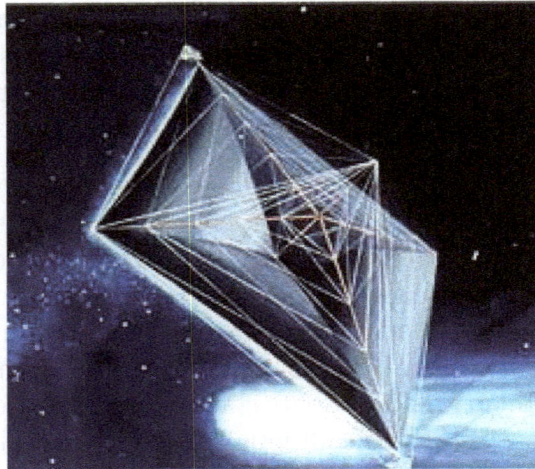

Artist's concept of a solar sail

Spacecraft for interstellar travel also need propulsion methods. No such spacecraft has yet been built, but many designs have been discussed. Because interstellar distances are very great, a tremendous velocity is needed to get a spacecraft to its destination in a reasonable amount of time. Acquiring such a velocity on launch and getting rid of it on arrival will be a formidable challenge for spacecraft designers.

Effectiveness

When in space, the purpose of a propulsion system is to change the velocity, or v, of a spacecraft. Because this is more difficult for more massive spacecraft, designers generally discuss momentum, mv. The amount of change in momentum is called impulse. So the goal of a propulsion method in space is to create an impulse.

When launching a spacecraft from Earth, a propulsion method must overcome a higher gravitational pull to provide a positive net acceleration. In orbit, any additional impulse, even very tiny, will result in a change in the orbit path.

The rate of change of velocity is called acceleration, and the rate of change of momentum is called force. To reach a given velocity, one can apply a small acceleration over a long period of time, or one can apply a large acceleration over a short time. Similarly, one can achieve a given impulse with a large force over a short time or a small force over a long time. This means that for maneuvering in space, a propulsion method that produces tiny accelerations but runs for a long time can produce the same impulse as a propulsion method that produces large accelerations for a short time. When launching from a planet, tiny accelerations cannot overcome the planet's gravitational pull and so cannot be used.

Earth's surface is situated fairly deep in a gravity well. The escape velocity required to get out of it is 11.2 kilometers/second. As human beings evolved in a gravitational field of 1g (9.8 m/s²), an ideal propulsion system would be one that provides a continuous acceleration of 1g (though human bodies can tolerate much larger accelerations over short periods). The occupants of a rocket or spaceship having such a propulsion system would be free from all the ill effects of free fall, such as nausea, muscular weakness, reduced sense of taste, or leaching of calcium from their bones.

The law of conservation of momentum means that in order for a propulsion method to change the momentum of a space craft it must change the momentum of something else as well. A few designs take advantage of things like magnetic fields or light pressure in order to change the spacecraft's momentum, but in free space the rocket must bring along some mass to accelerate away in order to push itself forward. Such mass is called reaction mass.

In order for a rocket to work, it needs two things: reaction mass and energy. The impulse provided by launching a particle of reaction mass having mass m at velocity v is mv. But this particle has kinetic energy $mv^2/2$, which must come from somewhere. In a conventional solid, liquid, or hybrid rocket, the fuel is burned, providing the energy, and the reaction products are allowed to flow out the back, providing the reaction mass. In an ion thruster, electricity is used to accelerate ions out the back. Here some other source must provide the electrical energy (perhaps a solar panel or a nuclear reactor), whereas the ions provide the reaction mass.

When discussing the efficiency of a propulsion system, designers often focus on effectively using the reaction mass. Reaction mass must be carried along with the rocket and is irretrievably consumed when used. One way of measuring the amount of impulse that can be obtained from a fixed amount of reaction mass is the specific impulse, the impulse per unit weight-on-Earth (typically designated by I_{sp}). The unit for this value is seconds. Because the weight on Earth of the reaction mass is often unimportant when discussing vehicles in space, specific impulse can also be discussed in terms of impulse per unit mass. This alternate form of specific impulse uses the same units as velocity (e.g. m/s), and in fact it is equal to the effective exhaust velocity of the engine (typically designated v_e). Confusingly, both values are sometimes called specific impulse. The two values differ by a factor of g_n, the standard acceleration due to gravity 9.80665 m/s² ($I_{sp}g_n = v_e$).

A rocket with a high exhaust velocity can achieve the same impulse with less reaction mass. However, the energy required for that impulse is proportional to the exhaust velocity, so that more mass-efficient engines require much more energy, and are typically less energy efficient. This is a problem if the engine is to provide a large amount of thrust. To generate a large amount of impulse per second, it must use a large amount of energy per second. So high-mass-efficient engines require enormous amounts of energy per second to produce high thrusts. As a result, most high-mass-efficient engine designs also provide lower thrust due to the unavailability of high amounts of energy.

Methods

Propulsion methods can be classified based on their means of accelerating the reaction mass. There are also some special methods for launches, planetary arrivals, and landings.

Reaction Engines

A reaction engine is an engine which provides propulsion by expelling reaction mass, in accordance with Newton's third law of motion. This law of motion is most commonly paraphrased as: "For every action force there is an equal, but opposite, reaction force".

Examples include both duct engines and rocket engines, and more uncommon variations such as

Hall effect thrusters, ion drives and mass drivers. Duct engines are obviously not used for space propulsion due to the lack of air; however some proposed spacecraft have these kinds of engines to assist takeoff and landing.

Delta-v and Propellant

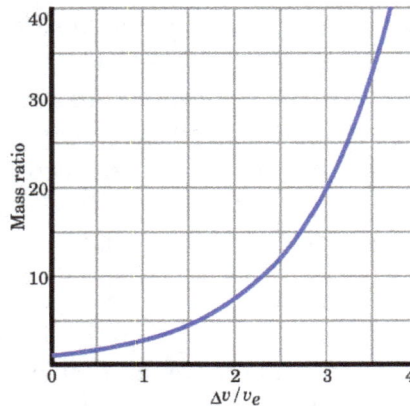

Rocket mass ratios versus final velocity, as calculated from the rocket equation

Exhausting the entire usable propellant of a spacecraft through the engines in a straight line in free space would produce a net velocity change to the vehicle; this number is termed 'delta-v' (Δv).

If the exhaust velocity is constant then the total Δv of a vehicle can be calculated using the rocket equation, where M is the mass of propellant, P is the mass of the payload (including the rocket structure), and v_e is the velocity of the rocket exhaust. This is known as the Tsiolkovsky rocket equation:

$$\Delta v = v_e \ln\left(\frac{M+P}{P}\right).$$

For historical reasons, as discussed above, v_e is sometimes written as

$$v_e = I_{sp} g_o$$

where I_{sp} is the specific impulse of the rocket, measured in seconds, and g_o is the gravitational acceleration at sea level.

For a high delta-v mission, the majority of the spacecraft's mass needs to be reaction mass. Because a rocket must carry all of its reaction mass, most of the initially-expended reaction mass goes towards accelerating reaction mass rather than payload. If the rocket has a payload of mass P, the spacecraft needs to change its velocity by Δv, and the rocket engine has exhaust velocity v_e, then the mass M of reaction mass which is needed can be calculated using the rocket equation and the formula for I_{sp}:

$$M = P\left(e^{\frac{\Delta v}{v_e}} - 1\right).$$

For Δv much smaller than v_e, this equation is roughly linear, and little reaction mass is needed. If Δv is comparable to v_e, then there needs to be about twice as much fuel as combined payload and structure (which includes engines, fuel tanks, and so on). Beyond this, the growth is exponential; speeds much higher than the exhaust velocity require very high ratios of fuel mass to payload and structural mass.

For a mission, for example, when launching from or landing on a planet, the effects of gravitational attraction and any atmospheric drag must be overcome by using fuel. It is typical to combine the effects of these and other effects into an effective mission delta-v. For example, a launch mission to low Earth orbit requires about 9.3–10 km/s delta-v. These mission delta-vs are typically numerically integrated on a computer.

Some effects such as Oberth effect can only be significantly utilised by high thrust engines such as rockets, i.e. engines that can produce a high g-force (thrust per unit mass, equal to delta-v per unit time).

Power use and Propulsive Efficiency

For all reaction engines (such as rockets and ion drives) some energy must go into accelerating the reaction mass. Every engine will waste some energy, but even assuming 100% efficiency, to accelerate an exhaust the engine will need energy amounting to

$$\frac{1}{2}\dot{m}v_e^2$$

This energy is not necessarily lost- some of it usually ends up as kinetic energy of the vehicle, and the rest is wasted in residual motion of the exhaust.

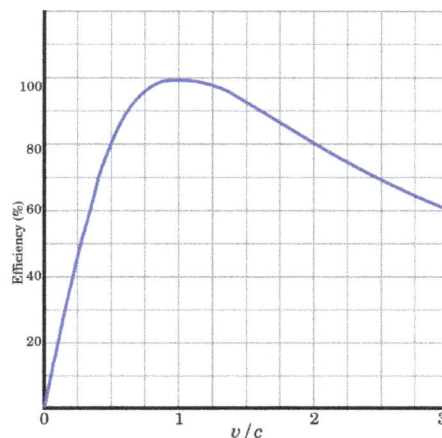

Due to energy carried away in the exhaust, the energy efficiency of a reaction engine varies with the speed of the exhaust relative to the speed of the vehicle, this is called propulsive efficiency

Comparing the rocket equation (which shows how much energy ends up in the final vehicle) and the above equation (which shows the total energy required) shows that even with 100% engine efficiency, certainly not all energy supplied ends up in the vehicle - some of it, indeed usually most of it, ends up as kinetic energy of the exhaust.

The exact amount depends on the design of the vehicle, and the mission. However, there are some useful fixed points:

- if the I_{sp} is fixed, for a mission delta-v, there is a particular I_{sp} that minimises the overall energy used by the rocket. This comes to an exhaust velocity of about ⅔ of the mission delta-v. Drives with a specific impulse that is both high and fixed such as Ion thrusters have exhaust velocities that can be enormously higher than this ideal for many missions.

- if the exhaust velocity can be made to vary so that at each instant it is equal and opposite to the vehicle velocity then the absolute minimum energy usage is achieved. When this is achieved, the exhaust stops in space and has no kinetic energy; and the propulsive efficiency is 100%- all the energy ends up in the vehicle (in principle such a drive would be 100% efficient, in practice there would be thermal losses from within the drive system and residual heat in the exhaust). However, in most cases this uses an impractical quantity of propellant, but is a useful theoretical consideration. Anyway, the vehicle has to move before the method can be applied.

Some drives (such as VASIMR or Electrodeless plasma thruster) actually can significantly vary their exhaust velocity. This can help reduce propellant usage or improve acceleration at different stages of the flight. However the best energetic performance and acceleration is still obtained when the exhaust velocity is close to the vehicle speed. Proposed ion and plasma drives usually have exhaust velocities enormously higher than that ideal (in the case of VASIMR the lowest quoted speed is around 15000 m/s compared to a mission delta-v from high Earth orbit to Mars of about 4000m/s).

It might be thought that adding power generation capacity is helpful, and although initially this can improve performance, this inevitably increases the weight of the power source, and eventually the mass of the power source and the associated engines and propellant dominates the weight of the vehicle, and then adding more power gives no significant improvement.

For, although solar power and nuclear power are virtually unlimited sources of energy, the maximum power they can supply is substantially proportional to the mass of the powerplant (i.e. specific power takes a largely constant value which is dependent on the particular powerplant technology). For any given specific power, with a large v_e which is desirable to save propellant mass, it turns out that the maximum acceleration is inversely proportional to v_e. Hence the time to reach a required delta-v is proportional to v_e. Thus the latter should not be too large.

Energy

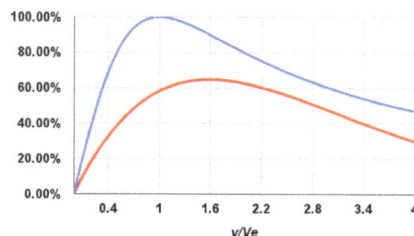

Plot of instantaneous propulsive efficiency (blue) and overall efficiency for a vehicle accelerating from rest (red) as percentages of the engine efficiency

In the ideal case m_1 is useful payload and $m_0 - m_1$ is reaction mass (this corresponds to empty tanks having no mass, etc.). The energy required can simply be computed as

$$\frac{1}{2}(m_0 - m_1)v_e^2$$

This corresponds to the kinetic energy the expelled reaction mass would have at a speed equal to the exhaust speed. If the reaction mass had to be accelerated from zero speed to the exhaust speed, all energy produced would go into the reaction mass and nothing would be left for kinetic energy gain by the rocket and payload. However, if the rocket already moves and accelerates (the reaction mass is expelled in the direction opposite to the direction in which the rocket moves).

Example

Suppose a 10,000 kg space probe will be sent to Mars. The required Äv from LEO is approximately 3000 m/s, using a Hohmann transfer orbit. For the sake of argument, assume the following thrusters are options to be used:

Engine	Effective exhaust velocity (km/s)	Specific impulse (s)	Fuel mass (kg)	Energy required (GJ)	Energy per kg of propellant	Minimum[a] power/thrust	Power generator mass/thrust[b]
Solid rocket	1	100	190,000	95	500 kJ	0.5 kW/N	N/A
Bipropellant rocket	5	500	8,200	103	12.6 MJ	2.5 kW/N	N/A
Ion thruster	50	5,000	620	775	1.25 GJ	25 kW/N	25 kg/N

1.   ^ Assuming 100% energetic efficiency; 50% is more typical in practice.

2.   ^ Assumes a specific power of 1 kW/kg

Observe that the more fuel-efficient engines can use far less fuel; their mass is almost negligible (relative to the mass of the payload and the engine itself) for some of the engines. However, note also that these require a large total amount of energy. For Earth launch, engines require a thrust to weight ratio of more than one. To do this with the ion or more theoretical electrical drives, the engine would have to be supplied with one to several gigawatts of power, equivalent to a major metropolitan generating station. From the table it can be seen that this is clearly impractical with current power sources.

Alternative approaches include some forms of laser propulsion, where the reaction mass does not provide the energy required to accelerate it, with the energy instead being provided from an external laser or other beam-powered propulsion system. Small models of some of these concepts have flown, although the engineering problems are complex and the ground based power systems are not a solved problem.

Instead, a much smaller, less powerful generator may be included which will take much longer to generate the total energy needed. This lower power is only sufficient to accelerate a tiny amount of fuel per second, and would be insufficient for launching from Earth. However, over long periods in orbit where there is no friction, the velocity will be finally achieved. For example, it took the

SMART-1 more than a year to reach the Moon, whereas with a chemical rocket it takes a few days. Because the ion drive needs much less fuel, the total launched mass is usually lower, which typically results in a lower overall cost, but the journey takes longer.

Mission planning therefore frequently involves adjusting and choosing the propulsion system so as to minimise the total cost of the project, and can involve trading off launch costs and mission duration against payload fraction.

Rocket Engines

SpaceX's Kestrel engine is tested

Most rocket engines are internal combustion heat engines (although non combusting forms exist). Rocket engines generally produce a high temperature reaction mass, as a hot gas. This is achieved by combusting a solid, liquid or gaseous fuel with an oxidiser within a combustion chamber. The extremely hot gas is then allowed to escape through a high-expansion ratio nozzle. This bell-shaped nozzle is what gives a rocket engine its characteristic shape. The effect of the nozzle is to dramatically accelerate the mass, converting most of the thermal energy into kinetic energy. Exhaust speed reaching as high as 10 times the speed of sound at sea level are common.

Rocket engines provide essentially the highest specific powers and high specific thrusts of any engine used for spacecraft propulsion.

Ion propulsion rockets can heat a plasma or charged gas inside a magnetic bottle and release it via a magnetic nozzle, so that no solid matter need come in contact with the plasma. Of course, the machinery to do this is complex, but research into nuclear fusion has developed methods, some of which have been proposed to be used in propulsion systems, and some have been tested in a lab.

Electromagnetic Propulsion

Rather than relying on high temperature and fluid dynamics to accelerate the reaction mass to high speeds, there are a variety of methods that use electrostatic or electromagnetic forces to accelerate the reaction mass directly. Usually the reaction mass is a stream of ions. Such an engine typically uses electric power, first to ionize atoms, and then to create a voltage gradient to accelerate the ions to high exhaust velocities.

This test engine accelerates ions using electrostatic forces

The idea of electric propulsion dates back to 1906, when Robert Goddard considered the possibility in his personal notebook. Konstantin Tsiolkovsky published the idea in 1911.

For these drives, at the highest exhaust speeds, energetic efficiency and thrust are all inversely proportional to exhaust velocity. Their very high exhaust velocity means they require huge amounts of energy and thus with practical power sources provide low thrust, but use hardly any fuel.

For some missions, particularly reasonably close to the Sun, solar energy may be sufficient, and has very often been used, but for others further out or at higher power, nuclear energy is necessary; engines drawing their power from a nuclear source are called nuclear electric rockets.

With any current source of electrical power, chemical, nuclear or solar, the maximum amount of power that can be generated limits the amount of thrust that can be produced to a small value. Power generation adds significant mass to the spacecraft, and ultimately the weight of the power source limits the performance of the vehicle.

Current nuclear power generators are approximately half the weight of solar panels per watt of energy supplied, at terrestrial distances from the Sun. Chemical power generators are not used due to the far lower total available energy. Beamed power to the spacecraft shows some potential.

6 kW Hall thruster in operation at the NASA Jet Propulsion Laboratory.

Some electromagnetic methods:

- Ion thrusters (accelerate ions first and later neutralize the ion beam with an electron stream emitted from a cathode called a neutralizer)

    - Electrostatic ion thruster

    - Field-emission electric propulsion

    - Hall effect thruster

    - Colloid thruster

- Electrothermal thrusters (electromagnetic fields are used to generate a plasma to increase the heat of the bulk propellant, the thermal energy imparted to the propellant gas is then converted into kinetic energy by a nozzle of either physical material construction or by magnetic means)

    - DC arcjet

    - microwave arcjet

    - Helicon Double Layer Thruster

- Electromagnetic thrusters (ions are accelerated either by the Lorentz Force or by the effect of electromagnetic fields where the electric field is not in the direction of the acceleration)

    - Magnetoplasmadynamic thruster

    - Electrodeless plasma thruster

    - Pulsed inductive thruster

    - Pulsed plasma thruster

    - Variable specific impulse magnetoplasma rocket (VASIMR)

- Mass drivers (for propulsion)

In electrothermal and electromagnetic thrusters, both ions and electrons are accelerated simultaneously, no neutralizer is required.

Without Internal Reaction Mass

The law of conservation of momentum is usually taken to imply that any engine which uses no reaction mass cannot accelerate the center of mass of a spaceship (changing orientation, on the other hand, is possible). But space is not empty, especially space inside the Solar System; there are gravitation fields, magnetic fields, electromagnetic waves, solar wind and solar radiation. Electromagnetic waves in particular are known to contain momentum, despite being massless; specifically the momentum flux density P of an EM wave is quantitatively $1/c$ times the Poynting vector S, i.e. $P = S/c$, where c is the velocity of light. Field propulsion methods which do not rely on reaction mass thus must try to take advantage of this fact by coupling to a momentum-bearing field such as an EM wave that exists in the vicinity of the craft. However, because many of these phenomena are diffuse in nature, corresponding propulsion structures need to be proportionately large.

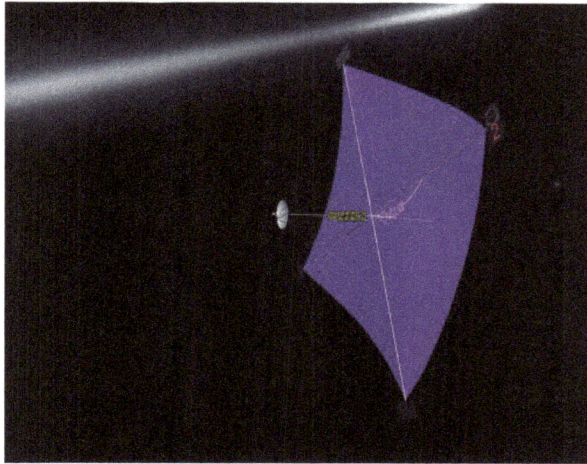

NASA study of a solar sail. The sail would be half a kilometer wide.

There are several different space drives that need little or no reaction mass to function. A tether propulsion system employs a long cable with a high tensile strength to change a spacecraft's orbit, such as by interaction with a planet's magnetic field or through momentum exchange with another object. Solar sails rely on radiation pressure from electromagnetic energy, but they require a large collection surface to function effectively. The magnetic sail deflects charged particles from the solar wind with a magnetic field, thereby imparting momentum to the spacecraft. A variant is the mini-magnetospheric plasma propulsion system, which uses a small cloud of plasma held in a magnetic field to deflect the Sun's charged particles. An E-sail would use very thin and lightweight wires holding an electric charge to deflect these particles, and may have more controllable directionality.

As a proof of concept, NanoSail-D became the first nanosatellite to orbit Earth. There are plans to add them to future Earth orbit satellites, enabling them to de-orbit and burn up once they are no longer needed. Cubesail will be the first mission to demonstrate solar sailing in low Earth orbit, and the first mission to demonstrate full three-axis attitude control of a solar sail.

Japan also launched its own solar sail powered spacecraft IKAROS in May 2010. IKAROS successfully demonstrated propulsion and guidance and is still flying today.

A satellite or other space vehicle is subject to the law of conservation of angular momentum, which constrains a body from a net change in angular velocity. Thus, for a vehicle to change its relative orientation without expending reaction mass, another part of the vehicle may rotate in the opposite direction. Non-conservative external forces, primarily gravitational and atmospheric, can contribute up to several degrees per day to angular momentum, so secondary systems are designed to "bleed off" undesired rotational energies built up over time. Accordingly, many spacecraft utilize reaction wheels or control moment gyroscopes to control orientation in space.

A gravitational slingshot can carry a space probe onward to other destinations without the expense of reaction mass. By harnessing the gravitational energy of other celestial objects, the spacecraft can pick up kinetic energy. However, even more energy can be obtained from the gravity assist if rockets are used.

Planetary and Atmospheric Propulsion

A successful proof of concept Lightcraft test, a subset of beam-powered propulsion.

Launch-Assist Mechanisms

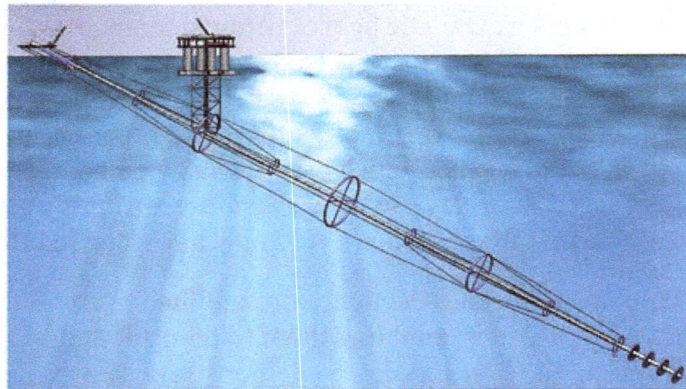

The conceptual ocean-located Quicklauncher, a light-gas gun–based space gun

There have been many ideas proposed for launch-assist mechanisms that have the potential of drastically reducing the cost of getting into orbit. Proposed non-rocket spacelaunch launch-assist mechanisms include:

- Skyhook (requires reusable suborbital launch vehicle, not engineeringly feasible using presently available materials)

- Space elevator (tether from Earth's surface to geostationary orbit, cannot be built with existing materials)

- Launch loop (a very fast enclosed rotating loop about 80 km tall)

- Space fountain (a very tall building held up by a stream of masses fired from its base)

- Orbital ring (a ring around Earth with spokes hanging down off bearings)

- Electromagnetic catapult (railgun, coilgun) (an electric gun)

- Rocket sled launch

- Space gun (Project HARP, ram accelerator) (a chemically powered gun)

- Beam-powered propulsion rockets and jets powered from the ground via a beam

- High-altitude platforms to assist initial stage

Airbreathing Engines

Studies generally show that conventional air-breathing engines, such as ramjets or turbojets are basically too heavy (have too low a thrust/weight ratio) to give any significant performance improvement when installed on a launch vehicle itself. However, launch vehicles can be air launched from separate lift vehicles (e.g. B-29, Pegasus Rocket and White Knight) which do use such propulsion systems. Jet engines mounted on a launch rail could also be so used.

On the other hand, very lightweight or very high speed engines have been proposed that take advantage of the air during ascent:

- SABRE - a lightweight hydrogen fuelled turbojet with precooler

- ATREX - a lightweight hydrogen fuelled turbojet with precooler

- Liquid air cycle engine - a hydrogen fuelled jet engine that liquifies the air before burning it in a rocket engine

- Scramjet - jet engines that use supersonic combustion

Normal rocket launch vehicles fly almost vertically before rolling over at an altitude of some tens of kilometers before burning sideways for orbit; this initial vertical climb wastes propellant but is optimal as it greatly reduces airdrag. Airbreathing engines burn propellant much more efficiently and this would permit a far flatter launch trajectory, the vehicles would typically fly approximately tangentially to Earth's surface until leaving the atmosphere then perform a rocket burn to bridge the final delta-v to orbital velocity.

Planetary Arrival and Landing

A test version of the MARS Pathfinder airbag system

When a vehicle is to enter orbit around its destination planet, or when it is to land, it must adjust its velocity. This can be done using all the methods listed above (provided they can generate a high enough thrust), but there are a few methods that can take advantage of planetary atmospheres and/or surfaces.

- Aerobraking allows a spacecraft to reduce the high point of an elliptical orbit by repeated brushes with the atmosphere at the low point of the orbit. This can save a considerable amount of fuel because it takes much less delta-V to enter an elliptical orbit compared to a low circular orbit. Because the braking is done over the course of many orbits, heating is comparatively minor, and a heat shield is not required. This has been done on several Mars missions such as Mars Global Surveyor, Mars Odyssey and Mars Reconnaissance Orbiter, and at least one Venus mission, Magellan.

- Aerocapture is a much more aggressive manoeuver, converting an incoming hyperbolic orbit to an elliptical orbit in one pass. This requires a heat shield and much trickier navigation, because it must be completed in one pass through the atmosphere, and unlike aerobraking no preview of the atmosphere is possible. If the intent is to remain in orbit, then at least one more propulsive maneuver is required after aerocapture—otherwise the low point of the resulting orbit will remain in the atmosphere, resulting in eventual re-entry. Aerocapture has not yet been tried on a planetary mission, but the re-entry skip by Zond 6 and Zond 7 upon lunar return were aerocapture maneuvers, because they turned a hyperbolic orbit into an elliptical orbit. On these missions, because there was no attempt to raise the perigee after the aerocapture, the resulting orbit still intersected the atmosphere, and re-entry occurred at the next perigee.

- A ballute is an inflatable drag device.

- Parachutes can land a probe on a planet or moon with an atmosphere, usually after the atmosphere has scrubbed off most of the velocity, using a heat shield.

- Airbags can soften the final landing.

- Lithobraking, or stopping by impacting the surface, is usually done by accident. However, it may be done deliberately with the probe expected to survive

Table of Methods

Below is a summary of some of the more popular, proven technologies, followed by increasingly speculative methods.

Four numbers are shown. The first is the effective exhaust velocity: the equivalent speed that the propellant leaves the vehicle. This is not necessarily the most important characteristic of the propulsion method; thrust and power consumption and other factors can be. However:

- if the delta-v is much more than the exhaust velocity, then exorbitant amounts of fuel are necessary.

- if it is much more than the delta-v, then, proportionally more energy is needed; if the power is limited, as with solar energy, this means that the journey takes a proportionally longer time

The second and third are the typical amounts of thrust and the typical burn times of the method. Outside a gravitational potential small amounts of thrust applied over a long period will give the same effect as large amounts of thrust over a short period. (This result does not apply when the object is significantly influenced by gravity.)

The fourth is the maximum delta-v this technique can give (without staging). For rocket-like propulsion systems this is a function of mass fraction and exhaust velocity. Mass fraction for rocket-like systems is usually limited by propulsion system weight and tankage weight. For a system to achieve this limit, typically the payload may need to be a negligible percentage of the vehicle, and so the practical limit on some systems can be much lower.

Propulsion methods					
Method	Effective exhaust velocity (km/s)	Thrust (N)	Firing duration	Maximum delta-v (km/s)	Technology readiness level
Solid-fuel rocket	<2.5	<10^7	Minutes	7	9: Flight proven
Hybrid rocket			Minutes	>3	9: Flight proven
Monopropellant rocket	1 – 3	0.1 – 100	Milliseconds – minutes	3	9: Flight proven
Liquid-fuel rocket	<4.4	<10^7	Minutes	9	9: Flight proven
Electrostatic ion thruster	15 – 210		Months – years	>100	9: Flight proven
Hall-effect thruster (HET)	8 – 50		Months – years	>100	9: Flight proven
Resistojet rocket	2 – 6	10^{-2} – 10	Minutes	?	8: Flight qualified
Arcjet rocket	4 – 16	10^{-2} – 10	Minutes	?	8: Flight qualified
Field emission electric propulsion (FEEP)	100 – 130	10^{-6} – 10^{-3}	Months – years	?	8: Flight qualified
Pulsed plasma thruster (PPT)	20	0.1	2,000 – 10,000 hours	?	7: Prototype demoed in space
Dual-mode propulsion rocket	1 – 4.7	0.1 – 10^7	Milliseconds – minutes	3 – 9	7: Prototype demoed in space

Solar sails	• 299792, light • 145 – 750, solar wind	• 9/km² at 1 AU • 230/km² at 0.2 AU • 10^{-10}/km² at 4 ly	Indefinite	>40	• 9: Light pressure atti-tude-con-trol flight proven • 6: De-ploy-only demoed in space • 5: Light-sail validated in lit vacuum
Tripropellant rocket	2.5 – 5.3	$0.1 – 10^7$	Minutes	9	6: Prototype demoed on ground
Magnetoplasmadynamic thruster (MPD)	20 – 100	100	Weeks	?	6: Model, 1 kW demoed in space
Nuclear–thermal rocket	9	10^7	Minutes	>20	6: Prototype demoed on ground
Propulsive mass drivers	0 – 30	$10^4 – 10^8$	Months	?	6: Model, 32 MJ demoed on ground
Tether propulsion	N/A	$1 – 10^{12}$	Minutes	7	6: Model, 31.7 km demoed in space
Air-augmented rocket	5 – 6	$0.1 – 10^7$	Seconds – minutes	>7?	6: Prototype demoed on ground
Liquid-air-cycle engine	4.5	$10^3 – 10^7$	Seconds – minutes	?	6: Prototype demoed on ground
Pulsed-inductive thruster (PIT)	10 – 80	20	Months	?	5: Component vali-dated in vacuum
Variable-specific-impulse mag-netoplasma rocket (VASIMR)	10 – 300	40 – 1,200	Days – months	>100	5: Component, 200 kW validated in vacuum
Magnetic-field oscillating am-plified thruster	10 – 130	0.1 – 1	Days – months	>100	5: Component vali-dated in vacuum
Solar–thermal rocket	7 – 12	1 – 100	Weeks	>20	4: Component validated in lab
Radioisotope rocket	7 – 8	1.3 – 1.5	Months	?	4: Component validated in lab
Nuclear–electric rocket	As electric propulsion method used				4: Component, 400 kW validated in lab
Orion Project (near-term nu-clear pulse propulsion)	20 – 100	$10^9 – 10^{12}$	Days	30 – 60	3: Validated, 900 kg proof-of-concept
Space elevator	N/A	N/A	Indefinite	>12	3: Validated proof-of-concept
Reaction Engines SABRE	30/4.5	$0.1 – 10^7$	Minutes	9.4	3: Validated proof-of-concept

Magnetic sails	145 − 750, solar wind	2/t	Indefinite	?	3: Validated proof-of-concept
Mini-magnetospheric plasma propulsion	200	1/kW	Months	?	3: Validated proof-of-concept
Beam-powered/laser	As propulsion method powered by beam				3: Validated, 71 m proof-of-concept
Launch loop/orbital ring	N/A	10^4	Minutes	11 − 30	2: Technology concept formulated
Nuclear pulse propulsion (Project Daedalus' drive)	20 − 1,000	$10^9 - 10^{12}$	Years	15,000	2: Technology concept formulated
Gas-core reactor rocket	10 − 20	$10^3 - 10^6$	?	?	2: Technology concept formulated
Nuclear salt-water rocket	100	$10^3 - 10^7$	Half-hour	?	2: Technology concept formulated
Fission sail	?	?	?	?	2: Technology concept formulated
Fission-fragment rocket	15,000	?	?	?	2: Technology concept formulated
Nuclear–photonic rocket	299,792	$10^{-5} - 1$	Years – decades	?	2: Technology concept formulated
Fusion rocket	100 − 1,000	?	?	?	2: Technology concept formulated
Antimatter-catalyzed nuclear pulse propulsion	200 − 4,000	?	Days – weeks	?	2: Technology concept formulated
Antimatter rocket	10,000 − 100,000	?	?	?	2: Technology concept formulated
Bussard ramjet	2.2 − 20,000	?	Indefinite	30,000	2: Technology concept formulated
Method	Effective exhaust velocity (km/s)	Thrust (N)	Firing duration	Maximum delta-v (km/s)	Technology readiness level

Testing

Spacecraft propulsion systems are often first statically tested on Earth's surface, within the atmosphere but many systems require a vacuum chamber to test fully. Rockets are usually tested at a rocket engine test facility well away from habitation and other buildings for safety reasons. Ion drives are far less dangerous and require much less stringent safety, usually only a large-ish vacuum chamber is needed.

Famous static test locations can be found at Rocket Ground Test Facilities

Some systems cannot be adequately tested on the ground and test launches may be employed at a Rocket Launch Site.

Speculative Methods

Artist's conception of a warp drive design

A variety of hypothetical propulsion techniques have been considered that mostly require a deeper understanding of the properties of space, particularly inertial frames and the quantum vacuum. To date, such methods are highly speculative and include:

- Abraham—Minkowski drive
- Alcubierre drive
- Diametric drive
- Differential sail
- Disjunction drive
- Gravitational shielding
- Heim drive
- Photon rocket
- Pitch drive & bias drive
- Quantum tunnelling
- Quantum vacuum thruster
- Reactionless drive
- Traversable wormholes
- Woodward effect

A NASA assessment of its Breakthrough Propulsion Physics Program divides such proposals into those that are non-viable for propulsion purposes, those that are of uncertain potential, and those that are theoretically possible.

Jet Propulsion

Jet propulsion is thrust produced by passing a jet of matter (typically air or water) in the opposite direction to the direction of motion. By Newton's third law, the moving body is propelled in the opposite direction to the jet. It is most commonly used in the jet engine, but is also the most common means of spacecraft propulsion.

A number of animals, including cephalopods, sea hares, arthropods, and fish have convergently evolved jet propulsion mechanisms.

Physics

Jet propulsion is most effective when the Reynolds number is high - that is, the object being propelled is relatively large and passing through a low-viscosity medium.

In biology, the most efficient jets are pulsed, rather than continuous: at least when the Reynolds number is greater than 6.

Jet Engine

A jet engine is a reaction engine that discharges a fast moving jet of fluid to generate thrust by jet propulsion and in accordance with Newton's laws of motion. This broad definition of jet engines includes turbojets, turbofans, rockets, ramjets, pulse jets and pump-jets. In general, most jet engines are internal combustion engines but non-combusting forms also exist.

Jet-propelled Animals

Jet propulsion in cephalopods is produced by water being exhaled through a siphon, which typically narrows to a small opening to produce the maximum exhalent velocity. The water passes through the gills prior to exhalation, fulfilling the dual purpose of respiration and locomotion. Sea hares (gastropod molluscs) employ a similar means of jet propulsion, but without the sophisticated neurological machinery of cephalopods they navigate somewhat more clumsily.

Some teleost fish have also developed jet propulsion, passing water through the gills to supplement fin-driven motion.

In some dragonfly larvae, jet propulsion is achieved by the expulsion of water from a specialised cavity through the anus. Given the small size of the organism, a great speed is achieved.

Scallops and cardiids, siphonophores, tunicates (such as salps), and some jellyfish also employ jet propulsion. The most efficient jet-propelled organisms are the salps, which use an order of magnitude less energy (per kilogram per metre) than squid.

References

- Wake, M.H. (1993). "The Skull as a Locomotor Organ". In Hanken, James. The Skull. University of Chicago Press. p. 460. ISBN 978-0-226-31573-7.

- Chamberlain Jr, John A. (1987). "32. Locomotion of Nautilus". In Saunders, W. B.; Landman, N. H. Nautilus: The Biology and Paleobiology of a Living Fossil. ISBN 9789048132980.

- Long, Hongtao (6 May 2016). "Trial operation of magnetic levitation line in Changsha to start". People's Daily Online. Retrieved 6 May 2015.

- "Japan's maglev train breaks world speed record with 600 km/h test run". The Guardian. United Kingdom: Guardian News and Media Limited. 21 April 2015. Retrieved 21 April 2015.

- "S. Korea unveils first domestically-developed urban maglev train". Global Times. Xinhua. 16 May 2014. Retrieved 9 January 2015.

- "Hover cars have arrived! Self-driving sky pods set to soar above traffic on high-speed magnetic tracks". MailOnline. 27 June 2014. Retrieved 8 January 2015.

- MICHAEL, GEBICKI (27 November 2014). "What's the world's fastest passenger train". Stuff.co.nz. Retrieved 24 December 2014.

- "JR Tokai unveils a model for the new high-speed maglev train 'L0'". Daily Onigiri. DailyOnigiri.com. 4 November 2010. Retrieved 17 January 2013.

- "Bullet train may connect Mysore-Bangalore in 1hr 30 mins Photos". Yahoo! India Finance. 20 April 2012. Retrieved 2012-11-04.

Permissions

Index

www.ingramcontent.com/pod-product-compliance
Lightning Source LLC
Chambersburg PA
CBHW061320190326

41458CB00011B/3848